Contemporary College Algebra

Data, Functions, Modeling
Fifth Edition

Don Small

The United States Military Academy
West Point, New York

Boston Burr Ridge, IL Dubuque, IA Madison, WI New York San Francisco St. Louis
Bangkok Bogotá Caracas Lisbon London Madrid
Mexico City Milan New Delhi Seoul Singapore Sydney Taipei Toronto

The McGraw-Hill Companies

Contemporary College Algebra

Data, Functions, Modeling

McGraw-Hill's Primis Custom Publishing consists of products that are produced from camera-ready copy. Peer review, class testing, and accuracy are primarily the responsibility of the author(s).

6 7 8 9 0 QSR QSR 0 9 8 7 6 5

ISBN 0-07-299383-9

Sponsoring Editor: Christine Bowie
Production Editor: Kathy Phelan
Printer/Binder: Quebecor World

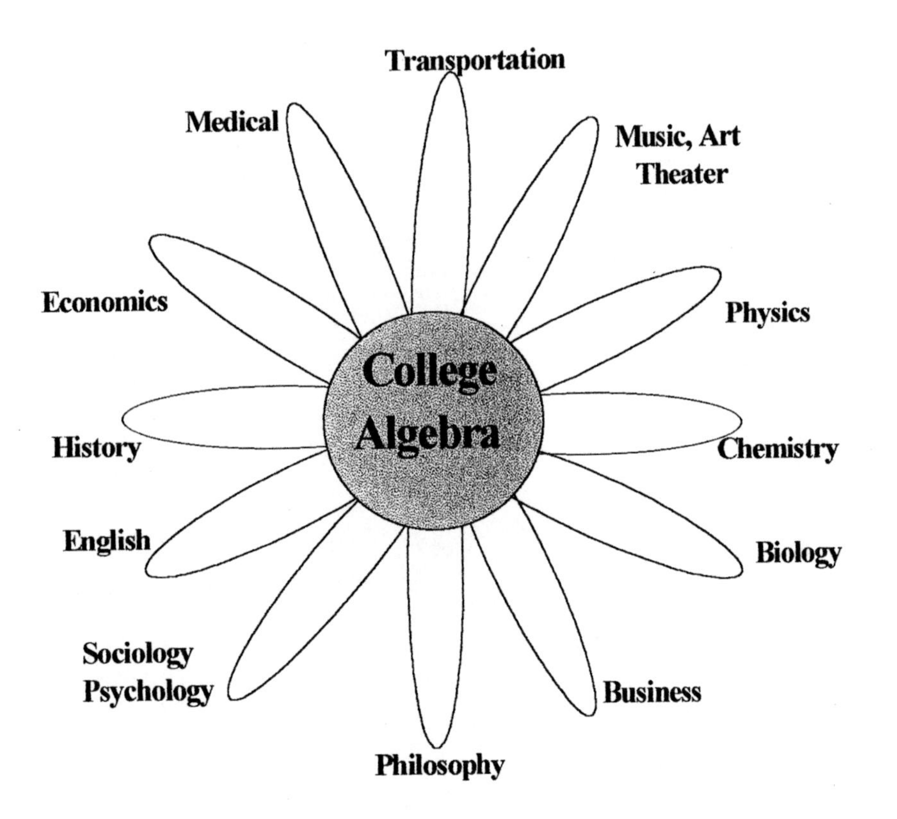

This material is based upon work supported by the National Science Foundation (DUE 97525919), The Exxon Education Foundation, and the Brown Foundation.

Preface

The mind is not a vessel to be filled, but a flame to be kindled.—Plutarch

Primary Goal

Our philosophy is to **educate students for the future rather than train them for the past**. We therefore have made a conscious effort to incorporate into our goals the common qualifications for entering the work force as enunciated by social, business, and industrial leaders.

The primary goal of this text is to empower students to become **exploratory learners**, not to master a list of algebraic rules. Each section contains *Queries* that engage students in questioning and exploring the material being presented. Exercises that explicitly ask students to explore, ask what-if type questions, make up examples, further investigate worked examples, iterate for the purpose of recognizing a pattern and developing a sense for the behavior of a solution, and graphically fit a curve to a data set are some of the means that are used to establish an exploratory environment for the students.

Other Goals of this Text

1. Improve communication skills—reading, writing, presenting, listening.

   The large majority of the exercises are presented in the story problem format to address the reading aspect of this goal. The story problem format also addresses the applicability aspect of college algebra, as real-life situations are usually described verbally or in written form rather than in terms of equations.
2. Small-group work—in-class group activities and out-of-class group projects. In-class activities culminate in student presentations to the class, and out-of-class projects culminate in both a written report and a student presentation.
3. Use of technology—every student is expected to have daily access to a graphing calculator and/or computer. The ability to use technology for plotting and computation is a very important skill.
4. Modeling—to empower students to use mathematics to quantify real-life situations.
5. Confidence—develop personal confidence as a problem solver. Develop confidence in the iterative process: "try something, note the errors, modify previous attempt to lessen the errors, and try again" until a satisfactory approximation has been obtained. The initial attempt is usually informed by sketching a picture.
6. Enjoy applying mathematics to meaningful situations.

This text is to be **read, studied, and annotated.** Students should study the worked examples for the purpose of understanding the concepts and reasoning involved. Students are expected to personalize their text by filling in missing details, making up examples and illustrations, and raising questions. The purpose of the exercises is to help clarify and expand the reasoning process. As such, working exercises is secondary in importance to studying the written material in the sections.

Real-World Contexts

Concepts and techniques are introduced and motivated by real-life situations. Computational techniques are introduced in response to the need to solve real-life situations. For example, the quadratic formula is introduced in Chapter 4 in order to solve motion problems that involve quadratic equations. The ability to understand elementary data analysis, to extract function relations from data, and to mathematically model real-life situations in different disciplines is fundamental to the liberal arts education of every student.

Fun Projects

Fun Projects are small-group (three to five students) out-of-class projects. The projects are designed for six to ten hours of work and culminate in a written report. Instructors are encouraged to assign two or three projects during the course. The purpose of the projects are to provide opportunities to

1. Mathematically model real-world situations
2. Research a topic (usually on the Internet)
3. Provide a writing assignment
4. Provide a small-group experience
5. Have fun exploring and creating solutions to meaningful problems

The Project Report should consist of

Cover Page (creative design by students)
Title Page (project name, date, instructor name, students' names)
Executive Summary (one-page abstract of the problem, approach used, results obtained)
Supporting Data (computations, labeled drawings, labeled computer plots, and/or printouts)
Group Log (time, date, location, and brief description of each meeting)
Evaluation Summary of the group's learning experience in working on the project
List of references consulted

All group members should be involved in answering each of the questions. In addition, each member of the group should be assigned a particular responsibility in connection with the project, such as one of the following.

Leader: Responsible for developing the group. Responsible for seeing that the project is completed in a satisfactory manner and on time.
Recorder: Arranges group meetings and records group activities.
Checker: Checks accuracy of all computations. Checks to see that all questions are answered.
Typist: Types Executive and Evaluation Summaries.
Reader: Responsible for proofreading and final assembly of the report.

A Few Suggestions to Students

1. Be an exploratory learner: sketch pictures, question, create what-if questions, make up examples, question the reasonableness of results, look for applications.
2. Read with a pencil in hand. Make your text useful to you by using the margins or additional paper

to write explanatory notes and questions, fill in missing computations, make up additional worked problems, and so on. When you personalize your text by augmenting it, you transform it into an effective learning tool for you.

3. Do not get bogged down in computations. The course is about applying mathematics to real-world situations, not about computations.
4. Be patient and persevere in your studying. Focus on understanding the reasoning in the worked examples.
5. Answer the *Queries*.
6. Make up examples and what-if exercises.
7. Work the activities in the accompanying CD-Rom.

Chapter Content

The analysis of data is the starting point for most of the topics in this text. The analysis of data motivates the concept of function for the purpose of drawing predictions from data. Just as data is displayed differently for varied purposes, functions are represented differently (graphically, symbolically, numerically, and verbally) to address varied concerns. The ability to graphically approximate a data set is a key skill in applying mathematics. A strong foundation in elementary data analysis and the function concept prepares the student to model real-life situations.

Chapter 1—Overview

Chapter 2—Data and Variables: We study how to read and display data: table, pie chart, scatter and line plots, and bar charts. We learn the meanings, use, and methods to compute the three principle summary measures of a set: average (mean), median, and mode. Our understanding that data is information about a variable introduces us to an understanding of the meaning of variable and its use as a mathematical pronoun. The exploration of relations between variables leads to the study of straight lines, a fundamental concept in the application of mathematics to real-life situations. Applications of linear equations lead naturally to systems of linear equations, linear inequalities, and their applications in linear programming.

Chapter 3—Functions: The concept of a function is one of the most important concepts in mathematics. The concept is developed informally through discussion of academic grades, modeling water level in a well, and warming a can of soda. Definitions of a function and related terms are clearly presented and illustrated. Graphically extracting functional relations from data introduces the shapes of the basic functions: power, radical, exponential, logarithmic, and periodic (sine, cosine). The skill to graphically fit a curve to a scatter plot is enhanced by studying the basic graph transformations of shifting and scaling. The algebra of functions (addition, multiplication, composition, and inverses) is developed graphically, symbolically, and numerically. The ability to display data and to graphically approximate numerical solutions of equations and zeros of functions is an important thread throughout the text. A key element in transforming data into information is the development of symbolic approximation of data (that is, regression analysis). Optimization of functions, finding maximum and minimum values, completes the chapter.

Chapter 4—Modeling: College algebra is a college or university program in the sense that it or an equivalent course is required by all disciplines. Therefore an appropriate goal of *Contemporary College Algebra* is to prepare students to mathematically model real-life situations arising in different disciplines. To illustrate the breadth of the applicability of college algebra, the focus of Chapter 4 is on modeling problems in business, physical and life sciences, and the arts. The primary modeling techniques are graphical approximations and recursive sequences. The recursive sequence model developed on the reasoning

(New Situation) = (Old Situation) + (Change)

is applicable across the disciplines. In particular, the recursive sequence model of the accumulation of money in a savings account serves as paradigm for most of the discrete models developed in *Contemporary College Algebra.*

Exercises

There is a rich assortment of exercises at the end of each section to augment the *Queries* and worked examples in the section. The purpose of the exercises is to support and expand the conceptual understanding of the material. As such, several of the exercises refer to worked examples in the section. Many of the exercises are suitable for small-group in-class activities or small-group out-of-class projects.

Labels

Queries are numbered consecutively within sections. Examples, figures, and tables have three place labels, the first location denotes the chapter, the second location denotes the section, and the third location denotes the example, figure, or table within the section. For example, Figure 3.4.6 is the sixth figure in Section 4 of Chapter 3 and Table 3.FunProject2.4 denotes the fourth table in Fun Project number 2 in Chapter 3.

Computational Skill

Development of computational skill aided by the calculator and/or computer is an expected outcome of studying this text. Computational skill, including approximation of answers and checking the reasonableness of answers, will develop by working on real-life problems. Each section's exercise set begins with three Computational Skill exercises, the third of which is to make up three additional exercises similar to the first two. The purpose is to highlight the important computational techniques used in the section and to encourage students to ask what-if type questions. Appendix A on Computational Skills and Basic Functions is included for reference. Time should not be spent on drilling to master hand computation skills, such as factoring. In educating students to become contributing and productive members of society, we need to engage them in the use of the computational tools used in society—graphing calculators and computers.

The following table indicates where important algebraic or arithmetic techniques are introduced in the text.

Percentage	Section 2.1	Systems of Linear Equations	Section 3.8
Fractions	Section 2.2	Iteration	Section 4.1
Radian and Degree Measure	Section 2.4	Geometric Series	Section 4.1
Linear Equations	Section 2.6	Financial Problems	Section 4.2
Inequalities, Absolute Value	Section 2.7	Exponential Functions	Section 4.2
Evaluation of Functions	Section 3.1	Logarithmic Functions	Section 4.2
Basic Functions	Section 3.3	Quadratic Formula	Section 4.3
Graph Transformation	Section 3.4	Parametric Functions	Section 4.3
Algebraic Operations	Section 3.5	Trigonometric Functions	Section 4.3
Factoring	Section 3.6	Logistic Function	Section 4.3

Let us begin an exciting journey through *Contemporary College Algebra*.
Fasten your seat belts, **you are driving**!

Acknowledgment

There are many people who have contributed to the development of the Contemporary College Algebra program, a product of the Historically Black College and University Consortium for College Algebra Reform. I appreciate their efforts and offer my sincere thanks for their support and contributions. Special thanks to the members of the Advisory Council. In addition, I would like to mention a few individuals who made special contributions. Mr. Bob Witte, Program Officer for the Exxon Education Foundation, was the first person to support our vision by providing funding for the initial year of the program. He renewed the funding for each of the following four years. Dr. Della Bell, Texas Southern University, chaired the Consortium through its first seven years, making countless arrangements and administering our grants. Her attention to detail was indispensable to the program. Dr. General Marshall, Huston-Tillotson College, Mr. Gene Taylor, Grambling State University, Bill Echols, and Dr. Victor Obot and Mr. Carrington Stewart both of Texas Southern University, all were instrumental in developing the program. I also want to thank Mr. Tom Dyson and Dr. Frank Hughes, both of Textronix Inc., for their work on the CD-Rom. I want to express grateful appreciation to Dr. David Arney and Dr. Gary Krahn, successive Chairs of the U.S. Military Academy's Department of Mathematical Sciences, who provided encouragement and funding from Project InterMath, a National Science Foundation funded consortium for interdisciplinary cooperation administered by the Department of Mathematical Sciences of the U.S. Military Academy. I also gratefully acknowledge the financial support received from the Brown Foundation and the National Science Foundation.

Don Small

Fun Projects are small-group (three to five students) projects designed to be completed outside of class. These projects compliment the small-group activities and projects incorporated into the individual sections. Sets of these projects follow each of the chapters 2, 3, and 4. See the Fun Project section of Chapter 2 for the purpose, organization, and responsibilities associated with these projects. The following list of Fun Projects is grouped by chapter.

Chapter 2.

1. Doubling Leads to Large Numbers
2. City and Country Populations
3. Pepsi, Cole, or ... ?
4. The Chain Letter
5. Packaging
6. Kicking Field Goals
7. The Cost of Driving
8. Daily Recommended Amount of Sodium in a 2,000 Calorie per day Diet

Chapter 3.

1. Comfort Function for Stairways
2. Determining the Dimensions of a Soda Can
3. Postage Stamps
4. The Bones Know
5. Homecoming Parade
6. Income Tax
7. Measuring Earthquakes
8. How Many Times Must I Pump?

Chapter 4.

1. Consumer Price Index (CPI)
2. Shingling a Roof
3. Pollution—Walton Lake Cleanup
4. Sweepstakes
5. Mercury in the Reservoirs: Water's OK, but Don't Eat the Fish
6. The Future of the World's Oil Supply
7. College Tuition Plans
8. Oxygen Levels in the Narraguagus River
9. Do Manatees have a Future?

Contemporary College Algebra

Table of Contents

Chapter 4 - Modeling

Appendix

Barbara Charline Jordan

"The first Black woman everything..." (Washington Post)

Barbara Charline Jordan (1936–1995), one of Texas Southern University's most distinguished alumna, was born in Houston, Texas. Her determination to excel led her into a career characterized by a succession of "firsts." In college, she was the first woman to travel with the University's Debate Team and was the first black woman to earn first place in a debate at Baylor College. Following her graduation magna cum laude from Texas Southern University, she earned her law degree from Boston College and became the first black woman to serve as Administrative Assistant to the County Judge of Harris County, Texas.

In 1966, she was the first black to be elected to the Texas Senate since 1883; authored Texas's first minimum wage bill; the first black Texan to serve in the U.S. Congress; the first black women in the U.S. Congress representing a southern state; and the first black woman to offer the keynote address to the Democratic National Convention (1976).

A member of the House Judiciary Committee Hearings on Watergate, Barbara Jordan declared: "My faith in the Constitution is whole. It is complete. It is total. I am not going to sit here and be an idle spectator to the diminution, the subversion, the destruction of the Constitution." She was a national symbol and spokesperson for the principles of good government and high ethical standards in public service. Her constant advocacy for fairness and equality under law remains a beacon for all Americans. Exemplifying the best traits of a public servant, Barbara Jordan stands as a role model for people everywhere. In 1994, President Clinton awarded Barbara Jordan the Presidential Medal of Freedom, the highest civilian award in the land.

Barbara Jordan joined the faculty of the Lyndon B. Johnson School of Public Affairs at the University of Texas in 1979. She is the recipient of honorary doctorate degrees from twenty-five colleges and universities including, Texas Southern, Princeton, and Howard Universities and the Tuskegee Institute.

Chapter 1 - Overview

Contemporary College Algebra is designed to **educate students for the future, rather than train them for the past**. The goal is to empower students to use mathematics to quantify real-life situations-that is, to mathematically model real-life situations. The process involves helping students become exploratory learners who will question, experiment, rely on technology for computation, and treat every mistake as a stepping-stone to greater understanding. The objective of the Queries, Exercises, Group Activities, and Group Projects in this text is to help students gain conceptual understanding, not computational mastery of rules and techniques. Two important components in any program to educate students for the future are the development of their communication skills (reading, writing, listening, presenting) and of their experience working as a team member in a small group. These components are addressed through exercises presented as word problems, several writing exercises, and numerous small-group activities and projects that involve written and verbal reports.

Technology in the form of graphing calculators and computer algebra systems have profoundly changed undergraduate mathematics programs. Graphing calculators and computer algebra systems have freed students from hand computations, while providing them with powerful visualization opportunities. This, in turn, is fueling a reform in the goals, content, and pedagogy associated with college algebra.

The content of this text is designed to aid the student in developing "mathematical life skills" to answer the following categorical questions.

How to present data?

How to transform data into information?

How to transform information into knowledge?

Technology introduced the Information Age by providing the means for storing huge quantities of data that could be quickly referenced and displayed in tables, charts, and graphs, as well as in written form. The rate of technological advancement for handling data is growing exponentially. For example, today's Internet offers possibilities for accessing data that most people would not have comprehended only a few years ago. The ability to *present and to interpret data* in tables, charts, and graphs are necessary skills for today's students and are the starting points for the *Contemporary College Algebra* course.

Data is transformed into information through the identification of relations linking data values. For example, linear relations (those that can be plotted on a straight line) are characterized by having a constant rate of change represented by the slope of the line. These relations are the simplest and most prevalent of all the relations used to extract information from data. Understanding that constant change can be represented by the slope of a straight line and that a straight line is completely characterized by a point and a slope is fundamental to the appreciation and application of mathematics. Sections 2.6, "Discovering Relations Between Variables," and 2.7, "Applications of Linear Equations," illustrate the universality of linear relations.

The most important relations in mathematics are called functions (the subject of Chapter 3). These are characterized as input - output relations in which the output is dependent on the input. Thus they are also referred to as dependency relations. For example, the amount of interest received on a given investment over a fixed period of time is dependent on (is a function of) the interest rate. The following graph shows the amount of interest received, for yearly interest rates from 0% to 20%, when $1,000 is invested for one year.

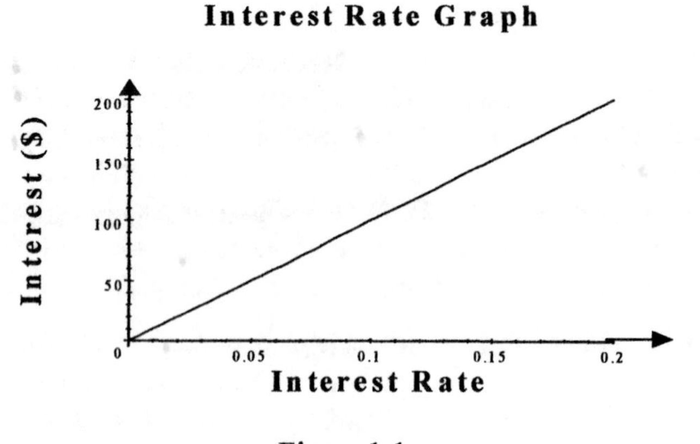

Figure 1.1

Functions are graphed with the input values measured along the horizontal axis and the corresponding output values measured along the vertical axis. There are only five basic functions used in this text: power (polynomial), radical, exponential, logarithm, and the periodic functions (sine and cosine). Of course, there are infinitely many combinations of these basic functions. The shapes of the graphs of these functions are called the basic shapes (see Sections 3.3–3.6).

Fitting a curve to a set of data points is a common method for *transforming data into information.* The functional description of the curve can then be used to make predictions based on the data. We primarily use the graphical "conjecture and test" method for fitting a curve to data points. Under this method, one plots the data points, recognizes the basic shape of the data points, then conjectures and plots a function whose graph has the same basic shape. Coefficients in the conjectured function expression are then adjusted in order to create a better fit. Usually four or more adjustments need to be made before an acceptable fit is obtained (see Section 3.6).

Modeling is used to *transform information into knowledge.* A mathematical model of a real-world situation is a mathematical description of the situation that enables one to analyze the situation under different scenarios. Modeling is a primary theme that runs through this text. In Chapter 2, the most basic form of a mathematical model, the straight line (linear model), is developed. In Chapter 3, real-life situations are modeled through the process of collecting and plotting data and then fitting a curve to the data plot. (The function whose graph fits the data is the model.) Three approaches to curve fitting are developed. The first is the iterative, graphical approach of "conjecture and test." Regression analysis is the second approach. This is a symbolic method that provides the best fit in the sense of minimizing the sum of the squares of the distances between the data points and the curve. Regression algorithms are preprogrammed into most graphing calculators. The third method gives a polynomial function whose graph contains the data points. This method is generally used only when there are few data points, say four or fewer (see Section 3.7). In Chapter 4, numerical models in the form of tables of iterations and symbolic models based on recursive sequences (also called discrete dynamical systems) are developed to model real-life situations in a broad spectrum of fields including business, the physical sciences, the life sciences, and the arts. The recursive models are developed on the reasoning paradigm

(New Situation) = (Old Situation) + (Change)

The following real-life situation, discussed in Chapter 4, illustrates the process of developing a recursive sequence model.

> Malcolm has "maxed out" his credit card with an accumulated $5,000 of debt on his card. The credit card company charges 1.5% monthly interest (18% annual rate). Malcolm resolves to pay off his debt by paying $100 a month and not charging any additional items to his credit card. He asks you to determine how many months it will take him to pay off his debt. You begin by creating a recursive sequence model of the situation based on the paradigm (**New Situation**) = (**Old Situation**) + (**Change**). Time is measured in months and the **Change** in Malcolm's situation during the nth month is (interest charged minus payment). Thus we have
>
> $$b(n) = b(n-1) + (\text{interest charged} - \text{payment})$$
>
> where $b(n)$ represents the balance of the debt after the nth month. Hence
>
> $$b(n) = b(n-1) + 0.015b(n-1) - 100$$
>
> or, collecting terms,
>
> $$b(n) = (1.015)b(n-1) - 100$$
>
> with $b(0) = 5,000$. In words, this model says that the balance at the end of the nth month is equal to the balance at the end of the $n-1^{st}$ month plus the interest charged during the nth month minus the $100 payment. We will see in Chapter 4 that an analytical solution to this model is
>
> $$b(k) = (1.015)^k(-1666.6667) + \tfrac{100}{0.015}.$$
>
> The debt is paid off when the balance is zero, that is when $b(n) = 0$. Using a calculator, we determine that $b(n) = 0$ when $n = 93.11105$. Thus it will take Malcolm 93.11105 months or about seven and three-quarter years to pay off his credit card debt, with his last payment being $11.17. ▲

Note that each of the important components in the previous problem—account balance, interest rate, monthly payment, and number of months—is explicitly shown in the model. This is characteristic of a model. One can investigate other scenarios by making appropriate changes in these components. For example, how long would it take Malcolm to pay off his debt if he paid $200 per month? Or, if Malcolm's grandmother agreed to pay off the balance of his debt after he kept his resolution for one year, how much would she have to pay?

Figure 1.2 illustrates the modeling process.

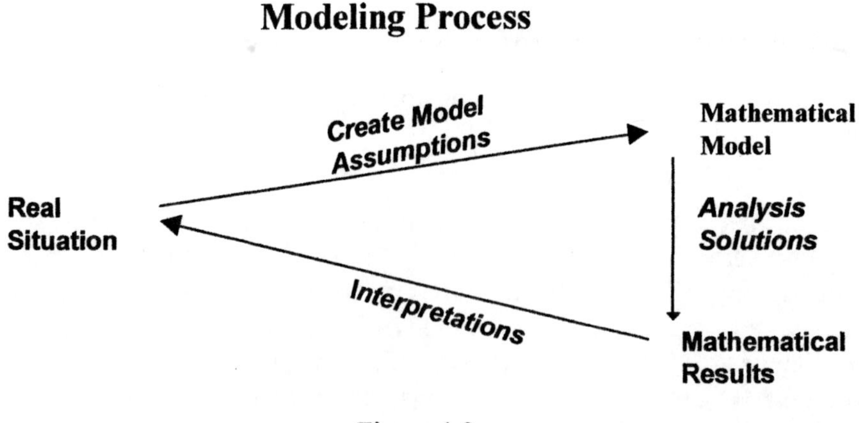

Figure 1.2

Real-world situations are usually too complicated to be modeled exactly. Thus simplifying assumptions are made in order to construct a model. For example, in Malcolm's situation we assumed that each month had the same number of days. Mathematical solutions need to be interpreted in light of the real-world situation. In Malcolm's situation, n represents the number of months. Thus even though $n = 293.11105$ is a correct mathematical solution to the model, it is not a real-world solution. Interpreting the mathematical solution in the real-world situation means that n must be rounded up to the next higher integer and the amount of the last payment recomputed.

Technology, in the form of a graphing calculator, has opened up numerous possibilities for the college algebra student to mathematically model and analyze real-life situations. The objective of *Contemporary College Algebra* is to prepare the student to do just that (**and to have a good time doing it**).

Jackie Robinson

Jackie Roosevelt Robinson, the first African American to play baseball in the major leagues, was born January 31, 1919, in Cario, Georgia. To cope with prejudice and gain respect, Jackie immersed himself in sports. A gifted athlete, he excelled in football, basketball, baseball, and track while in high school. He was the first student at the University of California at Los Angles to earn four varsity letters in one year.

Brooklyn Dodgers: 1947–1956
Baseball's Hall of Fame (1962)
Career batting average: .311,
19 Career Steals of Home,
Rookie Year: average .297,
12 homers, stole 29 bases,
"Rookie of the Year."
Most Valuable Player Award
(1949), Batting Title (1949),
Av/season: 75 walks,
Av/season: 29 strikeouts

After his discharge from the Army in 1945, Jackie coached basketball at Huston (now Huston-Tillotson) College. The following year he joined the Kansas City Monarchs of the National Negro League, and then the Brooklyn Dodgers', where he played for the Montreal Royals. Branch Rickey, President of the Brooklyn Dodgers, said that he chose Jackie for his fierce pride and determination as well as for his excellent athletic and ball-handling abilities. These were the attributes that characterized Jackie during all of his endeavors and were the source of his many successes on and off the field of athletic competition. Branch Rickey warned Jackie that he could expect abuse from both fans and players and that he would need to "turn the other cheek." When Jackie asked Branch if he wanted a ballplayer who lacked the courage to fight back, Branch responded, "No, I want a ballplayer who has the courage not to fight back."

Jackie Robinson overcame the extreme abuse and vindictiveness directed toward him by outstanding achievements on the ball field, achievements that eventually won him the respect of all persons. Competing and playing to win by the rules were his hallmarks, and his accomplishments have been symbols of inspiration and hope to millions of young people.

> "Life is not important except in its impact on the lives of others. The feeling that an individual who is committed and will persevere can make a difference is part of his legacy." Rachel Robinson

Chapter 2 Data and Variables

Technology has thrust us into the Information Age. The exploding volume of information that is easily accessible threatens to overwhelm every discipline, industry, and business. From matchmaking to population figures for myrid census categories, from the *Encyclopedia Britannica* to airline schedules, from recipes to sports statistics, we are increasingly exposed to data. How do we filter this data to recognize authentic data from bogus data, to determine relevent data from nonrelevant data? How do we display data?

2.1 Displaying Data

The tearing down of the Berlin Wall symbolized a major change in world economics as well as in politics. Dismantling trade barriers, promoting free trade, and linking economies through foreign investments form the basis for today's global economy. Over the past two decades foreign investment in the U.S. economy has reached the $800 billion level. The following table, plots, and charts illustrated the rise in foreign investment in Texas during the decade 1987–1996. The state of Texas ranks second, behind California, in terms of foreign investment.

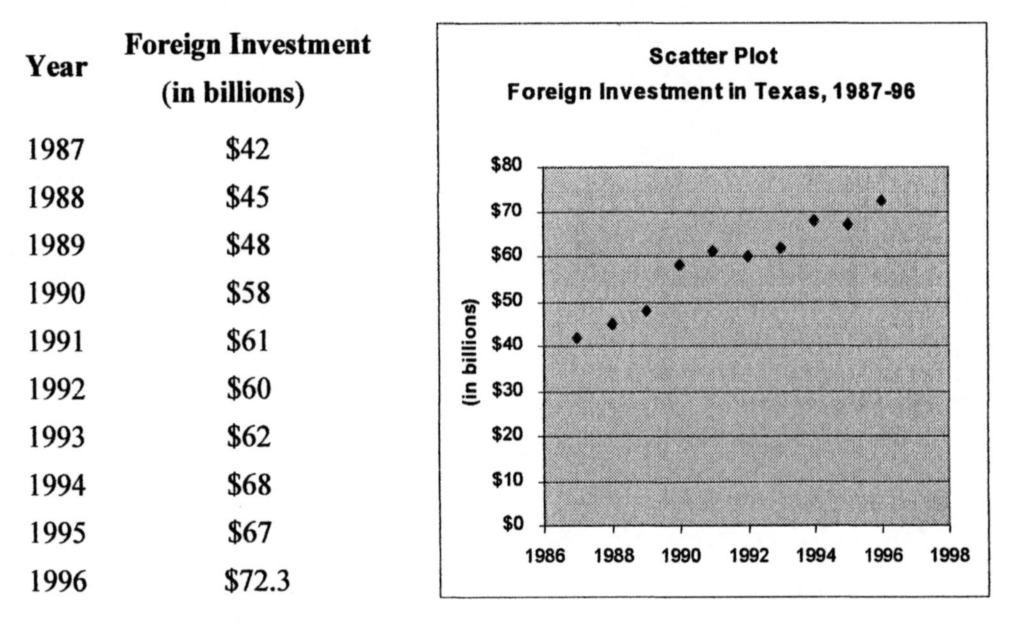

Year	Foreign Investment (in billions)
1987	$42
1988	$45
1989	$48
1990	$58
1991	$61
1992	$60
1993	$62
1994	$68
1995	$67
1996	$72.3

Table 2.1.1 Figure 2.1.1

(Source: Carole Keeton Rylander, Texas Comptroller of Public Accounts,Texas Department of Economic Development and United States Department of Commerce.)

The table, scatter plot, and bar chart present the data as though all of a year's investment occurred at the end of that year rather than throughout the year. Thus the data presents an approximation to the actual picture. The lines between the points in the following line chart suggest that the investments occur throughout the year and thus it would be reasonable to interpolate between data points.

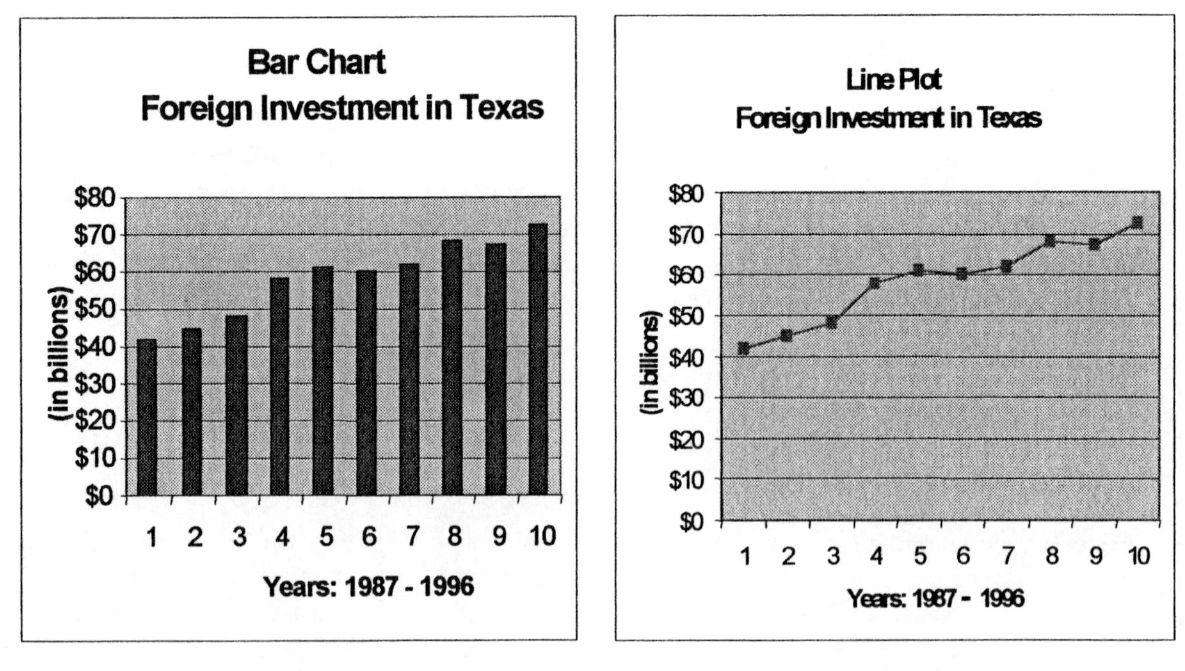

Figure 2.1.2 Figure 2.1.3

The table and pie chart show the affiliate employment for Texas in 1996. Affiliate employment is employment in a United States enterprise in which there is direct foreign investment. (The source is the same as that for the foreign investment in Texas.)

Enterprise	Affiliate Employees	Percent Enterprise Employees of Total Affiliate Employees
Petroleum	30,000	9.5
Wholesale Trade	31,000	9.8
Other Industries	34,000	10.7
Services	52,000	16.4
Manufacturing	128,000	40.4
Real Estate	2,000	0.6
Retail Trade	40,000	12.6
Total	317,000	100.0

Table 2.1.2

The percentage for each enterprise is computed by dividing the number of employees for that enterprise by the total number of affiliate enterprise employees.

For example, the percentage of affiliate services employees of the total affiliate employees is $\frac{52{,}000}{317{,}000} = 16.4\%$

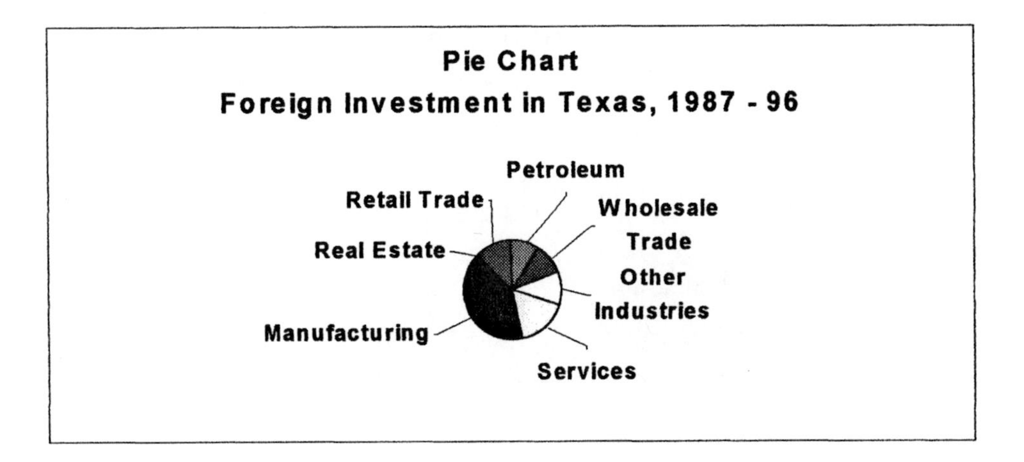

Figure 2.1.4

The tables, plots, and charts represent five different ways of displaying data.

Why do we need different ways of displaying data? The answer is that different forms of displaying data emphasize different aspects of the data. For example,

Tables display data in a factual format. Height-weight charts, tax tables, sports reports (e.g., league standings, batting averages), and census data are a few examples of how tables are used to display data.

Pie charts display data in pictorial form. This method is particularly useful when attempting to portray percentage comparison between the entries. The area of a pie slice represents that entry's percentage of the total amount. Thus the area of the pie slice representing service industry's affiliate employment is 16.4% of the total affiliate employment in Texas in 1996. Note that it would not make sense to portray the data on foreign investment by a pie chart because it would not be meaningful to speak of the investment for 1996 as a percentage. (Percentage of what?)

Pie charts are frequently used in the popular press and news magazines. For example, budgets or the percentage of food basket components are usually presented as pie charts. Market share percentages of competing companies are often displayed as pie charts. For example, the *New York Times* (February 13, 1999) estimated that in 1988 Coke held a 44.5% share and Pepsi held a 31.4% share of the total soft-drink market.

Shares of the Soft-Drink Market, 1988

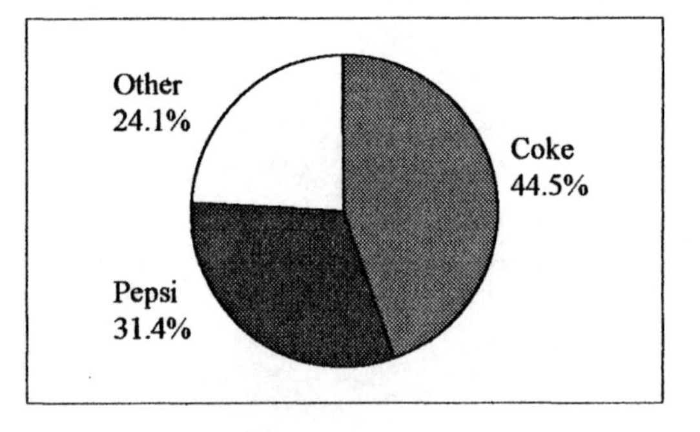

Figure 2.1.5

Query 1.

If the total United States soft-drink market in 1998 was valued at \$56.3 billion, what was the value of Coke's share?

Query 2.

In a pie chart, what can be said about the ratio of the length of the curved edge of a particular pie slice to the total circumference of the circle? (The length of the circumference of a circle of radius r is $2\pi r$.)

Bar charts display data in a manner that emphasizes comparison among the entries. The height or length of a bar in a bar chart represents the numerical value of the corresponding entry. The heights of the bars provide a visual comparison of the sizes of the corresponding entries. Bar charts, also called column charts, are frequently used in the popular press. For example, sizes of populations, amounts of exports, and numbers of drilling rigs are often displayed using bar charts.

Scatter plots display two variable data in a graphical manner. One variable is measured along the horizontal axis and the other variable is measured along the vertical axis. Scatter plots provide a graphical comparison of data. They are frequently used as a first step in statistical analysis of data relating two values. In Chapter 3, we will learn how to "fit a curve" to scatter plots that have unique data values.

Line plots are used when changes in the event in question are continuous, and thus a finite set of data points only approximates the true situation. In these cases, the lines connecting data points indicate the continuous nature of the event as well as indicating trends. For this reason, line plots are often used in presenting financial data covering a period of time. For example, the recent history of the Dow Jones stock market average is usually displayed as a line plot as are several other financial indices. Note that it would not have made sense to draw a line plot for the affiliate employment data as there are exactly seven distinct categories.

A line plot is obtained from a scatter plot by connecting pairs of points that have adjacent *first* components (in the set of first components). Thus the line plot of the three data points $\{(1,5),(2,1),(3,5)\}$ looks like a "V" with the points $(1,5)$ and $(2,1)$ joined (because 1 and 2 are adjacent first components) and the points $(2,1)$ and $(3,5)$ joined (because 2 and 3 are adjacent first components).

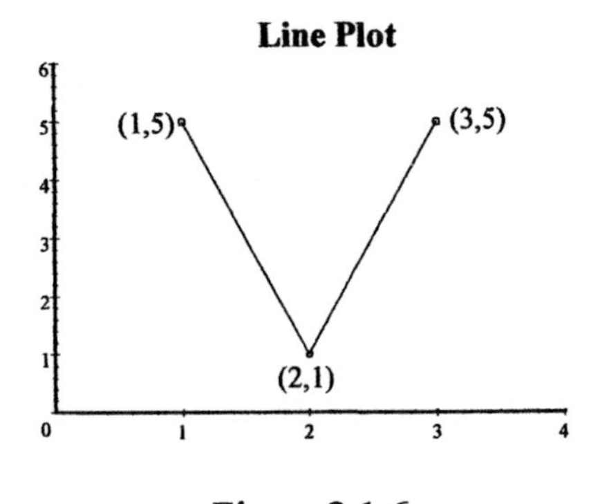

Figure 2.1.6

Note that the points $(1,5)$ and $(3,5)$ are not joined (because 1 and 3 are not adjacent in the set of first components) even though they are the closest pair of points. The first component often represents an *ordered* variable such as time.

For another example of displaying data, we consider the Top Ten Home Run Hitters of all time, the most prestigious grouping in baseball. What is the best way to display the membership in this club?

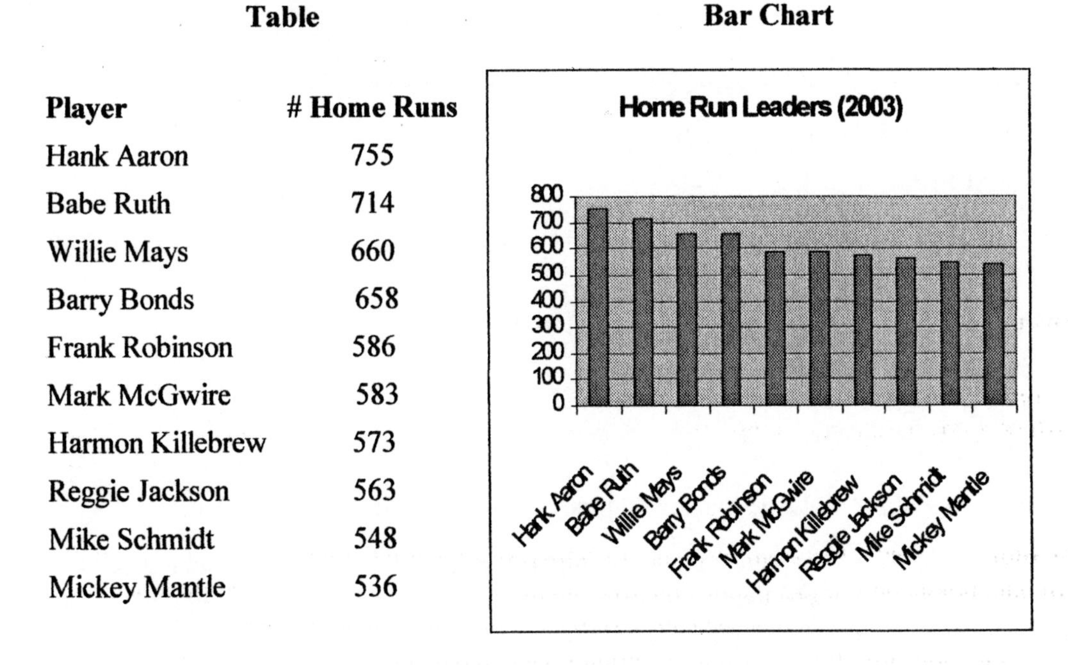

Table

Player	# Home Runs
Hank Aaron	755
Babe Ruth	714
Willie Mays	660
Barry Bonds	658
Frank Robinson	586
Mark McGwire	583
Harmon Killebrew	573
Reggie Jackson	563
Mike Schmidt	548
Mickey Mantle	536

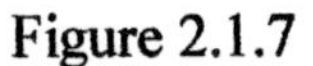

Figure 2.1.7

Note that it would not make sense to display this data in a pie chart because we would not be interested in showing that Hank Aaron's 755 home runs was 12% of the total number of home runs hit by the ten players that make up this prestigious group. Nor would it make sense to use a line plot to display this data because the ten players are distinct and thus a line joining two of the data points would have no meaning. Would it make sense to display this data in a scatter plot? If so, why?

The point we are making is that sometimes data represents discrete situations, such as the members in the Top Ten Hitters. Other times data represents a discrete approximation of a continuous situation, such as foreign investment in Texas. Data representing discrete situations is displayed using tables, scatter plots, and bar charts, but not line plots. (Why?) Data approximating continuous situations is displayed using tables, scatter plots, and bar charts, as well as line plots. Pie charts should only be used when it makes sense to speak of an event as a percentage of the sum of events.

Data and the analysis of data play a prominent role throughout this text. In this section, we will become comfortable displaying data with tables, bar charts, scatter plots, and line plots. In section 2.5, we will learn how to form pie charts when we revisit the topic of displaying data.

Query 3.

Which form of graphing is the best way to display the grades in a college algebra course? Explain, listing the assumptions you made.

Query 4.

Locate data relating to one of your interests, display the data, and make up a problem based on the data.

Graphing calculators have the capabilities for displaying tables, bar charts, scatter plots, and line plots. Spreadsheets also have these capabilities and, in addition, can display pie charts.

Exercises 2.1 (The use of graph paper will facilitate working exercises 4–10.)

1. (Computational Skill) Compute the following.

   a. 10% of 520

   b. 35% of 754

   c. 115% of 90

2. (Computational Skill) Determine the following percentages.

   a. 15 of 75

   b. 10 of 200

   c. 56 of 138

3. (Computational Skill) Make up and solve three exercises similar to those in Exercises 1a–c, 2a–c.

4. (Calculator Skill) Use a graphing calculator to plot the data on foreign investment given at the beginning of this section and then compare your plot against the plot given in the text. (Hint: Look up how to plot data in your calculator manual.)

5. Display the following data on the average value of homes in the United States using a bar chart, scatter plot, and line plot. The figures are corrected to 1990 dollars. The source is an Associated Press article in the *Bangor Daily News*, 26 June 1997.

Year	Average Value
1940	27,400
1950	39,900
1960	52,500
1970	57,300
1980	74,900
1990	79,100

Table 2.1.4

6. The average afternoon peak period freeway speed on freeways in Harris Country, Texas, for every third year between 1982 and 1994 is displayed in the following table (Source: 1996 *Houston Facts*). Display this data using a bar chart, scatter plot, and line plot.

Year	Miles per Hour
1982	38.3
1985	41.1
1988	45.6
1991	47.6
1994	49.0

Table 2.1.5

7. The data for 1997 Pell Grant Recipients, by income level, is displayed in the following pie chart. Display this data in both a table and a bar chart.

Pell Grant Recipants by Income Level, 1997

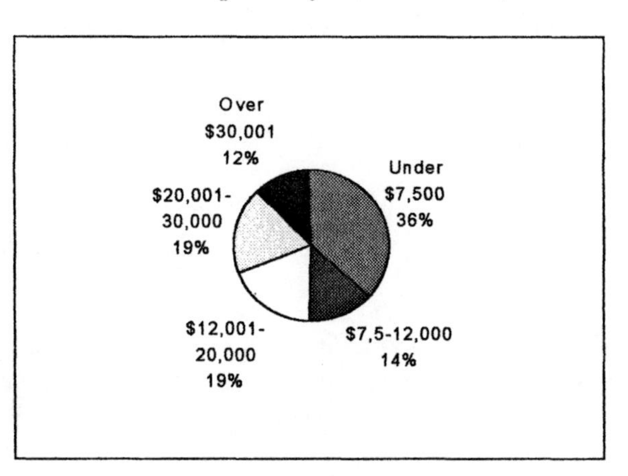

Figure 2.1.8

8. Form a data set whose entries are the chapters in this text and the corresponding number of pages in that chapter. Display this data set using a table, bar chart, and scatter plot.

9. Using newspapers and magazines as resources, find examples for each of the five forms of displaying data discussed in this section.

10. Form a table showing your typical day's activities and the amount of time spent on each activity (approximate your time in terms of 30-minute blocks). Display this data using a bar chart, pie chart, and scatter plot. Which of the five methods of displaying data would you use to describe to a friend how you spend your time? Explain.

11. The number of aircraft landings and takeoffs for the Houston, Texas Airport System in 1995 is given by category in the following table (Source: 1996 *Houston Facts*). Display this data using a bar chart and scatter plot

Houston, TX, Airport System

Category	**Landings & Takeoffs**
Air Carrier	408,855
General Aviation	195,247
Air Taxi	88,107
Military	25,606

Table 2.1.6

12. The following table gives the data on the heights of the six tallest buildings in Texas.

Building	**Height (feet)**
Chase Tower (Houston)	1,002
First Interstate Plaza (Houston)	972
Nations Bank Plaza (Dallas)	921
Transco Tower (Houston)	901
Renaissance Tower (Dallas)	886
Bank One Center (Dallas)	787

Table 2.1.7

For each of the five methods for displaying data that are discussed in this section, write an explanation of why that method would be appropriate or would be inappropriate for displaying this particular data.

13. Survey ten people for their height in inches and their weight in pounds. Use scatter plots to display the data obtained. Draw horizontal lines over your plots that show the average and the median of the data. (See Sections 2.2 and 2.3 for definitions of average and median.)

14. Collect temperature data for 30 days: mean, median, high, low. Draw a line plot for each of the four variables (mean, median, high, low) on the same set of axes. What observations can you draw from the four line plots? (See Sections 2.2 and 2.3 for definitions of average and median.)

2.2 Average (Mean)

Think about the various ways that people include mathematical ideas in their daily conversations. Some examples involving the concept of average are

- Your friends are talking about Jackie Robinson's batting average or maybe it is the average number of points the Rockets scored per game last year or maybe it is the average circulation of the *Atlanta Tribune*.
- On the TV, the weather forecaster is comparing today's temperature to the average temperature. The forecaster may also mention the average rainfall for the month.
- Mom is asking what your average is in math.
- Your sister is talking about the average lap speed at the Indianapolis Motor Speedway.
- Someone on the radio is talking about the Dow Jones average.
- A credit card official is explaining how to compute the average daily balance of an account.

Does the word average have the same meaning in all of the above instances?

The following bar chart displays the 1900–1990 population data by decades for Houston, Texas.

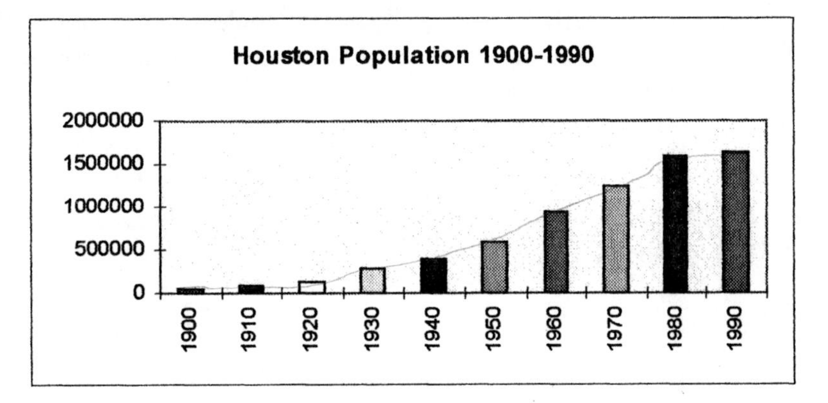

Figure 2.2.1

Draw a horizontal line through the bar chart to approximate the average population over this time span.

Query 1.

How does the weather forecaster compute the average temperature for a given day? How would you do it?

The **average of a numerical data set** is the number computed by adding all the data values together and dividing the resulting sum by the number of data entries.

$$\textbf{average} = \frac{\text{sum of data values}}{\text{number of entries}}$$

Example 2.2.1

Determine the average of the following set of test scores for a small class.

78, 75, 85, 63, 72, 88, 80, 92, 75, 94

Solution

The average test score is found by adding together the 10 test scores and then dividing by 10, the number of entries in this set: that is, the total number of students taking the test.

$$\frac{78+75+85+63+72+88+80+92+75+94}{10} = 80.2 \quad \blacktriangle$$

Note that the average value (80.2) is not one of the data values. This is usually the case. Note also that the number of data entries greater than the average is not equal to the number of data entries less than the average. This is also a typical result. One other interesting observation is that two students received the same grade (75), but this fact could not be determined by only knowing the average. The point is that although the average value is an important characteristic of a data set, it is not the only important characteristic.

Query 2.

Suppose in the preceding example, the class actually had 11 students, and one was sick and missed the test. What score would she have to earn on the makeup test for the class average to increase to 82? Explain your reasoning.

Query 3.

Hank Aaron had 3,771 hits in his 12,364 "at bats" during his 23 year baseball career. What was his lifetime batting average?

Query 4.

How is the average lap speed determined at a car race?

Query 5.

How is a student's grade point average determined?

Small-Group Activity

Develop a class presentation on generalizing the concept of average from a finite set of data to an infinite set. Include examples from at least four different settings (for example, average speed on a car trip). Answer the following two questions. (Hint: Distance = Rate x Time.)

a. On a 60-mile car trip, you average 40 mph over the first 30 miles and 60 mph over the last 30 miles. What is your average speed for the entire trip? (Answer: 48 mph.)

b. On a 60 mile car trip, you average 40 mph over the first 30 miles and 50 mph over the entire trip. What was your average speed over the last 30 miles? (Answer: 66.67 mph.)

Have you read or heard the word *mean* used in place of the word *average*? For example, the weather forecaster may say the *mean* temperature today was 76 degrees. Statisticians often use the term *mean* rather than *average*. The two terms have the same meaning, and we shall use the terms *average* and *mean* interchangeably.

Exercises 2.2

1. (Computational Skill) Determine the average for each of the following sets of numbers
   a. $\{3,7\}$
   b. $\{5,-2,10\}$
   c. $\{12,7,13,20\}$
2. (Computational Skill) Determine the value of x such that
   a. The average of 5 and x is 6.
   b. The average of $6, 10,$ and x is 5.
   c. The average of $5,4,6,7,$ and x is 6.
3. (Computational Skill) Make up and solve three exercises similar to Exercises 1a–c, 2a–c.
4. So far this semester your test scores have been 88, 75, and 80. If all of the tests are equally weighted, what score must you get on the fourth test to have a test average of 85?
5. This semester your (hour) test scores have been 72, 77, 81, and 86. If the weight of the Final Exam is twice the weight of an hour test, what score would you need to get on the Final Exam to have an 80 average for the course? An 85 average?
6. Suppose this semester you earn two As, one B, and two Cs. Assuming that all five courses were three credit courses, compute your grade point average. After answering, rework this exercise with the assumption that the B grade was earned in a five credit course and the other courses were four credits each.
7. Assume that you will have three one-hour tests this semester. Give five different examples of scores on these three tests that would yield a test average of 85.
8. So far this season you have hit safely 25 times out of 100 official "at bats." If you go "two for three" and then "three for four" in your next two games, what will be your batting average?
9. Annette drove 180 miles from Austin to Houston, Texas, in four hours. What was her average speed? Is it possible that Annette's driving speed never equaled her average speed? Explain.
10. Gene drove from Grambling, Louisiana, to Baltimore, Maryland, in eleven hours. If he drove 600 miles, what was his average speed? If Gene took four 20-minute rest stops, what was his average highway speed.?
11. The U.S. Postal Service published the following table of postal rates for a first-class letter on July 1, 1997. Determine the average cost of postage for a first-class letter within these six countries.

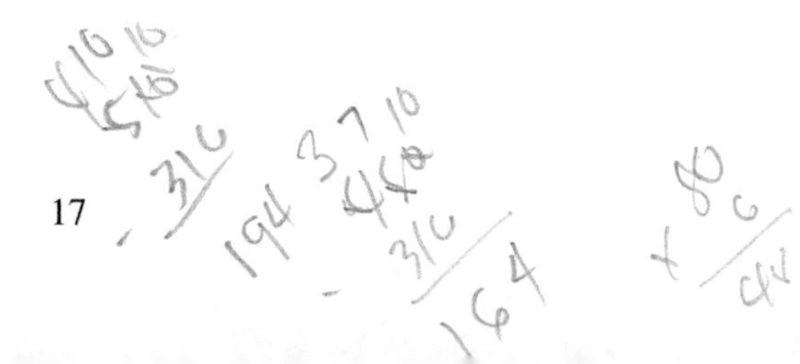

Country	**Postage (cents)**
Mexico	23
United States	32
Canada	33
Great Britain	43
Germany	57
Japan	70

Table 2.2.1

12. (*Small Group*) Develop a class presentation on the topic "weighted average." What does the phrase "weighted average" mean? For example, consider how your course grade may be determined by "averaging" the quizzes, homework, essays, hour tests, and final exam. Another example to consider is your grade point average. Is your grade point average an average of your grades or is it a weighted average of your grades? Make up an example of a grade report, including grade point average, of a student who takes five courses at your school. (That is, for each of the five courses give the course name, grade, number of credit hours, and number of grade points. Then compute the student's grade point average for that semester.) Make up two other examples of a weighted average.

13. (*Small-Group, Discovery*) Determine and verify the following.

a. Under what conditions is the average of a set of consecutive even integers a member of the set?

b. Under what conditions is the sum of a set of even integers equal to the average of the set times the number of elements in the set?

Hint: Experiment by summing and averaging several sets of even integers. Repeat the exercise for odd integers.

14. The following table gives the Decennial Census Population for Houston, Texas, from 1850 through 1990 (Source: 1996 Houston Facts).

Year	**Population**	**Year**	**Population**
1850	2,396	1930	292,352
1860	4,845	1940	384,514
1870	9,332	1950	596,163
1880	16,513	1960	938,219
1890	27,557	1970	1,233,605
1900	44,633	1980	1,595,138
1910	78,800	1990	1,630,533
1920	138,276		

Table 2.2.2

a. Compute the average population from 1900 through 1990. How does your computed average compare to the average you represented by drawing a horizontal line on the bar chart at the start of this section?

b. Compute the average population from 1850 through 1990. Explain why this number is less than the number you computed in Part a.

15. Some teachers omit the highest and lowest test scores before computing the test average. Write a paragraph explaining the rationale for such a practice and then give your reasons for supporting or objecting to this practice.

16. Answer each of the Queries in this section.

17. Two math classes take a test. The average is 70 in one class and 80 in the other class. What circumstance would guarantee that the average for both classes together is 75?

18. Alex bought desserts for 10 people, including himself. Four of the desserts cost $3 a piece, three desserts cost $2 a piece, and three desserts cost $1 apiece. He told his friends, "the desserts cost $1, $2, and $3. So the average cost is $(1+2+3)/3 = $2. If each of you gives me $2, we will be even."

    a. What is the average price of the desserts that were ordered?

    b. Did Alex gain, lose, or break even?

19. A fast-food restaurant employs ten people in addition to the manager. If the manager receives a salary of $75,000 and the average salary of the other ten workers is $17,000, what is the average salary for the eleven employees?

2.3 Median & Mode

The following two tables display apprehension data of the St. Louis Police Department, 1921–1948, by category of crime (Source: L. Haskins, K. Jeffery, *Understanding Quantitative History*, McGraw-Hill, 1990).

Average Annual Apprehension Numbers

Period	Disorder	Morals	Major Crimes	Suspicion	Other
1921–27	1292	2049	1235	1586	1157
1928–34	3437	4558	1668	2262	4249
1935–41	2840	2514	899	2422	1206
1942–48	1897	1311	725	1855	1431

Table 2.3.1

Average Annual Percentage of Total Apprehensions

Period	Disorder	Morals	Major Crimes	Suspicion	Other
1921–27	17.7	28.0	16.9	21.7	15.8
1928–34	21.3	28.2	10.3	14.0	26.3
1935–41	28.7	25.4	9.1	24.5	12.2
1942–48	26.3	18.2	10.0	25.7	19.8

Table 2.3.2

A striking observation from the first table is that the number of apprehensions in 1928–34 is much larger than in the other years in all of the categories except suspicion. However, this information cannot be seen in the second table because the row entries are percentages, not averages. Although both tables use the same underlying data, they present the data in different ways to emphasize two different aspects. The first table emphasizes the (column) comparisons of average apprehensions between year periods and the second table emphasizes (row) percentages. (Example. For the 1921-27 period there were 1292 people apprehended for Disorder. This number is 17.7% of the total number of persons apprehended in the 1921-27 period. This percentage, 17.7, is listed in the first row of the second table under the column heading Disorder.) This is an example of how a presentation of data to emphasize one aspect often disguises other aspects.

Another example of a data presentation that emphasizes one aspect at the cost of other aspects is given in the following table. We can determine that the average amount Texas received in crop subsidies for the period 1987–1996 was $585,400,000 per year (Sources: John Sharpe, Texas Comptroller of Public Accounts, United States Department of Agriculture and Environmental Working Groups).

Year	Payment(millions)	Year	Payments(millions)
1987	$932	1992	$529
1988	$782	1993	$801
1989	$706	1994	$454
1990	$490	1995	$192
1991	$483	1996	$481

Table 2.3.3

The mean or average value ($585 million) does not give very much information as to the distribution of the crop subsides over the ten-year period. Nor would only knowing that the smallest subsidy was $192 million and the largest subsidy was $932 million tell us that for five years the subsidies ranged from $454 million to $529 million. Knowing that the subsidies were more than $509.5 million for five years and less than that for the other five years does not tell us that in two years the subsidies were almost the same, $481 and $483 million. A question these comments raise is: How can a set of data be summarized?

How to summarize a data set is an important and difficult question. Suppose you had a listing of all of the ages of the students in your class. Could you then summarize that information with one numerical value? Would that number tell you how many students were the same age or would it tell you the average age or would it divide the class into two equal groups, one of whose ages was greater than your value and one whose ages was less than your value?

Large Group Activity: Summarizing Data

Represent the age data for the individuals in the group in the form of a table, scatter plot, and bar chart. Summarize your data with one numerical value. What information is gained when your summary value is used to represent the data? What information is lost when your summary value is used to represent the data? Develop a different set of data for which your method of summarizing the age data would not give an appropriate summary value. Develop an appropriate summary value for your new data. Present your results to the rest of the class explaining your reasoning in selecting your summary values.

Statisticians have developed several different ways of summarizing data. Each of the methods highlights an important characteristic of the data in question. However, each method often hides other characteristics. Thus it is important to clearly understand what characteristics a summary value highlights and what characteristics it hides. In this text, we shall restrict ourselves to three common summary values: *average* (*mean*), *median*, and *mode*.

The terms *average* and *mean* represent the same summary value, and the terms are used interchangeably. However, the *median* and *mode* summary values are very different.

The **median** of a set of numerical data is the middle number when the data is listed in numerical order. If the set contains an even number of entries, the median is the average of the two midmost numbers.

In order to compute the median, first arrange the data in numerical order (for example, smallest to largest). If the number of data entries is odd, then the median is the middle number in the ordered list. If the number of data entries is even, the median is the average of the two midmost numbers in the ordered list.

Example 2.3.1

To compute the median of the 10 test scores:

$$\{94, 75, 85, 75, 68, 92, 80, 75, 88, 78\},$$

we first arrange them in numerical order

$$\{68, 75, 75, 75, 78, 80, 85, 88, 92, 94\}$$

and then average the two midmost numbers, $\frac{78+80}{2} = 79$. Thus 79 is the median of this set of data. What is the average of the test scores? Is it the same as the median? ▲

The median is usually different from the average (mean). Note also that the median value (like the average value) does not indicate how many students scored 75 on the test.

Query 1.

Make up a set of ten numbers whose median is larger than its average.

A **mode** of a data set is an entry or entries that occur as many times or more times than any other entry in the set.

For the ten test scores listed earlier, 75 is the mode. Note that when the data is arranged in numerical order, it is easy to determine the mode.

Note that the definition of mode implies that if a data set has no repeated entries, then every element is a mode of the set. For example, each entry in the set $\{2, 4, 9\}$ is a mode of the set.

A mode for a data set does not provide any information on the average or median. In fact, the smallest number in the set could be a mode.

Query 2.

Determine the mode(s) for the fictitious data set of cereal sales illustrated with the following bar chart.

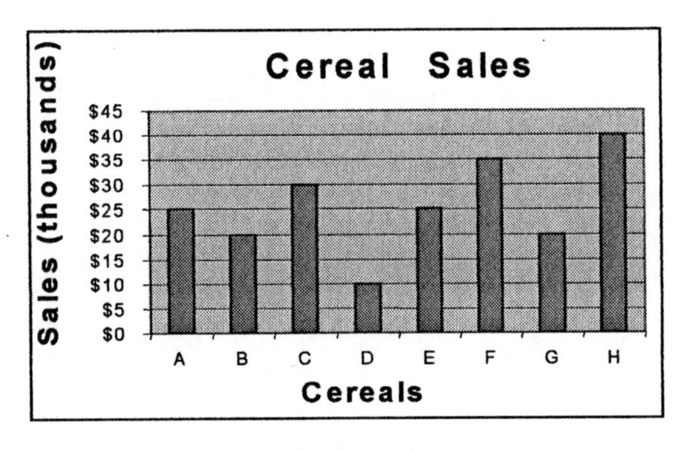

Figure 2.3.1

Query 3.

Can a data set have more than one mode? Why? If it can, make up a set of ten numbers that has two modes.

The average and median provide information concerning the center or equilibrium point of a data set. (Think of a set of data values distributed along a seesaw. If the fulcrum were at the average value, then the seesaw would be in balance. A corresponding image for median of a data set having at most one element at the median is the fulcrum of the Scale of Justice pan balance, with number of data elements less than the median on one pan and the number of data elements greater than the median on the other pan.) Does the word *median* in the highway traffic sign "Stay Off The Median" have the same meaning as the mathematical definition?

The mode(s), on the other hand, provide "clustering" information. If a data set represents views of several persons, a mode value could indicate a basis for building consensus. A retailer who surveyed potential customers for their preferences would be more interested in the mode(s) of the resulting data than in either the average or the median.

Query 4.

Why do reports on personal incomes of a population usually refer to the median income rather than the average income?

Query 5.

Does social pressure force one to conform toward the median, average, or mode behavior of the group? Explain

Query 6.

Is the average income in the United States greater than the median income? Why?

The summary values—average (mean), median, and mode—are often used to "represent" a set. For example, class averages are sometimes used to compare classes. One must be careful, however, to be aware of special circumstances that can distort a summary value such as an **outlier**, a data entry that lies unusually far from the main body of the data.

Exercises 2.3

To apply a "what-if" technique to an exercise means to modify the original exercise after it has been worked and then solve the modified exercise. The "what-if" technique is very effective in extending thinking and enforcing understanding. Apply the "what-if" technique to each of the exercises 4–14 by adding an eleventh data value while maintaining the stated restrictions.

1. (Computational Skill) Determine the median and mode for the following sets of numbers.
   a. $\{1,2,3,4,20\}$
   b. $\{5,-5,4,3,-2,5\}$
   c. $\{5,1,17,6,-2\}$
2. (Computational Skill) Determine a value of x such that the
   a. The median of $\{x,1,7,3,10\}$ is 5.
   b. The mode of $\{x,1,7,3,10\}$ is 7.
   c. The mean of $\{-4,5,-10,7,12,-4,\ x\}$ is 5.
3. (Computational Skill) Make up and solve three exercises similar to Exercises 1a–c, 2a–c.
4. Make up a data set of 10 entries for which the median and the mode are given by the same entry.
5. Make up a data set of 10 entries for which the median and the average are given by the same entry.
6. Make up a data set of 10 entries for which the mode and the average are given by the same entry.
7. Make up a data set of 10 entries for which the average, median, and mode are given by the same entry.
8. Make up a data set of 10 entries for which the mean is 4.
9. Make up a data set of 10 entries for which the median is 7.
10. Make up a data set of 10 entries for which the mode is 9.
11. Make up a data set of 10 entries for which the mean is 6 and the mode is 8.
12. Make up a data set of 10 entries for which the mean is 8 and the median is 5.
13. Make up a data set of 10 entries for which the median is 7 and the mode is 9.
14. Make up a data set of 10 entries for which the mean is 7, the median is 8, and the mode is 3.
15. For each of the 3 summary measures (average, median, mode), describe a situation in which that measure is the most appropriate one to use. Explain what you mean by "most appropriate."
16. Write a half-page essay on the limitations for summarizing data using a numerical average.
17. Write a half-page essay explaining the image of the average being the fulcrum or "balancing" point of a data set (seesaw) and the image of the median as the fulcrum of the Scale of Justice pan balance.

18. Write a paragraph explaining the purpose of a golfing or bowling handicap and how the handicap is computed.

19. Ask the manager of a shoe store to give you data showing the approximate number of pairs of shoes of the different sizes that the store stocks. Compile a bar chart and determine the median, mode, and average of the shoe sizes. (Consider only the lengths, not the widths.) Write a statement explaining what you think influenced the manager to maintain the described inventory.

20. An intuitive definition of an *outlier* in a data set is a data entry that lies unusually far from the main body of the data set. For example, 10 is an outlier in the data set $\{10, 50, 53, 54, 55, 60, 65\}$. Answers to the following questions illustrate the effect of outliers on averages.

    a. Compute the average of the data set $\{9, 13, 14, 18, 20\}$. Add an outlier to the data set in order to make the average 21.

    b. Compute the average of the data set $\{9, 10, 14, 25, 26\}$. Add an outlier to the data set in order to make the average 5.

    c. What effect, if any, does an outlier have on the mode or median of a data set? Explain.

21. A class shows that green is the class's favorite color. Is this an average, median, or mode? Explain.

22. The values of five houses in Cornwall are \$100,000, \$120,000, \$130,000, \$120,000, and \$500,000.

    a. What is the average price of these houses?

    b. What is the median price of these houses?

    c. Which price (average or median) is more representative of the group? Explain.

23. Which of the summary values (average, median, mode) should be used to represent the following data?

    a. The salary of four salesman and the owner of a small store.

    b. The height of all students in Jack Yates High School.

    c. The most popular dress size sold at Joan's apparel shop.

24. A study of a group of people with respect to weight, height, and age is summarized with the following statements. For each statement, draw one or more conclusions concerning the group and then write a paragraph justifying your conclusion(s).

    a. The average weight is less than the medium weight.

    b. The average height is larger than the medium height.

    c. The medium age is greater than the average age.

2.4 Variable Representation

The study of relations linking variables is at the heart of mathematics. Forming these types of relations is the primary method for extracting information from data. The most important type of these mathematical relations are dependency relations, which are called **functions.** These are relations in which one variable is dependent on the other variable(s), such as in a cause-effect relation. The understanding of functions is a central focus of this text. In a dependency relation, one variable is the dependent variable and the other variable(s) is (are) the independant variable(s). The analysis of the following questions illustrate the roles of the dependant and independant variables. Dependency relations or functions are often encountered in answering "what" or "why" or "how much" questions. Consider the following.

Question 1:

What determines the price of a pizza?

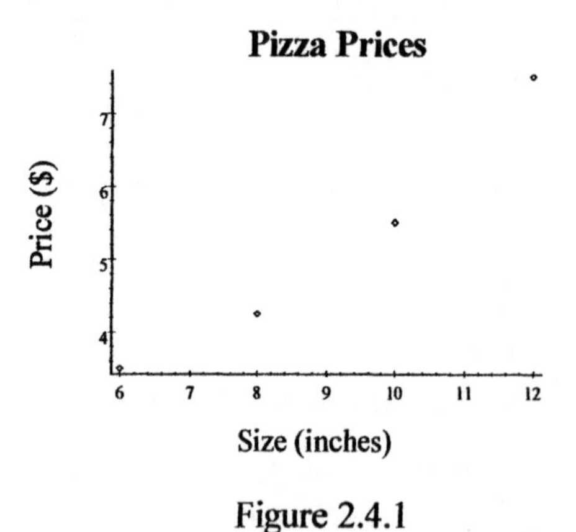

Figure 2.4.1

If our answer is size, then we would say that the price of a pizza is dependent on the size of the pizza. In this context, "size" and "price" are called variables. In the dependency relation linking these variables, size is the independent variable and price is the dependent variable.

Question 2:

What is your ideal weight (according to the following chart)?

Height/Weight Chart

Height (inches)	60	61	62	63	64	65	66	67	68	69	70	71	72	73
Weight (pounds)	112	116	120	124	127	130	133	137	141	146	149	155	157	159

Table 2.4.1

Height and weight are the variables for this question. If we designate height as the independent variable, this chart would define a dependency relation. Note that the chart would define a different relation if weight was designated as the independent variable.

Question 3:

Suppose Linda borrows $8,000 to buy a car and agrees to pay off her loan with equal monthly payments over 4 years. What determines the amount of her monthly payment?

If our answer is the interest rate, then we would say that the monthly payment is dependent on the interest rate. In this context, "interest rate" and "monthly payment" are called variables. In the dependency relation linking these variables, interest rate is the independent variable, and monthly payment is the dependent variable.

Question 4:

What are the characteristics of leadership?

If our answer is commitment, knowledge, and ability to communicate, then we would say that a person's leadership ability is dependent on the person's commitment, knowledge, and ability to communicate. In this context, "commitment, knowledge, ability to communicate, and leadership ability" are called variables. In the dependency relation linking these variables, commitment, knowledge, and ability to communicate are the independent variables, and leadership ability is the dependent variable.

The dependency relations in Questions 1–4 are examples of **functions.** This text is designed to provide students with opportunities to develop a conceptual understanding of variables and functions.

Activity

Make up an example of a dependency relation that has two independent variables and one dependent variable.
Hint: Consider a chart that has columns for age, average height, and average weight.

Remark

Arithmetic *is the study of numbers, operations on them, and relations between them. In contrast,* **algebra** *is the study of variables, operations on them, and dependency relations between them.*

Mathematics is the language of science. As such it must have objects that correspond to pronouns. What is a mathematical pronoun? (What is a pronoun?)

Webster's Unabridged Dictionary, 2nd edition, defines **pronoun** as "A word that may be used in place or as a substitute for a noun." Words like you, she, he, it, we, and they are pronouns. Identify the pronouns in the following quote about Jackie Robinson taken from Jules Tygiel's *Baseball's Great Experiment* (p. 190).

Robinson's impressive statistics revealed only a portion of the tale. "Never have records meant so little in discussing a player's value as they do in the case of Jackie Robinson," wrote Tom Meany. "His presence alone was enough to light a fire under his own team and unsettle his opponents." Sportswriter John Crosby asserts, "He was the greatest opportunist on any kind of playing field, seeing opportunities before they opened, pulling off plays lesser players can't even imagine." Robinson's intense competitiveness provided the crucial ingredient. A seasoned athlete, even in his rookie year, Robinson seemed to thrive on challenges and flourished before large audiences. At Montreal the preceding year, Dink Carroll had observed, "Robinson seems to have the same sense of the dramatic that characterized such great athletes as Babe Ruth, Red Grange, Jack Dempsey, Bobby Jones and others of that stamp. The bigger the occasion, the more they rose to it." Leo Durocher said "This guy didn't just come to play, he came to beat ya."

Do all of the pronouns refer to the same person or people? Does the use of pronouns in place of writing out the full names improve the readability of the writing?

Query 1.

Do you think that the following mathematical definition of variable qualifies variable to be called a mathematical pronoun? Why or why not?

Variables are symbols representing quantities that can take on two or more different values. The set of different possible values a variable can take on is called the **domain** of the variable.

In contrast to variable, a **constant** is a symbol that has exactly one value.

For example, in Question 2, the independent variable, height, can have any integer value between 60 and 73, thus the domain of the variable height is the set of integers from 60 to 73. In Question 4, the dependent variable, leadership ability, can have qualitative values like weak, strong, and forceful. Thus the domain of the variable, leadership ability, contains the terms weak, strong, and forceful.

Mathematics, like every other discipline, uses symbols as shorthand for communication. For instance, the statement

The sum of seven and three is ten.

is more efficiently written using symbolic shorthand as

$$7 + 3 = 10$$

In **algebra**, letters such as x are often used as symbolic shorthand to represent an unknown variable.

Note the use of x in the problem based on the following Associate Press report.

"Through August 9 (1996) of this year, 350 cases of measles have been reported in the United States, a 38 percent increase over the same period in 1995, officials from the Centers of Disease Control and Prevention said."

This report raises the interesting question (or problem):

Example 2.4.1

How many cases of measles have been reported in 1995 by August 9?

Solution:

We begin by defining the unknown in the problem.

Let x represent the number of cases of measles reported in 1995 by August 9.

The statement: "350 cases of measles (is) a 38 percent increase over the same period in 1995" means that

x plus 38 percent of x is equal to 350.

Thus we may describe the relationship in the number of cases of measles reported through August 9 in 1996 to the number in 1995 with the equation (symbolic shorthand)

$$x + .38x = 350$$

or

$$1.38x = 350.$$

Now dividing both sides of the equation by 1.38, we have

$$x = 253.623$$

which we round up to 254. Hence approximately 254 cases of measles were reported in 1995 by August 9. ▲

Query 2.

If the "38 percent increase" in the preceding problem had been a 38 percent decrease, how many cases of measles would have been reported in 1995 by August 9? (Answer: 565 cases.)

Query 3.

The label on a half-pint carton of 2% milk claims 38% less fat than whole milk. If a half-pint serving of 2% milk contains 5 grams of fat, how many grams of fat are there in a half-pint serving of whole milk?

The following problem provides a second example that illustrates using the letter x to represent an unknown quantity.

Example 2.4.2

Determine the length of the sides of a rectangular picture whose area is 100 square centimeters and one side is twice as long as the other side.

Solution:

We begin by defining the unknown for the problem. Because we are asked to find the length of the sides of the rectangle and we know that one side is twice as long as the other side, we let the length of the shorter side be our unknown quantity.

Let x represent the length of the shorter side.

Then $2x$ is the length of the longer side. Because the area of a rectangle is the product of the lengths of the 2 sides, we can construct the equation $100 = (x)(2x)$ and then solve for x.

100	$=$	$(x)(2x)$	
50	$=$	x^2	divide both sides by 2
$5\sqrt{2}$	$=$	x	take square root of both sides

The dimensions of the picture are $5\sqrt{2}$ cm by $10\sqrt{2}$ cm. ▲

Would it have made a difference if we had let x represent the length of the longer side? Explain.

Our third example that illustrates using the letter x to represent an unknown quantity involves a special type of sequence, called a recursive sequence. A **recursive sequence** is a listing of numbers such that each number, after the initial number(s), is determined by one or more of the preceding numbers.

Example 2.4.3 (recursive sequence):

The sequence 0, 1, 3, 7, 15, 31, _, _, is a recursive sequence. Each number, after the first, is equal to two times the previous number plus one. The first number, 0, is the initial number. What is the seventh number of this sequence? (Answer: 63) ▲

We define an addition sequence as a recursive sequence with the following properties.

- The first two numbers are the initial numbers.
- Each number after the initial numbers is the sum of the two preceding numbers.

Example 2.4.4 (addition sequence):

The four numbers 2, 3, 5, 8 start an addition sequence in which the first two numbers are the "initial" numbers, and each of the remaining numbers are determined by adding the two preceding numbers. (Thus the third number, 5, is determined by adding 2 and 3. The fourth number, 8, is determined by adding 3 and 5.) What is the fifth number? (Answer: 13) ▲

Example 2.4.5

Fill in the elements in the addition sequence in which the first number is 1 and the fourth number is 7.

Solution:

We begin by defining the unknown for the problem. Because the determination of the second number is the key to the solution, we let that number be the unknown.

Let x represent the second number.

We can now determine the numbers of the addition sequence in terms of the unknown x.

$$1, \quad x, \quad 1+x, \quad x+(1+x), \quad \_, \quad \_, \quad \_ \, .$$

Because the fourth number is 7, we have

$$\begin{aligned} x+(1+x) &= 7 \\ 1+2x &= 7 \\ 2x &= 6 \\ x &= 3 \end{aligned}$$

Thus $x = 3$ and therefore the first four elements of the addition sequence are

$$1, \quad 3, \quad 4, \quad 7 \, . \qquad \blacktriangle$$

Query 4.

How would you define a recursive sequence in which the numbers change by some multiplication formula? Give an example.

Unknowns are not always denoted by the letter x. We will sometimes use symbols that are suggestive of the quantities they represent. For example in Question 1, we might let p denote the price variable and c the cost variable. In Question 3, the letter r might be used to represent the interest rate variable and the letter p to represent the monthly payment. Note that the same letter can be used to represent different unknowns in different problems. To avoid confusion, the problem solver should always

Define the unknowns at the beginning of each problem.

Exercises 2.4

1. (Computational Skill) In the following, solve for the value of x that satisfies the equation.
   a. $2(1+x) = 3x$
   b. $(x+1)(x-1) = x^2 - x$
   c. $(x+1)^2 = x^2$

2. (Computational Skill) In the following, determine the value of b that satisfies the equation for the given value of x.
   a. $2(x+b) = 5x, \; x = 4$
   b. $(x+1)^2 = 2b, \; x = 3$
   c. $(x+b)(x-b) = -8, \; x = 1$

$a_1 = 2(a_1)+1$ – recursive formula

3. (Computational Skill) Make up and solve three exercises similar to Exercises 1a–c, 2a–c.

4. (Calculator Skill) Consider the equations (i) $x^2 - 6x + 5 = 0$, (ii) $2x^2 + 3x - 5 = 0$, and (iii) $p^3 - 2p = 5$.

   a. Use the Equation Solver in your calculator to solve for the value of the variable in each of the equations.
      Hint: Look up "solver" in your calculator manual. (For equation (iii), subtract 5 from both sides of the equation.)

   b. Plot each of the three equations. Use the Zoom feature on your calculator to approximate the points where the plots touch the x-axis. Compare these values to the solutions found using the "solver" function.

   c. Write a paragraph discussing whether or not the graphical method of part b is a legitimate method for approximating the solution of an equation.

5. (Small Group) Collect 5 sets of data from newspapers or magazines. For each set, define the variables involved, represent the data in two different ways, and list three questions concerning the data.

6. Solve *Query* 3.

7. Determine the missing elements in the following addition sequences. Is there only one answer for each part? (See Example 2.2.4.)

   a. 1, 2, _, _, _

   b. 1, _, _, 3, _, _, _

   c. 2, _, _, _, 7, _, _, _

   d. _, 3, _, _, _, 14, _, _, _

   e. _, _, _, _, 1, 2, _, _, _

8. Define a multiplication sequence as a recursive sequence in which the first two numbers are the initial numbers and each of the succeeding numbers is the product of the two preceding numbers. Determine the missing elements in the following multiplication sequences.

   a. 1, 2, _, _, _

   b. 2, _, _, 8, _, _, _

   c. _, 2, _, _, 9, _, _, _

   d. _, _, 2, _, _, 6, _, _, _

   e. 0, _, _, _, _, _, _, 0, _, _, _

9. (Fibonacci Sequence) The thirteenth century mathematician named Leonardo Fibonacci studied the population growth of rabbits. His rabbit problem that he included in his text *Liber abbaci* has become one of the most famous recursive sequences in the mathematics literature. The problem is as follows:

   Suppose rabbits breed in a way that each pair of rabbits, older than one month, in the population produces a pair of baby rabbits each month. No rabbits die or leave the population. A single pair of rabbits is born at the beginning of the year to start the population. Determine the number of pairs of rabbits at the end of year n for $n = 1,2,3,4,5,6$.

(Note that the pair that gave birth to the initial pair in the population is not counted in the population.)

10. Bertha and her husband have 3 children aged 2, 4, and 6. In 4 years, Bertha will be twice as old as the sum of the ages of her children. How old is Bertha now? (Clearly identify the unknown.)

11. Sandra can wash the windows in her house in half the time it takes Terry to wash them. It takes Doris one and one-half times as long to wash the windows as it does Terry. If it takes Doris four hours to wash the windows, how long would it take Sandra to wash them? (Clearly identify the unknown.)

12. Copy a short passage from a paper or magazine (e.g., *Sports Illustrated*) that uses the pronoun "she" to refer to different persons in the article. Write a half-page essay on how the pronoun "she" is used as a mathematical pronoun.

13. (Refer to Exercise 4.) Approximate the values of x that satisfy each of the equations shown in the following plots. Explain your reasoning. Check your results by substituting them into the equations.

Plot of $x^2 + 3 = 4x$ 　　　　　 Plot of $3x - 2^x = 0$

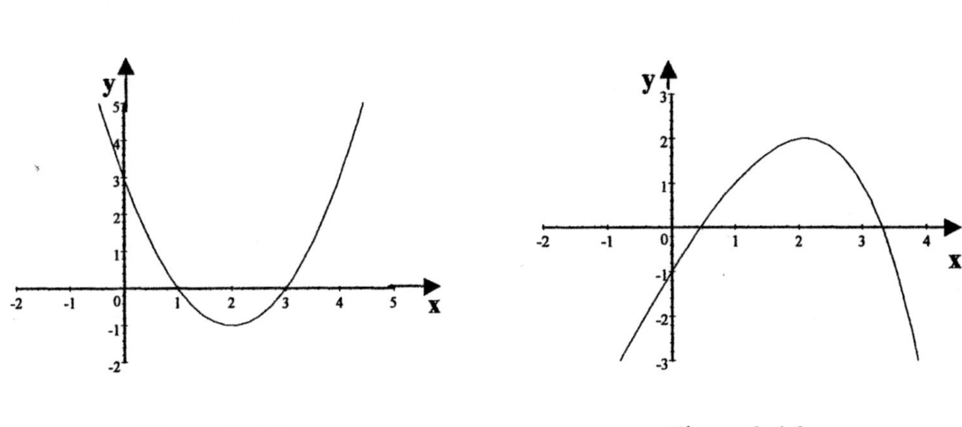

Figure 2.4.2

Figure 2.4.3

Historical Note: Henry Aaron said, as he and Frank Robinson were being inducted into the Hall of Fame in 1982,

> I'm proud to be standing where Jackie Robinson, Roy Campanella and others made it possible for players like Frank Robinson and myself to prove that a man's ability is limited only by his lack of opportunity.
>
> *Baseball's Great Experiment* by Jules Tygiel

2.5 Circle Properties and Pie Charts

Data: Data is information about a variable.

Data may be displayed in several ways as we saw in Section 2.1. The choice of how to display data depends on the aspect of the data to be emphasized. For example, numerical grades are usually presented in terms of percentages (or decimal equivalents) rather than the raw scores in order to make it easy to corollate the data with standard letter grades. Thus we often want to know the percentage associated with an entry. Before pursuing this, we pause and consider some terminology.

Terminology

Understanding terminology is crucial to understanding the topic in question. Because developing terminology is an evolutionary process with sources in different cultures and with different applications being considered, different words have evolved to refer to the same thing. This is unfortunate, but we accept it as a fact of life and will overcome the difficulties that new terminology presents through frequent and intentional use of the terminology.

- The words *average* and *mean* are used interchangeably.
- The words *entry* and *event* are used interchangeably.
- The **frequency** of an event is the number of times the event occurs. The phrases "frequency of an event" and "size of an event" are used interchangeably.
- The **relative frequency** of an event is the ratio of the frequency of that event to the total number of occurrences in the data set. The terms **relative frequency**, **percentage**, and **proportion** are used interchangeably. These can be expressed in decimal form such as 0.25 or in percentage form such as 25% or as fractions such as 1/4.

We illustrate these terms using data from a math contest that draws 80 youths from 10 different schools. The individual schools are the entries or events, the frequency or size of an entry is the number of youths attending from that school, and the relative frequency is the percentage or proportion of youths attending from a given school compared to the total number of youth attending the contest.

School	Frequency (# of Students)	Relative Frequency (Percentage, Proportion)
School #1	10	12.50%
School #2	5	6.25%
School #3	5	6.25%
School #4	10	12.50%
School #5	5	6.25%
School #6	8	10.00%
School #7	14	17.50%
School #8	10	12.50%
School #9	8	10.00%
School #10	5	6.25%
Total	**80**	**100.00%**

Table 2.5.1

Note that the vertical axis in the following bar chart represents the frequency or size of the groups attending from each school. In the pie chart, the area of each pie slice is equal to the relative frequency times the area of the circle (πr^2).

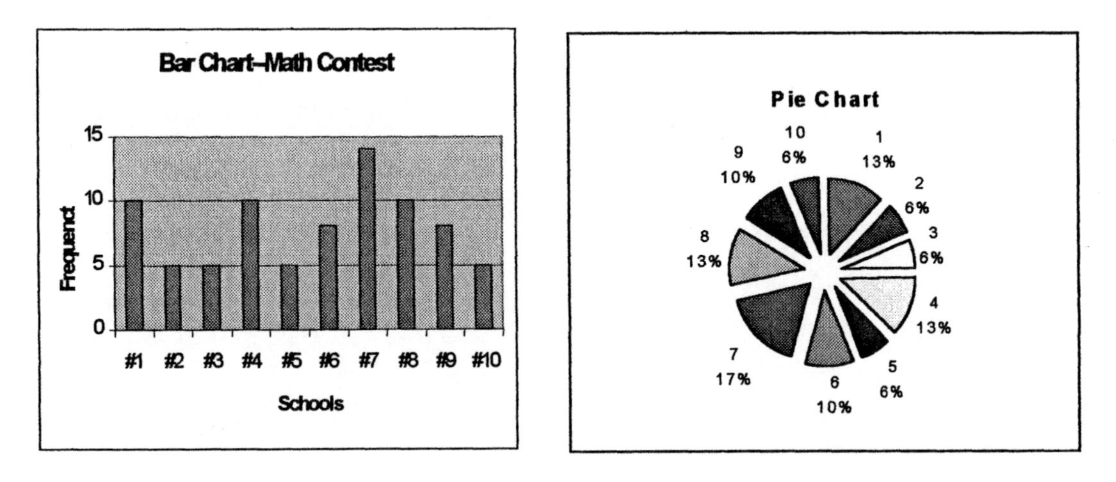

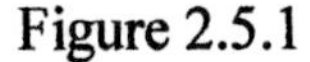

Figure 2.5.1

Figure 2.5.2

Constructing pie charts requires knowing how the measures of the central angles of the pie slices or the measures of the arcs of the pie slices relate to the relative frequencies. Understanding these relationships involves understanding the relationships between central angles, circumference arcs, and areas of pie slices. These relationships result from the following facts and proportions. (The central angles are measured in radians, not degrees.)

- *Area* of a circle of radius r is πr^2.
- Total central angle = 2π radians.
- The length of the *circumference* of a circle of radius r equals $2\pi r$.
- $$\frac{\text{area of pie slice}}{\text{area of circle}} = \frac{\text{central angle}}{\text{total angle}} = \frac{\text{arc length}}{\text{total circumference}}.$$
- The area of a pie slice of central angle θ and radius r equals $\frac{1}{2}r^2\theta$ because (from the first equality)

$$\frac{\text{area of pie slice}}{\text{area of circle}} = \frac{\text{central angle}}{\text{total angle}} = \frac{\theta}{2\pi}$$

so

$$\text{area of pie slice} = \frac{(\pi r^2)\theta}{2\pi} = \frac{1}{2}r^2\theta \quad \text{(cross multiply, simplify)}$$

- The arc length of a pie slice of central angle θ and radius r is $r\theta$ because (from the second equality)

$$\frac{\text{arc length}}{\text{total circumference}} = \frac{\text{area of pie slice}}{\text{area of circle}}$$

$$\frac{\text{arclength}}{2\pi r} = \frac{\frac{1}{2}r^2\theta}{\pi r^2}$$

$$\text{arc length} = r\theta \quad \text{(cross multiply, simplify).}$$

The relative frequency of an event is brought into the preceding results in the following manner. Because

$$\text{the relative frequency of an event} = \frac{\text{area of its pie slice}}{\text{total area of the circle}} = \frac{\frac{1}{2}r^2\theta}{\pi r^2} = \frac{\theta}{2\pi},$$

$$\theta = 2\pi(\text{relative frequency}).$$

For example, the central angle of the pie slice for School 6 in Table 2.5.1 is $2\pi(.10) \sim 0.63$ radians, and the arc length of this pie slice is $\sim 0.63r$. Why?

Example 2.5.1

USA Today in its December 21, 1999, issue gave the following data on the number of high school female athletes who participated in wrestling, football, and ice hockey in the 1998–99 school year

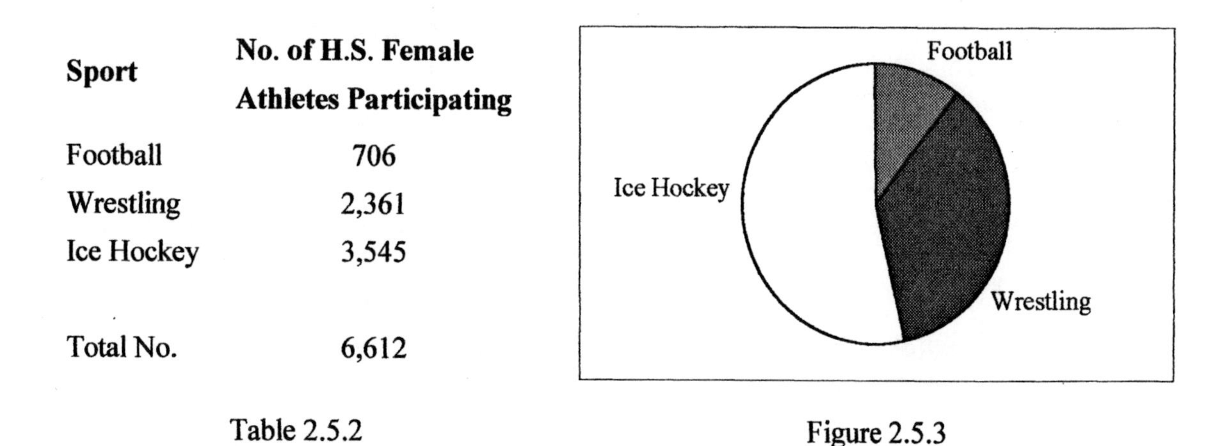

Sport	No. of H.S. Female Athletes Participating
Football	706
Wrestling	2,361
Ice Hockey	3,545
Total No.	6,612

Table 2.5.2 Figure 2.5.3

Assume that the radius of the circle in the pie chart is 1/2 inch. Determine the central angle, arc length, and area of the pie slice representing the number of high school female athletes who participated in wrestling during the 1998–99 school year.

Solution:

Because the central angle, θ, is equal to 2π times the relative frequency, the central angle for the "wrestling pie slice" is

$$\theta = 2\pi\left(\tfrac{2{,}361}{6{,}612}\right) = 0.7142\pi \text{ radians or } 128.55^\circ \text{ (degrees)}. \quad (1 \text{ radian} = \tfrac{180}{\pi} \text{ degrees})$$

Because the arc length of a pie slice is the radius of the circle times the central angle, the arc length of the "wrestling pie slice" is

$$\left(\tfrac{1}{2}\right)(0.7142\pi) = 0.3571\pi \sim 1.1219 \text{ inches}. \qquad (\pi \sim 3.1416)$$

Because the area of a pie slice is one-half the square of the radius of the circle times the central angle, the area of the "wrestling pie slice" is

$$\left(\tfrac{1}{2}\right)\left(\tfrac{1}{2}\right)^2(0.7142) = 0.8926 \text{ square inches.} \quad \blacktriangle$$

Conversion Between Radian and Degree Measure

There are two common ways of measuring the size of an angle: **degree measure** and **radian measure**. Consider the central angle of a circle. The amount of rotation required to move one side of the angle onto the other side is the **degree measure** of the angle. There are 360 degrees in the central angle of a full circle. The **radian measure** of the angle is the length of the subtended arc (portion of the circumference cut off by the angle) divided by the radius of the circle. Because the circumference of a circle of radius r is $2\pi r$ (as previously stated), the radian measure of the central angle of a full circle is $\dfrac{\text{length of circumference}}{\text{radius}} = \dfrac{2\pi r}{r} = 2\pi$. Thus there are 2π radians in the central angle of a circle and there are 360 degrees in the central angle of a circle, so

$$\mathbf{2\pi \text{ radians} = 360 \text{ degrees}}$$

or

$$\mathbf{1 \text{ radian} = \tfrac{180}{\pi} \text{ degrees and } 1 \text{ degree} = \tfrac{\pi}{180} \text{ radians}}$$

Note there are no units attached to radian measure because the units of measure of arc length and radius cancel. Thus radian is a "unitless" number. For this reason, when using radian measure we usually omit the words "radian measure."

Query 1.

How many radians are there in 45 degrees? How many degrees are there in π radians?

Query 2.

What effect does the length of the radius of a circle have on the measure (radian or degree) of a central angle? Explain.

Example 2.5.2

The number of aircraft landings and takeofs for the Houston, Texas, airport system in 1995 is given by category in the following table (source: 1996 *Houston Facts*). Display this data in a pie chart.

Houston, Texas Airport System

Category	**Landings & Takeoffs**
Air Carrier	408,855
General Aviation	195,247
Air Taxi	88,107
Military	25,606

Table 2.5.3

Solution:

We begin by computing the relative frequency or percentage for each category. Because the total number of landings and takeoffs is 717,815, the relative frequencies are

Category	**Relative Frequency**	**Pie Slice Angle:** $\theta = 2\pi$**(relative Frequency)**
Air Carrier	$\frac{408{,}855}{717{,}815} = 0.57$	$2\pi(0.57) = 3.581$ radians or 205.2 degrees
General Aviation	$\frac{195{,}247}{717{,}815} = 0.272$	$2\pi(0.272) = 1.709$ radians or 97.92 degrees
Air Taxi	$\frac{88{,}107}{717{,}815} = 0.123$	$2\pi(0.123) = 0.773$ radians or 44.28 degrees
Military	$\frac{25{,}606}{717{,}815} = 0.036$	$2\pi(0.036) = 0.226$ radians or 12.96 degrees

Table 2.5.4

The pie chart for landings and takeoffs at the Houston, Texas, Airport System is

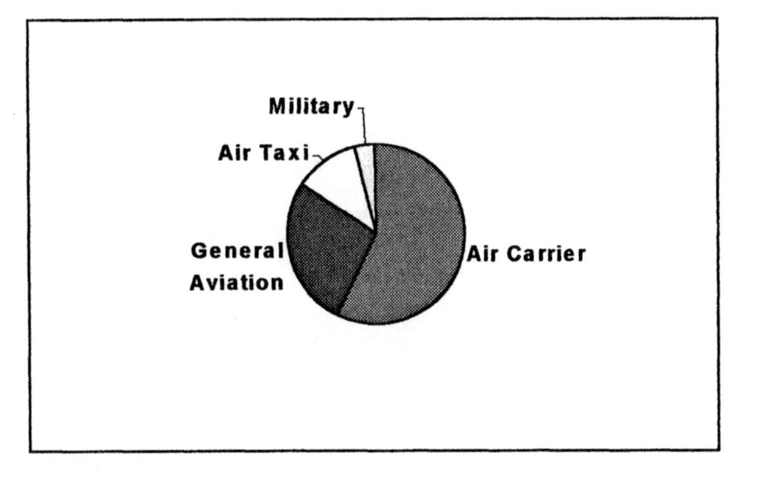

Figure 2.5.4

Small-Group Activity:

Instructor Jay's grade report for her algebra class listed 10 As, 25 Bs, 30 Cs, 20 Ds, and 10 Fs. Display this data in a three column table showing the grade, number of students receiving that grade, and the relative frequency of that grade.

Data on pairs of independent and dependent variables in a relationship are often represented by a table, as in a height/weight table, or by means of a graph. In a graph, the independent variable is measured along the horizontal axis and the dependent variable is measured along the vertical axis. A point on the graph represents an **ordered** pair of numbers. The first number in the ordered pair is the value of the independent variable and the second number is the value of the corresponding dependent variable. Thus to plot the ordered pair (2,3), move along the horizontal axis 2 units to the right from the **origin**, the point where the horizontal and vertical axes cross, and then move 3 units vertically (upward).

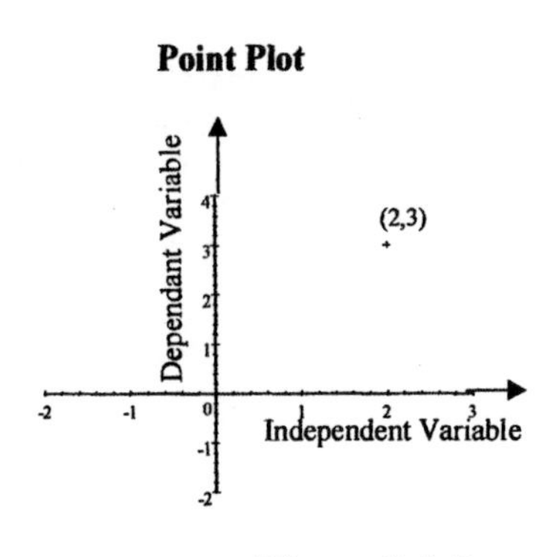

Figure 2.5.5

Thus the first component of a point on a graph represents the distance of the point from the vertical axis and the second component represents the distance of the point from the horizontal axis.

Small-Group Activity:

In Figure 2.5.5, plot and label the ordered pairs (1,2), (-1,2), and (3,-2).

The Family Economic Research Group, United States Department of Agriculture, estimate of the median cost of raising a child (in thousands of dollars) per year is described in tabular and graphical form as follows:

Age	Cost
1	6.1
2	6.5
3	6.9
4	7.8
5	8.3
6	8.8
7	9.2
8	9.8
9	10.3
10	10.7

Table 2.5.5

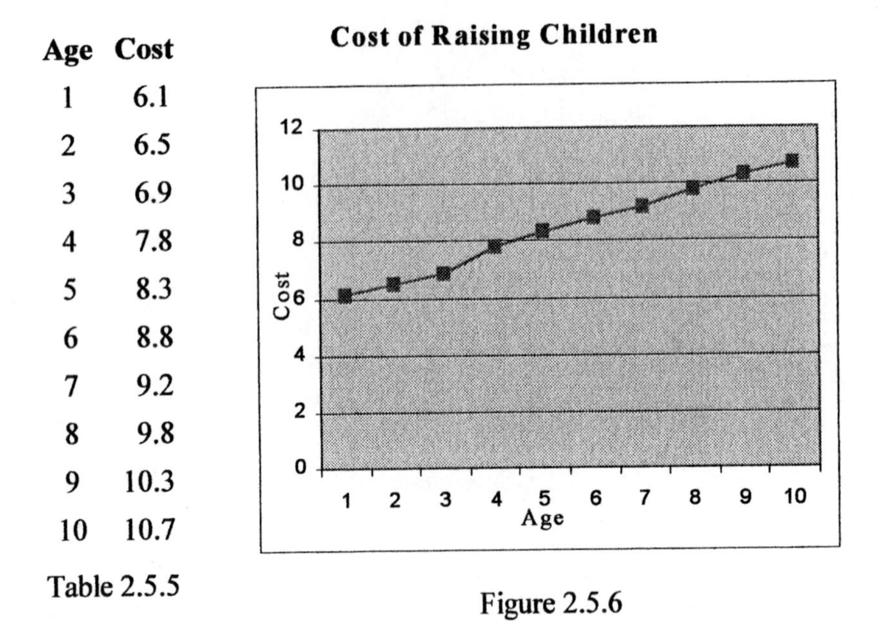

Figure 2.5.6

Small-Group Activity:

The coordinates of the last data point in Figure 2.5.6 are (10, 10.7). Using the tabular display of the data, label the coordinates of the other points in the graph.

Large-Group Activity:

List four different flavors of ice cream. Let each member of the group select his or her favorite from among the four flavors. Represent the data of the member selections in a pie chart and also in a bar chart.

Query 3.

What questions do you have concerning how data can be displayed?

Exercises 2.5

1. (Computational Skill)

a. Convert 90° to radian measure.

b. Convert 180° to radian measure.

c. Convert $\frac{2}{3}\pi$ to degree measure.

2. (Computational Skill)
   a. Determine the area of the "pie slice" whose central angle is 30° and whose radius is 3 in.
   b. Determine the area of the "pie slice" whose arc length is 4 in and whose radius is 3 in.
   c. If the area of the "pie slice" is 10 in^2 and its radius is 3 in, determine its central angle measured in radians and then in degrees.

3. (Computational Skill) Make up and solve three exercises similar to Exercises 1a–c, 2a–c.

4. (Calculator Skill) Form both a scatter plot and a line plot for the data given in Figure 2.5.6 on the cost of raising a child.
   Hint: Look up "plotting stat data" in your calculator manual.

5. The following figures refer to the number of vehicles per household in Houston in 1990. The data is from the United States Bureau of the Census, 1990 Census of Population, Summary Tape File 3A. Display this data using a bar chart, pie chart, and scatter plot

#Vehicles	#Households
0	75,590
1	270,866
2	204,429
≥ 3	65,902

Table 2.5.6

6. The adult age distribution in Austin, Texas in 1995 is given in the following table (source: Greater Austin Chamber of Commerce). Display this data in a pie chart.

Age	Percentage
18–24	20
25–34	29
35–44	22
45–54	11
55–64	8
65 +	10

Table 2.5.7

7. The "Austin Community Profile" published by the Greater Austin Chamber of Commerce (1996) gives the following data on household income. Display this data in a pie chart.

Household Income (dollars)	Percentage
< 20,000	36
20,000–25,000	9.5
25,001–35,000	16
35,001–50,000	16.5
50,001–75,000	13
75,001 +	9

Table 2.5.8

8. Central Pizza sells three sizes of pizzas: small, medium, and large. The small pizza is 6 inches in diameter and sells for $2.75. The medium pizza is 8 inches in diameter and sells for $4.30. The large pizza is 14 inches in diameter and sells for $9.00. Display this data in a two column table and then plot the data using size as the independent variable.

9. A 7-Eleven convenience store sells coffee in 8-, 12-, and 16-ounce cups. The respective prices are 65 cents, 85 cents, and one dollar. Display this data in a two column table and then plot the data using cup size as the independent variable.

10. The American Bagel Association reported (1996) the retail sales of bagels in millions from 1993 through 1996 as: $429 for 1993, $670 for 1994, $1,600 for 1995, and (estimated) $2,300 for 1996. Display this data in a bar chart and also in a line chart. Which method of displaying the data is best for you? Explain.

11. (Small Group) Survey your class to determine the eye color of each student. Display the resulting data in terms of a table, bar chart, and pie chart.

12. Determine the percent increase in 1996 in the number of bagel stores for each of the four companies listed in the following chart and then fill in the "percent change" column. The percent change in a data value measured at two different times is given by the quotient

$$\frac{\text{(New Value - Old Value)}}{\text{Old Value}} \text{ x } 100.$$

Company	December 1995	December 1996	Percent Change
Bruegger's	140 Stores	450 Stores	
Einstein/Noah	50 Stores	300 Stores	
Manhattan Bagel	146 Stores	287 Stores	
Don's Bagels	90 Stores	20 Stores	

Table 2.5.9

13. The Texas Department of Transportation (1996) listed the following information on the number of hours drivers waste each year sitting in traffic in the nation's most congested cities. Display this data in a bar chart.

City	**Hours Lost**
Washington, D.C.	70
San Francisco	66
Los Angeles	65
Houston	60
Detroit	57
Atlanta	53
Boston	44
New York	39
Chicago	34
Philadelphia	23

Table 2.5.10

14. Add a third column to the table in Exercise 13 showing the relative frequency.

15. The *New York Times* displayed the estimated gross receipts (in millions) for the 10 top films during the week of December 16–22, 1996, by the following chart. Add a third column to the chart showing the relative frequency and also display the data in a bar chart.

Title	**Gross**
Michael	17.8
Jerry Maguire	14.3
101 *Dalmations*	11.8
Beavis & Butt-Head	10.2
Scream	8.5
One Fine Day	8.1
The Preacher's Wife	7.7
Mars Attacks	5.1
My Fellow Americans	4.2
The Evening Star	3.3

Table 2.5.11

2.6 Discovering Relations Between Variables

(Only variables related linearly will be discussed in this section. Other types of relations, for example, parabolic, exponential, and trigonometric, will be discussed in Chapter 3.)

Dan is 2 years older than Travis. Therefore their ages are related by the equation

$$\text{Dan's age} = \text{Travis's age} + 2$$

If we define the variables, "D" to represent Dan's age and "T" to represent Travis's age, we can express the relationship between their ages by writing the equation (symbolic shorthand):

$$D = T + 2.$$

Knowing the relationship between their ages, we can easily construct a table of data on their ages.

Travis's Age	Dan's Age
1	3
2	4
3	5
4	6
5	7
6	8

Table 2.6.1

A more challenging problem is to start with a table of data and develop an equation describing the relationship between the variables. For example, develop an equation describing the relationship between the ages of Jonathan and Mike based on the following table of values.

Jonathan's Age	Mike's Age
2	5
4	7
5	8
7	10
10	13

Table 2.6.2

When given data on two related variables, it is always useful to plot the data. We do this, considering Jonathan's age to be the independent variable whose values will be measured along the horizontal axis and Mike's age to be the dependent variable whose values will be measured along the vertical axis. We first extend the table of age values to include a column of ordered pairs of age values, which we call **data points.**

Jonathan's Age	Mike's Age	Data Points
2	5	(2,5)
4	7	(4,7)
5	8	(5,8)
7	10	(7,10)
10	13	(10,13)

Note: The first element of a data point is a value of the independent variable and the second element is the value of the corresponding dependent variable.

Table 2.6.3

We now plot the data points.

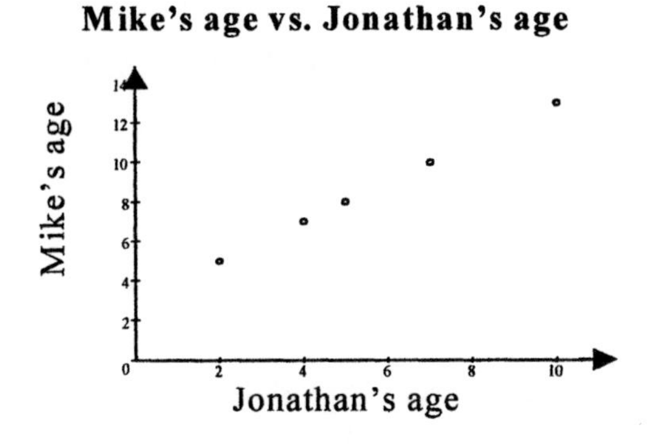

Figure 2.6.1

The points in the plot representing the data points appear to lie on a straight line. This suggests that the equation we want is the equation of the straight line passing through the points in the plot.

How do we determine the equation of a straight line?

A straight line is characterized by its direction (the measure of its rising or falling) and a point on the line. The direction of a line is called its **slope.** Two lines passing through a common point with different slopes are different lines. Likewise, two lines having the same slope but with no points in common are different lines. (Lines having the same slope are called parallel.)

Two Lines: Different Slopes, Common Point **Two Lines: Same Slope, No Common Point**

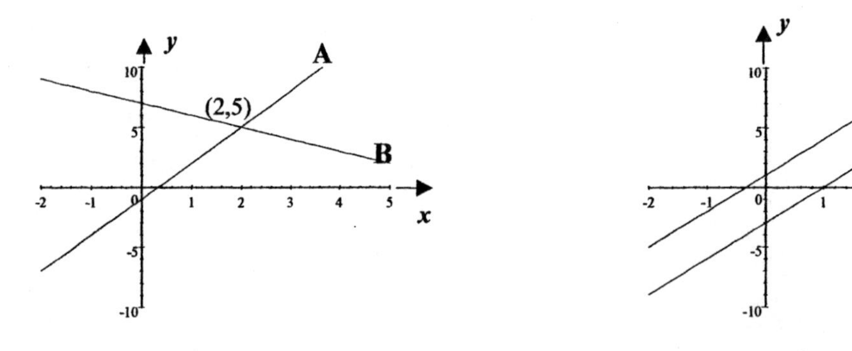

Figure 2.6.2

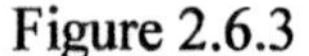

Figure 2.6.3

Query 1.

If two lines have two or more points in common, must they be the same line? Explain

Characterization of a Straight Line

A straight line is completely determined by its slope and a point on the line.

The **slope** of a straight line is also defined to be the **rate of change** of the dependent variable with respect to the independent variable. Thus the slope of a straight line is computed by dividing the change in the dependent variable by the corresponding change in the independent variable. Because the dependent variable is plotted on the vertical axis, the change in the dependent variable is a change in the **rise** of the line. Similarly because the independent variable is plotted on the horizontal axis, its change is a change in the **run** of the line. As a result, several texts describe the slope of a line by saying it is the **rise over the run**. That is, slope represents the steepness of a straight line.

The slope of a straight line is a constant (Why?). Thus any two points on a straight line can be used to compute the slope as shown in the following example.

Example 2.6.1

The slope of the straight line passing through the points (1,2) and (3,7) is

Slope Determined by Two Points

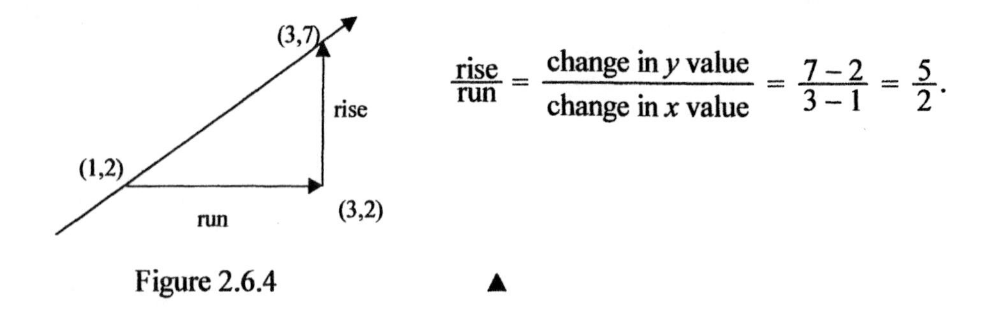

$$\frac{\text{rise}}{\text{run}} = \frac{\text{change in } y \text{ value}}{\text{change in } x \text{ value}} = \frac{7-2}{3-1} = \frac{5}{2}.$$

Figure 2.6.4

▲

The concept of slope occurs in several occupations, although with different names. The roofer uses the term *pitch* to describe the slope or steepness of a roof. The carpenter also uses the term *pitch* to describe the slope or steepness of a stairway. The highway engineer uses the term *grade* to describe the slope or steepness of a road. The physicist uses the term *average velocity* to represent the slope or average rate of change of a moving object over an interval. Economists describe slope as **marginal value**. These examples of how the mathematical concept of slope is applied in different fields illustrates the universality and applicability of mathematics.

Query 2.

When the ramp for a moving van is in place, the top of the ramp is level with the bed of the van and the bottom end rests on the road. Suppose a moving van is sitting on a level road and the bed of the van is 3 feet above the road. If the ramp extending out the back of the van is in place and the pitch (slope) of the ramp is $-\frac{1}{4}$, how far is the bottom end of the ramp from the van? (Answer: 12 feet) How long is the ramp?

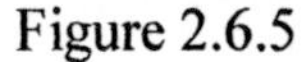

Figure 2.6.5

(See Appendix, Skill 8, for the formula of the distance between two points.) If the ramp were 10 feet long, what would be the pitch (slope)? (See Appendix, Skill 7, for the relationships between the lengths of the sides and the length of the hypotenuse of a right triangle.)

Query 3.

Federal regulations state that the maximum rise of a wheelchair ramp is one inch per (horizontal) running foot. What is the maximum allowable slope for a wheelchair ramp?

Returning to the relationship between Jonathan's and Mike's ages, we note that because each boy's age increases by one each year, the

$$\frac{\text{change in Jonathan's age}}{\text{change in Mike's age}} = 1.$$

Thus the slope of the line passing through their data points must be one.

We now compute the equation of the straight line representing the relation between Jonathan's age and Mike's age. First we define the variables. Let

$$\begin{aligned} \text{J} &= \text{Jonathan's age} \\ \text{M} &= \text{Mike's age.} \end{aligned}$$

We know that the slope can be computed using any two points on the line. Thus we let (J, M) denote an arbitrary point on the line and take any data point for the second point, say $(5, 8)$. The slope of the line can then be expressed as

$$\text{slope} = \frac{M-8}{J-5}.$$

Because we know from physical considerations that the slope is one, we can now solve for the equation of the line by setting the two different expressions for slope equal to each other.

$$\begin{aligned} \frac{M-8}{J-5} &= 1 & & \\ \text{So } M-8 &= (1)(J-5) & & \text{Multiply both sides by } J-5 \\ \text{or, } M &= (1)(J-5)+8 & & \text{Add 8 to both sides} \\ \text{or, } M &= J-5+8 & & \text{Multiply 1 times } J-5 \\ \text{or, } M &= J+3 & & \text{Add } -5 \text{ and } 8. \end{aligned}$$

The linear relationship between variables J and M determined by their table of values is $M = J + 3$.

Example 2.6.2

Determine the equation of the line with slope 3 that passes through the point (2,4).

Solution:

Let (x,y) denote an arbitrary point on the line. The slope is $\frac{y-4}{x-2}$ and the slope is also given to be 3. Thus

$$\begin{aligned} \frac{y-4}{x-2} &= 3. & & \\ \text{So } y-4 &= 3(x-2) & & \text{multiply both sides by } x-2 \\ \text{or, } y &= 3(x-2)+4 & & \text{add 4 to both sides} \\ \text{or, } y &= 3x-2 & & \text{simplify.} \end{aligned}$$

▲

Note that the **coefficient** of x, the number multiplying x, is the slope. Whenever the equation of a line is written in the **slope-intercept form**

$$y=mx+b$$

the coefficient m is the slope and the constant b is the y–intercept of the line. (The y-intercept is the y coordinate of the point where the line crosses the y-axis.)

Example 2.6.3

The line with equation $y = 3x + 2$ has a slope of 3 and a y-intercept of 2. ▲

Query 4.

What is the x-intercept of a straight line? What is the x-intercept of the line $y = 3x + 2$?

Example 2.6.4

The line with equation $2y = 5x - 6$ has a slope of $\frac{5}{2}$ because when the equation is written in slope-intercept form (that is, $y = \frac{5}{2}x - 3$), the coefficient of x is $\frac{5}{2}$. The y-intercept is -3 and the x-intercept is $\frac{6}{5}$. Why? ▲

We now reverse the question and ask, given the equation of a straight line, say $y = 2x - 1$, how do we obtain its graph?

<u>Method 1</u>. "Two-point method" for determining the graph of the line $y = 2x - 1$

Because a straight line is completely determined by two points, we will solve for the coordinates of two points on the line, plot those two points, and then draw a straight line through the two points. Let A be the point determined by setting $x = 2$ and then solving for y. That is, $y = 2(2) - 1 = 3$. Thus the coordinates of A are (2,3). In a similar manner, let B be the point with $x = 4$ and thus $y = 2(4) - 1 = 7$. So the coordinates of B are (4,7). We now plot the points A and B and draw a straight line through them.

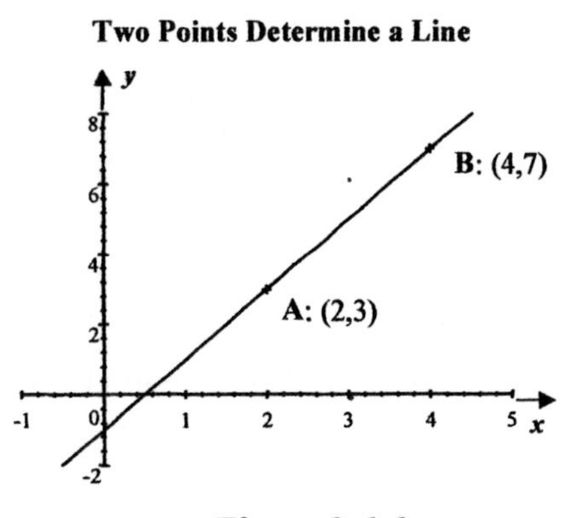

Figure 2.6.6

<u>Method 2</u>. "Point-slope method" for determining the graph of the line $y = 2x - 1$

Plot a point, say A, as in the previous method. Now determine a second point using the slope. Because the slope is 2 and 2 can be written as $\frac{2}{1}$, we move 2 units vertically upward (rise) from point A and then 1 unit to the right horizontally (run). The resulting point, (3,5), is on the desired line, which can now be drawn through these two points.

Line Determined by Point and Slope

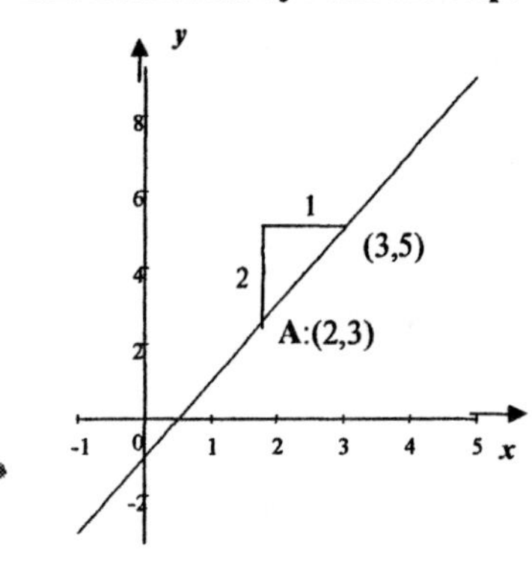

Figure 2.6.7

Explain why the slope = 2.

Query 5.

Redo the example illustrating Method 2, replacing the slope of positive 2 with a slope of negative 2. That is, $y = -2x - 1$. Does it make a difference if you write the slope $-2 = \frac{-2}{1}$ and thus move vertically downward 2 units (down is the negative direction) and then move horizontally 1 unit to the right, or write the slope $-2 = \frac{2}{-1}$ and thus move vertically upward 2 units and then move horizontally1 unit to the left (left is the negative direction)? Explain.

Example 2.6.5

Determine the equation of the straight line that passes through the points (1,2) and (5,3).

Solution:

Let (x,y) represent any point on the desired line. Because the slope of a line is determined by any two distinct points on the line, we have, using the points (x,y) and $(1,2)$,

$$\text{slope} = \frac{\text{rise}}{\text{run}} = \frac{\text{change in } y \text{ value}}{\text{change in } x \text{ value}} = \frac{y-2}{x-1}.$$

Now using the points $(1,2)$ and $(5,3)$, we compute

$$\text{slope} = \frac{\text{rise}}{\text{run}} = \frac{\text{change in } y \text{ value}}{\text{change in } x \text{ value}} = \frac{3-2}{5-1} = \frac{1}{4}.$$

Because there is one and only one slope for a given line, the two expressions for the slope of the desired line must be equal to each other. Therefore,

$$\frac{y-2}{x-1} = \frac{1}{4}.$$

Thus (cross multiplying) we have $4y - 8 = x - 1$ or $4y = x + 7$. Hence the equation of the desired line is

$$4y = x + 7 \text{ or } y = \tfrac{1}{4}x + \tfrac{7}{4}. \qquad \blacktriangle$$

Query 6.

If, in the previous example, the first pair of points were chosen to be (x, y) and the second pair to be $(5, 3)$, would the same equation for the line have been obtained? Why? If the second pair of points were chosen in the opposite order, would the same equation for the line have been obtained? Why?

Query 7.

A guy wire supports the telephone pole next to Don's house. One end of the wire is attached to the pole 2 feet from the top and the other end of the wire is attached to a ground anchor that is 12 feet from the pole. The pitch or slope of the wire is 2.

a. Draw a picture showing the pole, ground, and guy wire.
b. Assign variables to the appropriate parts of your picture.
c. Determine how far the top of the pole is from the ground. (Answer: 26 feet.)

Query 8.

Let (x_1, y_1) and (x_2, y_2) represent two points on a straight line and let m denote the slope of the line. How can you determine a formula that expresses m in terms of the coordinates of the two points?

Observations on a Straight Line

1. A straight line is completely determined by its slope and a point on the line. Any 2 points completely determine a straight line. (Why?)
2. The slope of a straight line is the average rate of change of the dependent variable with respect to the independent variable.
3. The **slope-intercept form** for the equation of a straight line is $y = mx + b$ where m is the slope and b is a constant.
4. Two variables are *linearly related* if their data points fall on a straight line.

Small-Group Activity (equipment: thermometer with Fahrenheit and centigrade markings):

Discover if there is a linear relationship between the Fahrenheit and centigrade (Celsius) temperature scales by forming a table of data values and plotting the data points. (If the points fall on a straight line, then the two scales are linearly related.) If the two temperature scales are linearly related, determine the equation of the line that passes through the data points.

Small -Group Activity (graphing calculator)

A. Discover what happens to the graph of a straight line when the value of the slope is changed. Write a short description and explanation and then prepare a five-minute class presentation. Hint: Select the equation of a straight line, say $y = 3x - 2$. Plot the graph. Superimpose on the same axes, the plots of four other lines obtained by changing only the slope value. The resulting plot is called a **multiplot**, that is, a plot made up of several plots. What can you say about the multiplot of the five lines? Repeat the hint with another line.

B. Discover what happens to the graph of a straight line when the value of the constant term (in the slope-intercept form) is changed. Write a short description and explanation and then prepare a 5 minute class presentation. Hint: Select the equation of a straight line, say $y = 3x - 2$. Plot the graph. Superimpose on the same axes, the plots of four other lines obtained by changing only the value of the constant term. What can you say about the multiplot of the five lines? Repeat the hint with another line.

Small-Group Activity "Line of Sight"

Designing locations for communication (relay) towers, aerial tramways, ski lifts, and air lanes all depend on establishing the equation of a line of sight. Suppose there is a section of the Rocky Mountains where the skyline is described by the graph of $y = -x^4 - 2x^3 + 12x^2 + x - 10$ for $-4 < x < 3$. The group's task is to determine the equation of the line (of sight) that just grazes the two mountain peaks. The group should designate one of its members to be the Plotter. This person plots the equation of the skyline. Then the other members take turns "guessing" the equation of the desired line until a suitable line is determined. After each guess, the Plotter superimposes the plot of the "guessed" line on top of the skyline plot. The next person describes the error(s) in the previous guess and then explains how to formulate a new guess that decreases the error(s). The Plotter then plots the new guess. This process continues until the group agrees that they have determined a suitable line. (The group must agree what "suitable" means.)

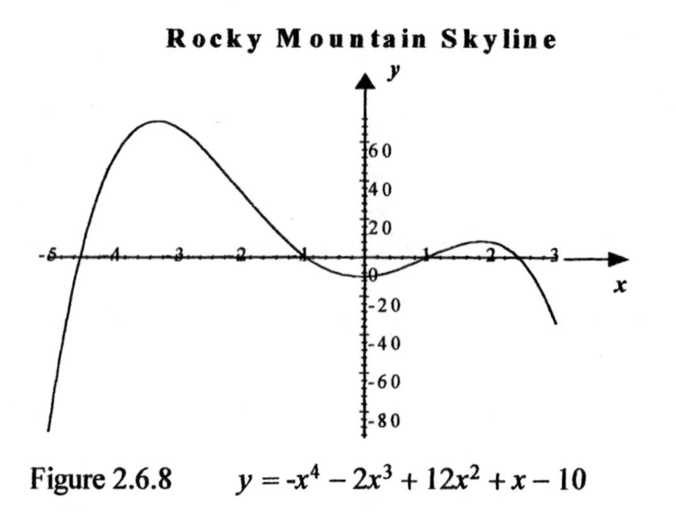

Figure 2.6.8 $\quad y = -x^4 - 2x^3 + 12x^2 + x - 10$

Query 9.

Is the following statement correct? Explain.
"Every point on the line $y = 2x$ has the property that its distance from the x-axis is twice its distance from the y-axis."

Query 10.

Does the point (3,5) lie on the line passing through the points (1,1) and (-1,-3)? Explain.

Exercises 2.6

1. (Computational Skill) Compute the following by hand and then check results using a calculator
   - **a.** $\frac{2}{3} + \frac{3}{4}$
   - **b.** $\frac{5}{2} + \frac{7}{3} - \frac{3}{4}$
   - **c.** $\dfrac{\frac{2}{5} + \frac{1}{2}}{\frac{1}{3} + \frac{1}{7}} - 4$
2. (Computational Skill) In the following, determine the value of x such that
   - **a.** $\frac{1}{2} - \frac{x}{5} = 2$
   - **b.** $\frac{3}{4} + \frac{x}{3} = 6$
   - **c.** $3 - \frac{2x}{5} + \frac{3x}{4} = 2$
3. (Computation Skill) Make up and solve three exercises similar to Exercises 1a–c, 2a–c.
4. (Calculator Skill) Using a graphing calculator, plot the following three equations on the same set of axes. Hint: look up "function graphing" in your calculator manual.

   $y = 2x + 2 \qquad y = -2x + 2 \qquad y = 5x + 2$
5. Let variables y and x be linearly related by the equation $y = 3x + 2$.
   - **a.** Plot the graph of the equation using y as the dependent variable and x as the independent variable.
   - **b.** Determine the rate of change of y with respect to x.
   - **c.** If x changes by 4 units, how much does y change?
   - **d.** If y changes by 2 units, how much does x change?
6. Determine the slope-intercept form equation for parts a and b and then answer part c.
   - **a.** A straight line that passes through the point (2,3) and has a slope of 4
   - **b.** A straight line that passes through the point (2,3) and has an average rate of change of 4
   - **c.** What conclusion can you draw from parts a and b? Explain.
7. Determine the slope-intercept form of the equation of the line that passes through the points (-2, -3) and (4, 1). What is the y-intercept of this line?
8. Determine the equations of four different straight lines that pass through the point (2,3). Plot the four lines. What can you say about different lines all passing through the same point? For example, can two different lines passing through the point (2,3) have the same slope? Why?
9. Discover the geometric significance of the constant term in the slope-intercept form equation of a straight line. Write a short description and explanation of the geometric significance. Hint: select the equation of a straight line, say $y = 2x + 3$. Plot the graph. Superimpose on the same axes, plots of a few (at least four) other lines obtained by changing only the constant value. Continue doing this until the geometric significance of the constant term is understood.
10. Match each of the following equations with its graph provided its graph is one of the four graphs shown. If there is no match for an equation, explain why there is no match.

a. $y = 5x + 2$ b. $y = x + 2$ c. $y = -3x + 4$ d. $y = -x - 1$

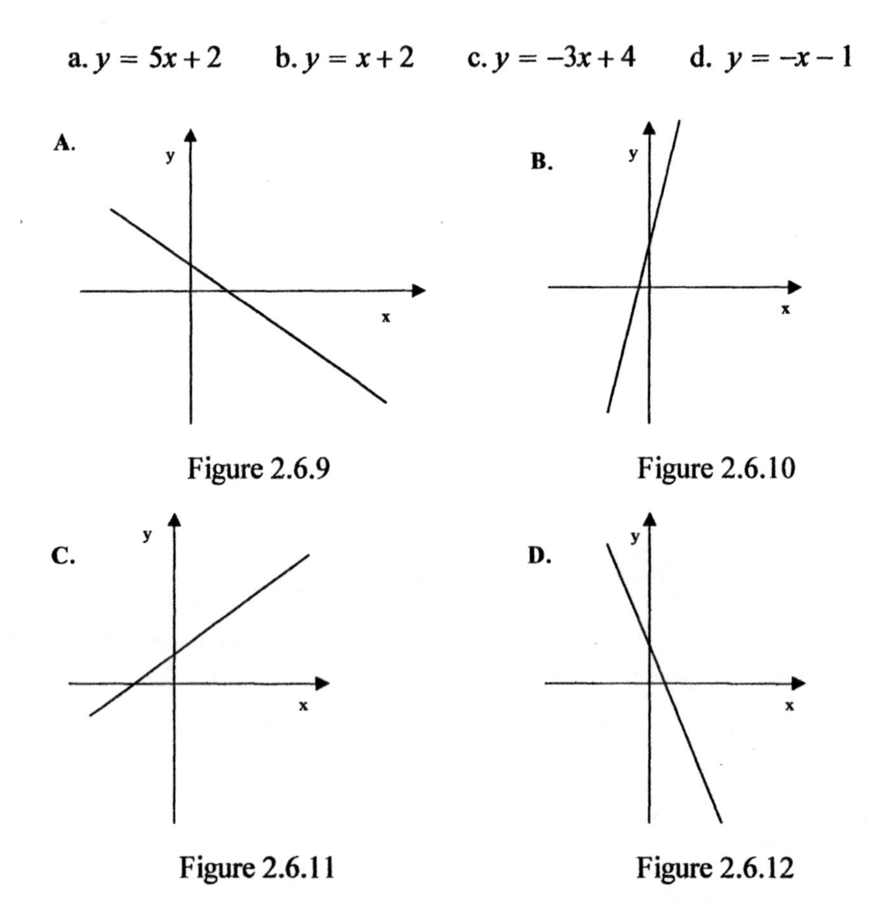

Figure 2.6.9 Figure 2.6.10

Figure 2.6.11 Figure 2.6.12

11. The following table gives the 1980 and 1990 census data for the five counties that constitute the Austin-San Marcus Metropolitan region (source: 1980, 1990 U.S. Census Bureau).

Country	1980	1990	Growth Rate 1980–1990
Bastrop	24,726	38,263	
Caldwell	17,803	26,392	
Hays	40,594	65,614	
Travis	419,573	576,407	
Williamson	76,521	139,551	

Table 2.6.4

a. Compute the annual (average) growth rate from 1980 to 1990 for each of the five counties.

b. Determine the average and median growth rates for the five-county region.

12. Determine the equation of the straight line in the following graph.

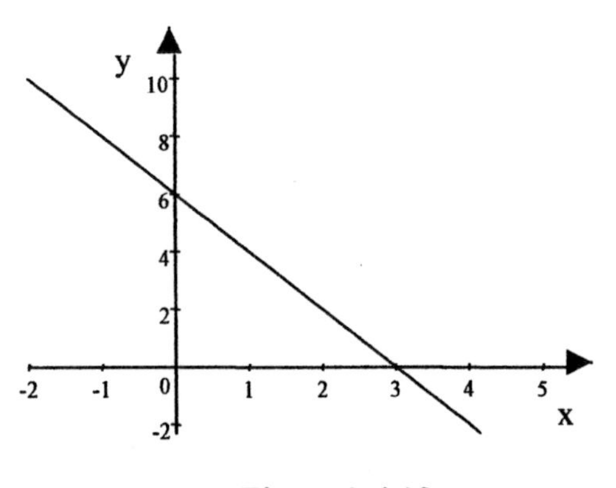

Figure 2.6.13

13. (*Small-Group Project*) "The Austin Community Profile" published by the Greater Austin Chamber of Commerce (1996) includes the following bar chart with the statement, "Austin's population has doubled about every 20 years."

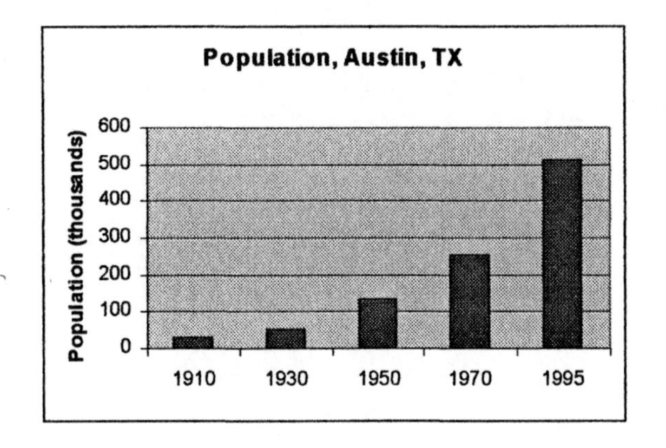

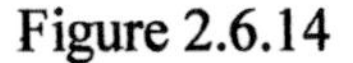

Figure 2.6.14

a. What is meant by the statement, "Austin's population has doubled about every 20 years?"

b. Can the statement, "Austin's population has doubled every 20 years" be converted into a linear equation? Why?

c. Plot the data given by the bar chart. (Let the year be the independent variable and the population be the dependent variable.)

d. Do the points in your plot lie on a straight line? If so, determine the equation of the line.

e. Write a short paragraph explaining the Greater Austin Chamber of Commerce's statement that "Austin's population has doubled about every 20 years."

14. The family income eligibility for free school lunch programs July 1, 1996, through June 30, 1997, is represented in the following table. The income levels are the maximum allowed to qualify children for free school lunch.

No. in Household	Income/Week
1	$194
2	$259
3	$325
4	$390
5	$456
6	$521
7	$587
8	$652

Table 2.6.5

a. Plot the data represented in the table using the number of persons in the household as the independent variable.

b. Do the points in your plot lie on a straight line? If so, determine the equation of the line. What would be a "fair" eligibility income level if there are nine persons in the household? Explain your reasoning.

15. Chicken eggs are classified by weight according to the following chart (e.g., the weight of a small egg is at least 1 ounce and less than 2 ounces):

Classification	Weight(oz.)
Small	1–2
Medium	2–5
Large	5–8
Extra large	8–11
Jumbo	11–14

Table 2.6.6

A supermarket advertised small eggs for $1.20 a dozen, large eggs for $1.50 a dozen, and jumbo eggs for $1.74 a dozen. Assuming the eggs all weigh the average weight for their classification, plot the data and then determine what "fair" prices would be for medium and extra large eggs. Explain your reasoning.

16. Irvin's parents allow him to use the family car provided he pays for the gas, which averages $1.60 a gallon. The car averages 20 miles a gallon. To help Irvin earn gas money, his parents pay him $10 a week to clean and wash the car, mow the yard, and take out the trash. Determine and then plot a weekly savings equation for Irvin that represents earnings minus money spent for gas. Write a physical interpretation for both intercepts of the savings equation.

17. Plotting the graph of a straight line, changing the window values for x and y, and then replotting can give the illusion of a different graph. Which of the following four plots could not be the plot of $y = -x + 8$?

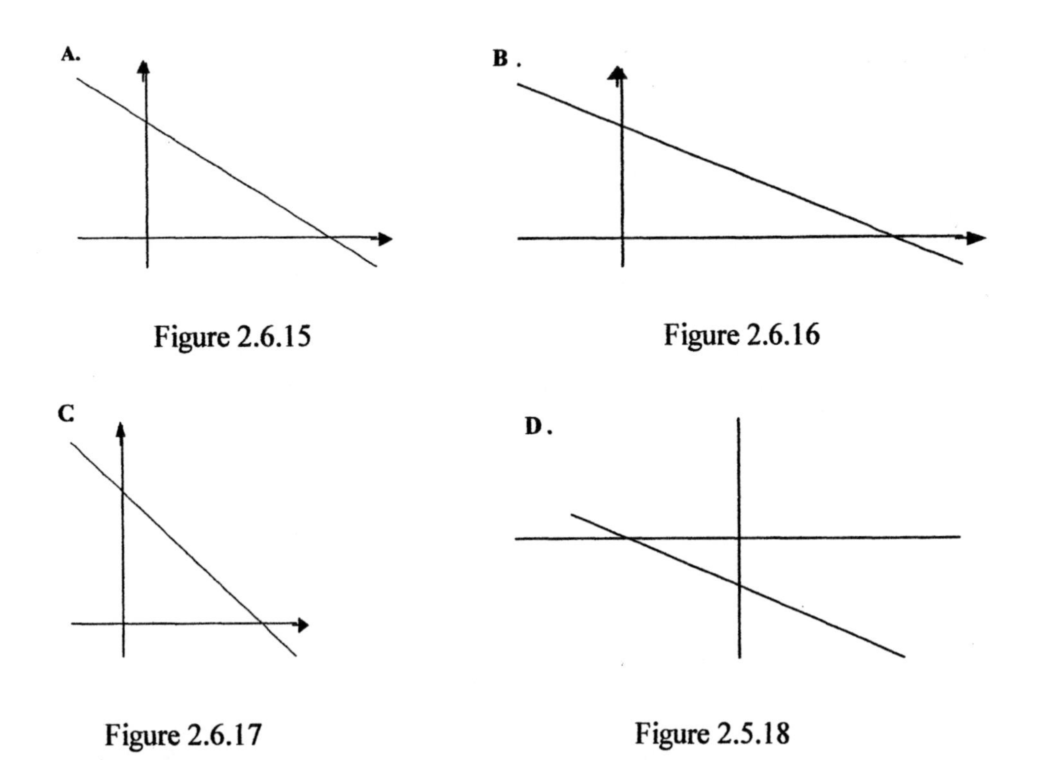

Figure 2.6.15

Figure 2.6.16

Figure 2.6.17

Figure 2.5.18

18. Determine a linear equation that represents the set of points whose distance from the x-axis is four times its distance from the y-axis.

19. Write a one-page essay, based on the following article, on the topic: Explaining the meaning of data by:

  a. The use of a summary value (e.g., mean, median, mode, see Section 1.2)

  b. The use of a plot to show trends

Historical Note

Since Ted Williams batted .406 for the Boston Red Sox in 1941, no hitter has passed the .400 mark in a full season. Baseball scholars and statisticians have tried to attribute this failure to better gloves for fielding, more-scientific managing, or tougher playing conditions, like the preponderance of night games. But for Dr. Gould, the disappearance of .400 hitting is the result of a general improvement in the caliber of play, which has resulted in shrinking variation in batting averages. The curve that describes the averages has pulled in tightly around the middle, and the extreme that used to represent those over .400 has disappeared. Focusing this way on a single number, like the .400 batting average, can make it impossible to spot a trend.
(From David Wheeler's comments on Dr. Gould's book *Full House: The Spread of Excellence from Plato to Darwin* in the article "An Eclectic Biologist Argues That Humans Are Not Evolution's Most Important Result: Bacteria Are," *The Chronicle of Higher Education*, September 6, 1996.)

2.7 Applications of Linear Equations

A **linear equation** in one independent variable is an equation whose graph is a straight line. (The graph of a linear equation in two independent variables is a plane in three-dimensional space.) The solution of a linear equation in one variable is the numerical value of the variable that reduces the equation to an identity. (That is, the equation has the same numerical value on both sides of the equal sign.) The solution of a linear equation in one variable can be found exactly using algebraic means or approximated using graphical means. To find the exact value, isolate the variable on one side of the equal sign. To approximate the solution graphically, plot both sides of the equation and note where the two plots intersect. Or, transfer all the terms to one side of the equal sign, plot that side of the equation, and note where the plot touches the horizontal axis.

The growth of a savings account over one interest period is modeled by a linear equation over that period.

Example 2.7.1

Wanda has \$1000 to invest. She decides to put part of her money into a money market account that pays 6.5% simple interest annually and the balance into a regular savings account that pays 4% simple interest annually. If at the end of the year Wanda's combined investments have earned her \$50 in interest, how much of her \$1000 did she invest in a regular savings account?

Solution:

We begin by identifying the unknown quantity the problem is asking us to find, which is, how much money did Wanda invest in a regular savings account? We let

$$x = \text{amount Wanda invested in a regular savings account.}$$

Therefore

$$1000 - x = \text{amount Wanda invested in a money market account.}$$

Because Wanda's combined investments have earned her \$50 of interest at the end of the year, we form an equation by writing

(interest from savings account) plus (interest from money market account) equals 50.

The interest from the savings account is 4% of x or $.04x$ and the interest from the money market is 6.5% of $(1000 - x)$ or $.065(1000 - x)$. Thus we have the interest equation (in mathematical shorthand)

$$.04x + .065(1000 - x) = 50$$

We can graphically approximate the solution of this equation by plotting both sides of the equation and noting where the two plots intersect.

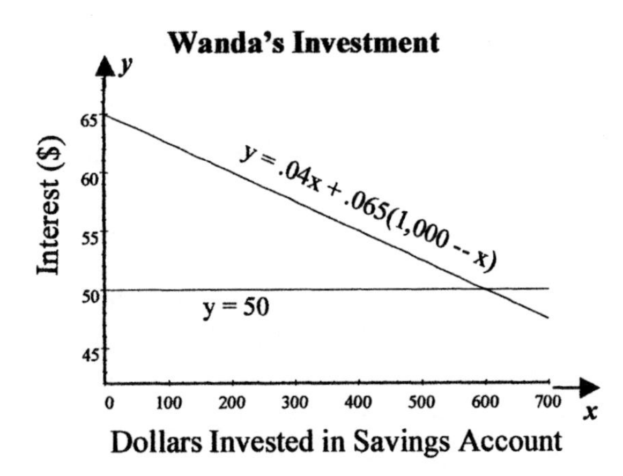

Figure 2.7.1

What is your approximation of the coordinates of the point where the two plots cross?

We now solve for x symbolically

$.04x + .065(1000 - x)$	$=$	50	
$.04x + 65 - .065x$	$=$	50	multiply: .065 times $(1000 - x)$
$-0.025x + 65$	$=$	50	add the two x terms together
$-.025x$	$=$	-15	subtract 65 from both sides
x	$=$	$\frac{-15}{-.025} = 600$	divide both sides by $= -.025$

Thus Wanda invested \$600 in the regular savings account and \$400 in the money market account. ▲

Query 1.

Referring to the previous exercise, how much interest did Wanda receive from each of her two investments?

Straight-line depreciation is another example of the use of linear equations in business.

Example 2.7.2

Oscar, a professional saxophone player, purchased a saxophone for \$1,100. He figures that he will use the instrument for 6 years and then sell it for \$300. For federal income tax purposes, he would like to determine a straight-line depreciation equation by which he could tell the re-sale value of the instrument at any time during the 6-year period.

Solution:

We begin by identifying what the problem is asking us to find. The problem asks for the equation of a straight line that represents the depreciated cost of Oscar's saxophone at any time during the 6-year period. We let

t = time in years that Oscar has owned the saxophone

c = cost of the saxophone at time t.

If we let t be the independent variable and c be the dependent variable, the form of the equation for the depreciation line is

$$c = mt + b.$$

There are two unknown coefficients, m and b, in this equation, and therefore we need to know two conditions that the line must satisfy. These two conditions are given by the initial price and the resale price of the saxophone.

At time $t = 0$, $c = 1100$.
At time $t = 6$, $c = 300$.

Thus the line must pass through the point (0,1100) representing the initial cost and the point (6,300) representing the resale value after 6 years. This gives us a system of two equations in the two unknown coefficients m and b.

$$1100 = m * 0 + b$$
$$300 = m * 6 + b$$

We solve this system of equations for the unknown coefficients m and b by solving the first equation for b, $b = 1100$ and then substituting this value of b into the second equation. This gives the following:

300	$=$	$6m + 1100$	write $m * 6$ as $6m$
-800	$=$	$6m$	subtract 1100 from both sides
$\frac{-800}{6}$	$=$	m	divide both sides by 6.

The desired equation of the straight line is

$$c = \frac{-800}{6}t + 1100.$$

To check our computations, we substitute the coordinates of each of the two known points into the equation. Because both points lie on the desired line, their coordinates must satisfy the equation of the line. This means that when the coordinates of the points are substituted into the equation an equality should result.

Substituting the coordinates of the point (0,1100) into the equation of the line gives

$$1100 = \frac{-800}{6}(0) + 1100$$

or

$$1100 = 1100, \text{ an equality.}$$

Substituting the coordinates of the point (6,300) into the equation of the line gives

$$300 = \frac{-800}{6}6 + 1100$$

or

$$300 = -800 + 1100 \text{ or } 300 = 300, \text{ an equality.}$$

Thus our computations are correct, and $c = \frac{-800}{6}t + 1100$ is the correct equation for the depreciation line. ▲

Query 2.

1. If Oscar depreciated his saxophone on his federal income tax according to the preceding straight-line depreciation, what is the depreciated cost of the saxophone to him after 4.5 years?
2. For how many years could Oscar claim a depreciation credit for his saxophone on his federal income tax if he never sold his instrument?

Query 3.

1. Determine the equation of the depreciation line in Example 2.7.2 by applying the 2-point method for determining the equation of a straight line (as illustrated in Section 2.5). Do you get the same result?
2. Plot the graph of the depreciation line.
3. Compare the computed depreciated cost of the saxophone after 4.5 years with the value given in your calculator's table of values for the depreciation line.

A third example of using linear equations comes from driving and involves determining distance given a rate and a time. The basic relationship involving distance, rate, and time is

distance = (rate) x (time).

Example 2.7.3

Ruby is driving along Interstate 20 traveling from Dallas to Marshall. She has her cruise control set at 65 miles per hour. As she passes a sign showing there are 50 miles to Marshall, she wonders how many minutes it will take her to get to Marshall if she maintains her present speed.

Solution:

We begin by identifying the unknown quantity the problem is asking us to find, which is how many minutes it will take Ruby to go 50 miles traveling at a constant rate of 65 miles per hour. We let

t = time in minutes to go 50 miles traveling at 65 miles per hour.

Because the question is asking for the time in minutes, we transform the rate of 65 miles per hour into miles per minute

$$\text{rate} = (65\ \tfrac{\text{miles}}{\text{hour}})\,(\tfrac{1}{60}\ \tfrac{\text{hour}}{\text{minutes}}) = \tfrac{65}{60}\ \tfrac{\text{miles}}{\text{hour}}\ \tfrac{\text{hour}}{\text{minutes}} = \tfrac{65}{60}\text{miles per minute.}$$

We now substitute into the basic distance-rate-time relation and then solve for time (in minutes).

distance	=	**(rate) x (time)**	
50	=	$\frac{65}{60}t$	substitute
3000	=	$65t$	multiply both sides by 60
$\frac{3000}{65}$	=	t	divide both sides by 65

Our answer is that it will take Ruby $t = \frac{3000 \text{ miles, minutes}}{65 \text{ miles}}$ or $t = 46.15$ minutes. ▲

How can you determine if the answer is reasonable?

Example 2.7.4

Resolve Example 2.7.3 using a graphical analysis.

Solution:

Display the information graphically on a distance versus time plot. Because Ruby is driving at a constant rate, the distance versus time plot should be a straight line. (Why?) In order to determine the plot, we need two points. If we measure both time and distance from the sign indicating 50 miles to Marshall, then the sign represents the point $(0,0)$. That is, zero distance and zero time. Because Ruby is driving 65 mph, a second point is $(60,65)$, which represents a time of 60 minutes and a distance of 65 miles. The line determined by these two points is shown in the following plot along with a horizontal line representing 50 miles (the distance from the sign to Marshall) and a vertical line segment determined by the intersection of the other two lines. The point where the vertical line segment meets the time axis (horizontal axis) is the time it will take Ruby to go 50 miles.

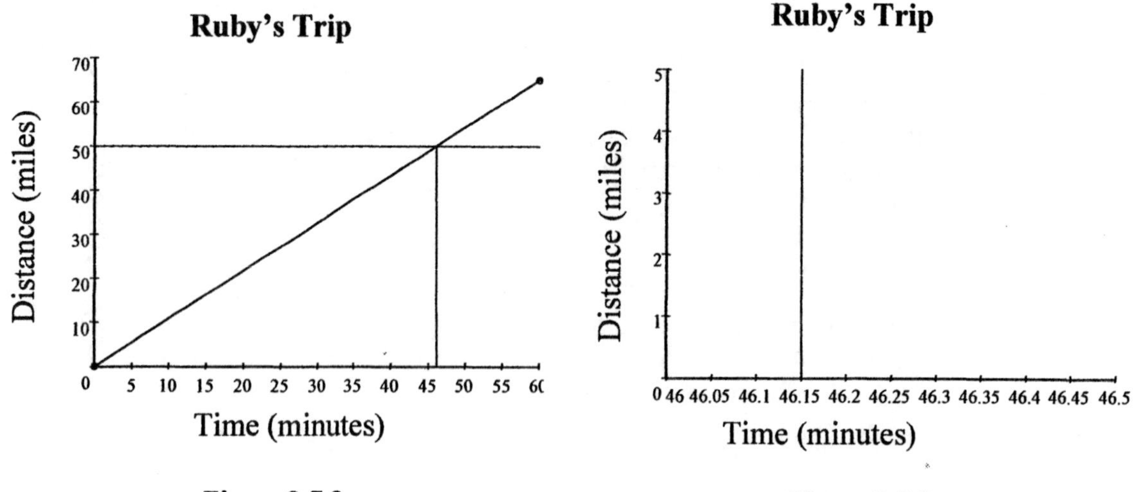

Figure 2.7.2

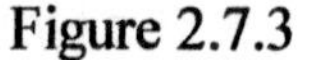
Figure 2.7.3

Zooming in on this point gives 46.15, the answer that was obtained previously using a symbolic method. ▲

Note that the preceding plot illustrates that the slope of the line is the rate of change of the dependent variable (distance) with respect to the independent variable (time).

Query 4.

A softball pitcher throws a fastball at 50 miles per hour. If the pitcher releases the ball 60 feet from home plate, how long does it take for the ball to get to the plate? (1 mile = 5280 feet)

The preceding problems have illustrated arithmetic rules for working with equations.

Arithmetic Rules for Equations

1. A number may be added or subtracted to both sides of an equation without changing the equality. (For example, $2x = 5y \Rightarrow 2x + 4 = 5y + 4$.)
2. Both sides of an equation may be multiplied or divided by a nonzero number without changing the equality. (For example, $2x = 5y \Rightarrow (3)2x = (3)5y$.)
3. Two (or more) equations can be combined by adding or subtracting their left sides and their right sides ("equals plus equals give equals").
   (For example, $2x = 5y$ "plus" $x + 2 = 2y - 3 \Rightarrow 3x + 2 = 7y - 3$.)

We repeat comments from Section 2.6 for emphasis.

The **slope** of the line represents the average **rate of change** of the dependent variable with respect to the independent variable. This is also the *unitary* rate of change in the dependent variable. That is, for every one unit change in the independent variable, the slope gives the corresponding change in the dependent variable.

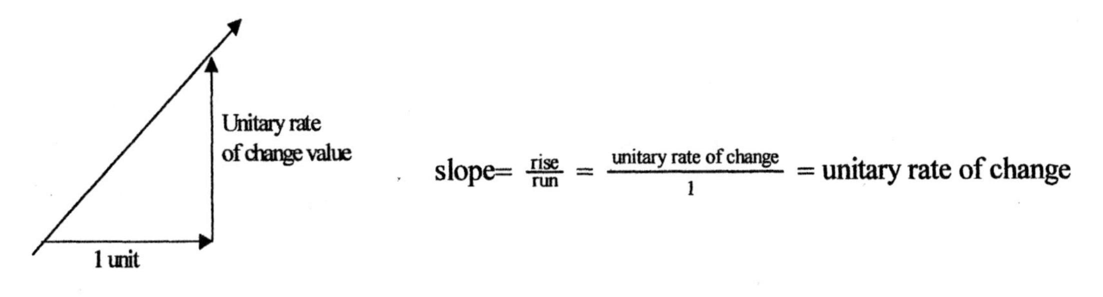

Figure 2.7.4

Economist use the term *marginal value* to describe the unitary rate of change.

Because we can compute the rate of change (slope) from any two points on a straight line, we can determine the equation of the line from any two points on the line. The ability to compute the equation of a straight line and to be able to determine the slope (rate of change) from the equation of a straight line is fundamentally important in the study of mathematics.

Exercises 2.7

1. (Computational Skill) For the following, given the basic formula: **distance = (rate) x (time)**, compute:
   a. Distance(miles), given: rate = 60 miles per hour, time = 2 hours
   b. Rate(miles per hour), given: distance = 100 miles, time = 90 minutes
   c. Time(hours), given: distance = 90 miles, rate = 40 miles per hour.
2. (Computational Skill) In the following, determine the equation of the line satisfying the given conditions.
   a. The line passes through the points $(1,2)$ and $(5,-3)$.
   b. The line passes through the point $(4,1)$ and has a slope of 2.

c. The line has a y-intercept of -3 and a slope of -2.

3. (Computational Skill) Make up and solve three exercises similar to Exercise 1a–c, 2a–c.

4. Alvin is going to claim a depreciation tax credit on his business car. The car cost $15,000 when new, and Alvin anticipates it will be worth $7,000 three years later. If he uses a straight-line depreciation model, what is the rate of depreciation? That is, what is the slope of the depreciation line?

5. Louis invests $4,000, part in a money market account that pays 5% interest yearly and the remainder in a savings account that pays 3% interest yearly. If, at the end of one year, Louis's combined investments have earned him $175.00, how much did he invest in the money market account?

6. If the sum of 3 consecutive integers is 27, determine the integers. Is is possible for the sum of 3 consecutive integers to be 28? Why or why not?

7. If the sum of 2 integers is 27 and one of the integers is twice as large as the other, what are the integers?

8. Determine 3 different odd integers that sum to 15. Is your answer unique?

9. Cora likes to shop at a special liquidator's discount store because the store offers a 10% discount on the amount of the bill that is over $100. Plot a graph that shows the amount of the discount for purchases that cost between $0 and $350.

10. Cora likes to shop at a special liquidator's discount store because the store offers a 10% discount on the amount of the bill that is over $100. If Cora pays $150, which includes a 6% sales tax, for her purchases, how much was her bill before the sales tax and discount were applied?

11. Doris is driving east on Interstate 20 with her cruise control set on 65 miles per hour. If Doris's car is 16 feet long, how long does it take her to pass a 70-foot tractor-trailer rig if the truck is traveling at 60 miles per hour? (1 mile = 5,280 feet)

12. Doris is driving east on Interstate 20. If Doris's car is 16 feet long, and it takes her 10 seconds to pass a 70-foot tractor-trailer rig that is traveling at 60 miles per hour, how fast is Doris traveling? (1 mile = 5,280 feet)

13. Doris is driving east on Interstate 20 with her cruise control set on 65 miles per hour. If Doris's car is 16 feet long and it takes her 7 seconds to pass a 70-foot tractor-trailer rig, how fast is the truck traveling? (1 mile = 5,280 feet)

14. Doris is driving east on Interstate 20 with her cruise control set on 65 mph. Doris's car is 16 feet long. If she takes 0.2 seconds to pass the same type of car going in the opposite direction, how fast is the other car traveling? (1 mile = 5,280 feet)

15. Gene drove 1,200 miles on a trip from Grambling, Louisiana, to Baltimore, Maryland, and back. If he averaged 27 miles per gallon and the average cost of gas was $1.189 per gallon, how much did Gene spend on gas for his trip?

16. Fred walks along a level path directly away from a streetlight. If Fred is 6 feet tall and the streetlight is 15 feet high, how long is Fred's shadow when he is 5 feet from the street light? (Hint: Draw a picture and then use similar triangles.)

17. Fred wants to measure the height of a flagpole. Fred is 6 feet tall, and when he stands 10 feet from the flagpole, the tip of his shadow is at the tip of the flagpole shadow. If his shadow is four feet long, how tall is the flagpole? (Hint: Draw a picture and then use similar triangles.)

18. Consider a system of two linear equations in two unknowns and the corresponding graphs (lines) of the two equations. Describe all of the different possible graphical situations. For example, the two lines could be parallel. For each situation, give an example and describe what is true about the solution(s) of your example.

19. Consider a system of three linear equations in two unknowns and the corresponding graphs (lines) of the three equations. Describe all of the different possible graphical situations. For example, the three lines could all be parallel or two could be parallel and not the third or there could be three intersections. Give an example for each situation and describe what is true about solution(s) of your example.

20. Write an algorithm (recipe) for solving a system of two linear equations in two unknowns. Illustrate the steps of your algorithm with an example.

21. (Small Group) Use at least two different methods to measure the height of the flagpole on campus. (Asking someone to tell you the height is not an acceptable method.)

22. (Small Group) Develop three different designs for a swimming pool that will hold 6,000 gallons of water and whose depth will change from 3 feet at one end to 8 feet at the other end. For each design, include two drawings: one showing the surface shape and the other showing the cross-section shape with the depths indicated.

23. A lunch counter that is open from 10:00 A.M. to 2:00 P.M. averages sales of $810. The counter is serviced by two waitresses and one busboy. The tips, which average 10% of the bill, are shared equally by the three employees. How much should the owner pay each employee in order to guarantee that each one of them averages at least $7 per hour?

24. Write a paragraph discussing why the following graphical method is a legitimate method for approximating the solution of a linear equation. Why is the solution obtained approximate rather than exact? Illustrate your points.

    Graphical Method for Approximating the Solution of an Equation:

    a. Transform the equation by moving all terms to the left-hand side of the equal sign.
    b. Plot the expression on the left-hand side of the equal sign.
    c. Zoom on the point where the plot touches the x-axis.
    d. The x coordinate of this point is an approximate solution of the equation.

25. A person standing on the equator makes one full revolution around the center of the earth every day. Assuming that the earth's equator is a circle of radius 3,960 miles, how fast is the person moving? (That is, what is the person's orbital velocity?)

26. Communication satellites orbit the earth in *geosynchronous* orbits, that is, they appear to be fixed above a specific point on the earth's surface. If a communication satellite is in orbit 2,000 miles above a point on the equator, how fast is it moving? (See Exercise 25.)

2.8 Systems of Equations

Systems of equations occur very frequently in the mathematical descriptions (modeling) of situations involving two or more variables. How to solve systems is the subject of this section. We illustrate solving systems of linear equations using substitution, graphical, elimination, and matrix methods. The algorithmic nature of the elimination method provides a structure for developing a matrix method. This matrix method, accomplished by using a calculator, will be our principle method of solving systems involving three or more equations.

Example 2.8.1

Students in two study groups agree to coordinate their breaks. Evelyn spends \$2.60 buying 1 can of soda and 4 bottles of water for one group and Sam spends \$2.80 for 3 cans of soda and 2 bottles of water for the other group. In order for each person to pay their proper amount, they need to determine the cost of a can of soda and a bottle of water. Mathematically model this situation and then determine the cost of a can of soda and a bottle of water.

Solution.

We begin by identifying variables. Let

s = cost of a can of soda
w = cost of a bottle of water.

Using these variables, Evelyn's purchase is described by $s + 4w = \$2.60$ and Sam's purchase is described by $3s + 2w = \$2.80$. We model the situation with the system of 2 equations

$$\begin{cases} s + 4w & = & \$2.60 \\ 3s + 2w & = & \$2.80. \end{cases}$$

We illustrate the substitution, graphical, and elimination solution methods.

1. Substitution method. (Solve for a variable in one equation and substitute for that variable in the other equation.)

Directions	**Equation**
Solve the first equation for s in terms of w	$s = \$2.60 - 4w$
Substitute for s in the second equation	$3(\$2.60 - 4w) + 2w = \$2.80$
Simplify and isolate w	$\$7.80 - 12w + 2w = \$2.80$
Implies	$-10w = -\$5.00$
Implies	$w = \$0.50$
Substitute for w in the first equation and solve for s	$s = \$2.60 - 4(\$0.50) = \$0.60.$

Thus a can of soda costs \$0.60, and a bottle of water costs \$0.50. ▲

2. Graphical method. (Plot both equations and determine their point of intersection.)

Water, Soda Cost Functions

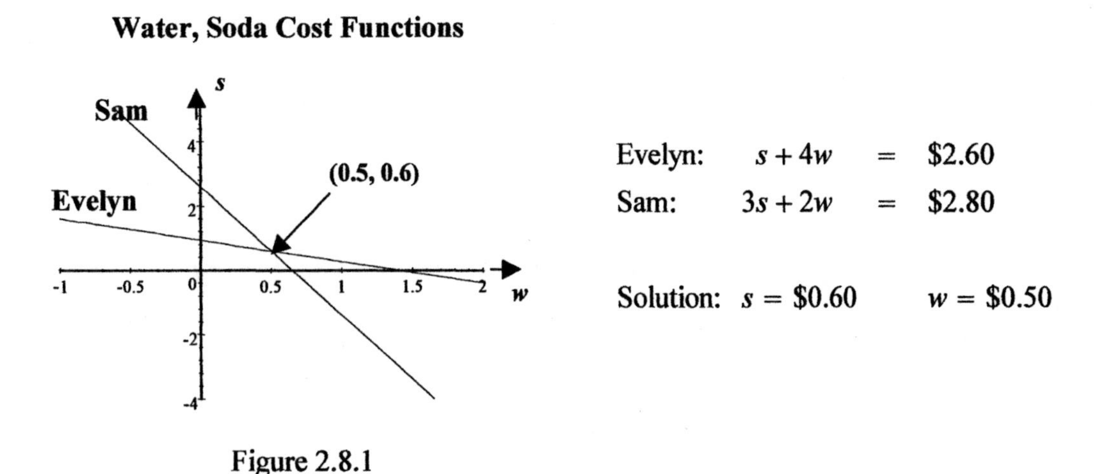

Evelyn:	$s + 4w$	=	\$2.60
Sam:	$3s + 2w$	=	\$2.80

Solution: s = \$0.60 $\quad$ w = \$0.50

Figure 2.8.1

Use the zoom and trace facilities of your calculator to approximate the coordinates of the intersection point. ▲

3. Elimination method: (Multiply one equation by an appropriate number and add it to the other equation in order to eliminate a variable. Repeat this process until a solution is determined. This method uses rules 2 and 3 from the **Arithmetic Rules for Equations** stated in Section 2.7.) Two systems with the same number of equations are said to be **equivalent** provided they have the same solutions.

Original system: $\begin{cases} s + 4w = \$2.60 \\ 3s + 2w = \$2.80 \end{cases}$

Multiply the first equation by –3 and add it to the second equation.

New equivalent system: $\begin{cases} s + 4w = \$2.60 \\ \quad - 10w = -\$5.00 \end{cases}$

Multiply second equation by $-\frac{1}{10}$.

New equivalent system: $\begin{cases} s + 4w = \$2.60 \\ \quad w = \$0.50 \end{cases}$

Multiply the second equation by –4 and add it to the first equation.

New equivalent system: $\begin{cases} s = \$0.60 \\ \quad w = \$5.00 \end{cases}$.

Thus soda costs \$0.60 per can and water costs \$.50 per bottle. ▲

The elimination method is governed by the following three rules for producing equivalent systems of linear equations:

1. Two equations may be interchanged.
2. An equation may be multiplied by a nonzero constant.
3. A constant multiple of one equation may be added to a second equation.

Example 2.8.1 illustrated a case where a system of equations had a unique solution. Is it possible to have a system that has no solutions? Or, a system that has infinitely many solutions?

Query 1.

Is it possible to have two lines that do not intersect? If so, what does that imply about the solution of the corresponding system of equations?

Query 2.

Can a system of two equations have infinitely many solutions? If so, what is true of the plots of the two equations?

Example 2.8.2.

Prices for a public supper were \$3 for children and \$6 for adults. If the 100 people who attended paid a total of \$540, how many of them were children and how many were adults?

Solution:

To develop a model, we begin by identifying variables. Let

c = number of children
a = number of adults.

Using these variables, the number of people is described by $c + a = 100$ and the amount collected is described by $3c + 6a = 540$. We want to solve the system of equations

$$\begin{cases} c + a = 100 \\ 3c + 6a = 540 \end{cases}$$

for c and a.

Substitution method. (Solve for a variable in one equation and substitute for that variable in the other equation.)

Directions	**Equation**
Solve the first equation for c in terms of a	$c = 100 - a$
Substitute for c in the second equation	$3(100 - a) + 6a = 540$
Simplify and isolate a	$300 - 3a + 6a = 540$
Implies	$3a = 240$
Implies	$a = 80$
Substitute for a in the first step and solve for c	$c = 100 - 80 = 20.$

Thus 80 adults and 20 children attended the public supper.

Graphical method. (Plot both equations and determine their point of intersection.)

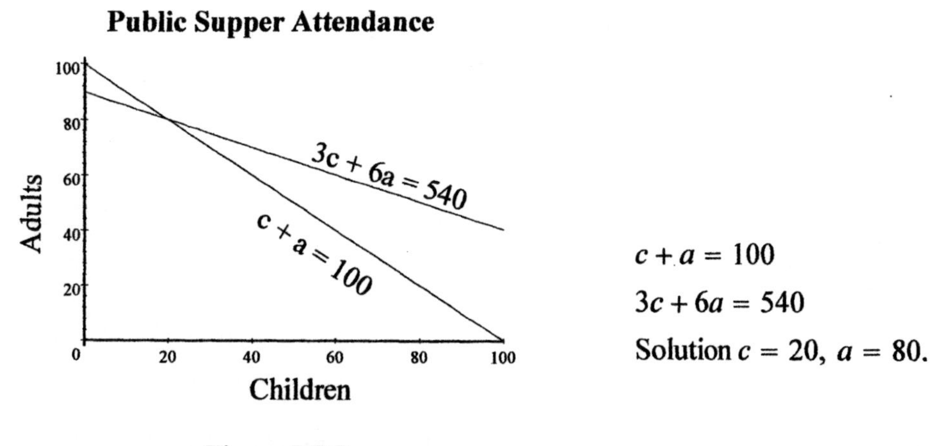

$c + a = 100$

$3c + 6a = 540$

Solution $c = 20,\ a = 80.$

Figure 2.8.2

Elimination method. (Only the sequence of equivalent systems of equations are presented. The reader is asked to fill in the explanation of how each system is obtained from the previous system. See Example 2.8.1.)

$$\begin{cases} c + a = 100 \\ 3c + 6a = 540 \end{cases}$$

$$\begin{cases} c + a = 100 \\ 0 + 3a = 240 \end{cases}$$

$$\begin{cases} c + a = 100 \\ a = 80 \end{cases}$$

$$\begin{cases} c = 20 \\ a = 80 \end{cases}$$

Thus 80 adults and 20 children attended the public supper. ▲

Supply and demand are the primary factors in a market-based economy. The relation between these two concepts is generally modeled by a plot similar to that in Figure 2.8.3. Quantity is measured on the horizontal axis and price on the vertical axis.

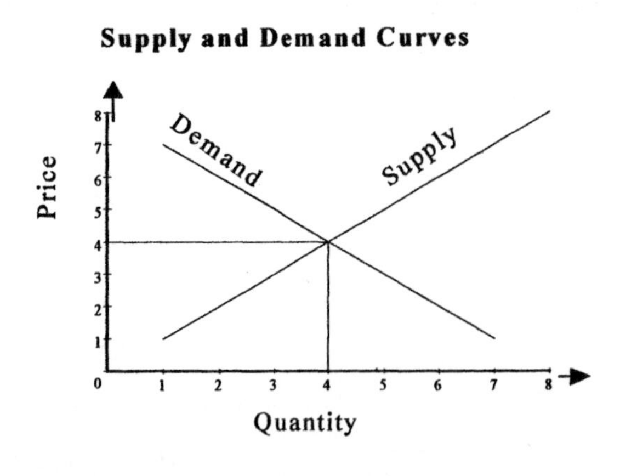

Figure 2.8.3

In an actual scenario, the curves would probably not be straight lines although the trends of the curves would be similar to those in Figure 2.8.3. That is, the supply curve has a positive slope reflecting the general rule that an increase in demand, supply being constant, will raise prices whereas the demand curve has a negative slope reflecting the general rule that an increase in supply, demand being constant, will decrease prices. The point of intersection of the supply and demand curves is called the **equilibrium point**, the point where supply equals demand. In Figure 2.8.3, the equilibrium point is (4,4).

Example 2.8.3.

Suppose Figure 2.8.3 represents the supply and demand curves for wheat. Quantity is measured in millions of bushels, and price is in dollars per bushel. Suppose also that the government decides to impose a $1 tax per bushel on the producers in an attempt to reduce production of wheat. How does the imposition of the tax effect the equilibrium point? Does the consumer share any of the tax burden?

Solution:

We assume that the demand curve does not change. However, the supply curve shifts upward to the left reflecting the $1 increase in price due to the tax. The new situation is shown in Figure 2.8.4.

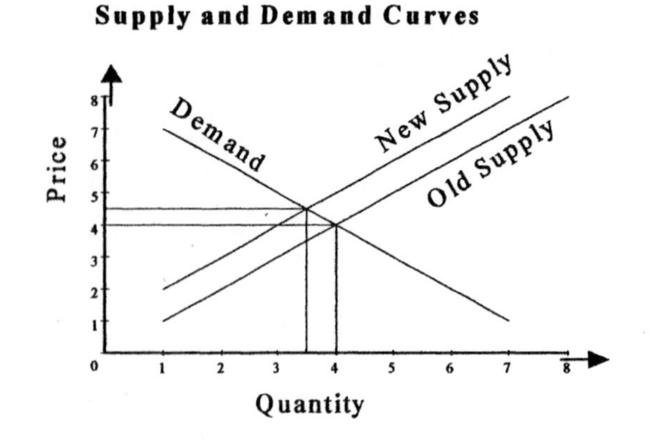

Figure 2.8.4

The equilibrium point has shifted from (4,4) to (3.5,4.5) and the new equilibrium output has dropped from 4 million bushels to 3.5 million bushels.

At the new equilibrium level, the producers receive $4.50 – $1 = $3.50 per bushel after paying the tax. Thus the producer actually pays $0.50 of the tax and the consumer pays the other $0.50. (The new equilibrium point and the division of the tax between producer and consumer is based on the slopes of the supply and demand curves.) ▲

Query 3.

Suppose in Example 2.8.3 the government, instead of imposing a $1 per bushel tax, had granted a $1 per bushel subsidy to the producers to allow them to reduce the price $1 to meet foreign competition. How would the subsidy effect the equilibrium point?

We now develop a **matrix** method for solving a system of linear equations. A matrix is a rectangular array of elements. For example, a table can be viewed as a matrix. A matrix is usually denoted by enclosing the rectangular array in brackets. The dimensions of a matrix are denoted by an ordered pair of numbers in which the first number is the number of rows and the second number is the number of columns. For instance,

$$\begin{bmatrix} 1 & 3 & 4 \\ 2 & -5 & 1 \end{bmatrix}$$

is a 2 x 3 matrix.

Query 4.

Give an example of a 3 x 2 matrix and a 2 x 5 matrix.

Our matrix method is derived from the elimination method. We will illustrate the process with the system of linear equations from Example 2.8.1 after stating some definitions and establishing a few matrix results.

In a system of equations, the roles of the unknowns, the equal signs, and the dollar signs are organizational. These symbols keep the information organized. For instance in the system

$$\begin{cases} s + 4w = \$2.60 \\ 3s + 2w = \$2.80 \end{cases},$$

the important information is contained within the coefficients and constant terms. Thus by agreeing to write the s terms first, the w terms second, and the dollar amounts third, we can display the information in our system of equations by the matrix

$$\begin{bmatrix} 1 & 4 & 2.6 \\ 3 & 2 & 2.8 \end{bmatrix}.$$

The matrix of the first two columns is called the **coefficient matrix** of the system. When the coefficient matrix is augmented by including the column of constants as the right-most column, the matrix is called the **augmented matrix** of the system. Each system of linear equations has a corresponding augmented matrix, and every matrix can be considered to be the augmented matrix for a corresponding system of linear equations.

Query 5.

What is the corresponding system of equations for the matrix $\begin{bmatrix} 1 & 3 & 2 & 4 \\ 5 & 1 & 2 & 2 \\ 2 & 4 & 1 & 2 \end{bmatrix}$?

We defined two systems to be equivalent provided they had the same solutions. We now define two augmented matrices to be **row-equivalent** provided their corresponding systems are equivalent.

The process of the matrix method is to transform the augmented matrix of the system using the following set of three row operations for producing row-equivalent matrices (note the similarity between these operations and the rules for producing equivalent systems of linear equations).

1. Two rows may be interchanged.
2. The elements of a row may be multiplied by a nonzero constant.
3. A constant multiple of one row may be added to a second row.

The process for a system of two equations in two unknowns continues until one of the following three augmented matrix forms is obtained (a, b, and c are nonzero constants):

$\begin{bmatrix} 1 & 0 & a \\ 0 & 1 & b \end{bmatrix}$: Unique solution, $x = a,\ y = b$, different intersecting lines.
Coefficient matrix is an identity matrix: 1s on the diagonal and 0s elsewhere

$\begin{bmatrix} 1 & a & b \\ 0 & 0 & 0 \end{bmatrix}$: Infinitely many solutions, $x + ay = b$, 2 equations for the same line.
The number of nonzero rows is less than the number of unknowns.

$\begin{bmatrix} 1 & a & b \\ 0 & 0 & c \end{bmatrix}$: No solutions, parallel lines.
A row of 0s except for the last entry, which is nonzero.

The reader is asked to generalize these three matrix forms to systems of more than two equations in two unknowns. Examples 2.8.5 and 2.8.6 illustrate the situation for infinitely many solutions in the cases of two equations with three unknowns and three equations with four unknowns.

Small-Group Activity.

A. Write the equations for two different lines that intersect. Form a system of these two equations and then transform the corresponding augmented matrix. Explain how can you tell from the transformed matrix that there is a unique solution?

B. Write two equations for the same line. Form a system of these two equations and then transform the corresponding augmented matrix. Explain how can you tell from the transformed matrix that there are infinitely many solutions.

C. Write the equations for two parallel lines. Form a system of these two equations and then transform the corresponding augmented matrix. Explain how can you tell from the transformed matrix that there is no solution.

We now illustrate how the matrix method is derived from the elimination method by applying each step of the elimination method to the augmented matrix of the system of Example 2.8.1 and expressing the results in matrix form.

The augmented matrix is $\begin{bmatrix} 1 & 4 & 2.6 \\ 3 & 2 & 2.8 \end{bmatrix}$.

Multiply the first equation by –3 and add it to the second equation.

Augmented matrix for new equivalent system: $\begin{bmatrix} 1 & 4 & 2.6 \\ 0 & -10 & -5.0 \end{bmatrix}$

Multiply second row by $-\frac{1}{10}$.

Augmented matrix for new equivalent system: $\begin{bmatrix} 1 & 4 & 2.6 \\ 0 & 1 & 0.5 \end{bmatrix}$

Multiply the second equation by –4 and add it to the first equation.

Augmented matrix for new equivalent system: $\begin{bmatrix} 1 & 0 & 0.6 \\ 0 & 1 & 0.5 \end{bmatrix}$

Thus an equivalent system is

$$\begin{cases} s = \$0.60 \\ w = \$0.50 \end{cases}$$

and so soda costs 60 cents a can and water costs 50 cents a bottle.

This last matrix is said to be in the **row reduced echelon-form** (rref) of the augmented matrix. Note that the system corresponding to the rref matrix gives the answer to the problem. Most graphing calculators have a rref program built into the calculator.

Which of the four methods is the best one to use for solving a system of linear equations? If the system consists of only two equations in two unknowns, the substitution or graphical method may be best. However, when there are more than two unknowns, using the rref method with a calculator is the most efficient and most accurate.

Example 2.8.4

Solve the system
$$\begin{cases} 2x + y - 3z + 2w + z = 2 \\ x - y - 2z + w + 3z = 0 \\ 3x + 4y + 2z - 3w + 2z = 1 \\ x - y - z + 2w + 3z = 0 \\ 4x + 2y - 3z + w - z = 3 \end{cases}.$$

Solution:

We enter the augmented matrix

$$\begin{bmatrix} 2 & 1 & -3 & 2 & 1 & 2 \\ 1 & -1 & -2 & 1 & 3 & 0 \\ 3 & 4 & 2 & -3 & 2 & 1 \\ 1 & -1 & -1 & 2 & 3 & 0 \\ 4 & 2 & -3 & 1 & -1 & 3 \end{bmatrix}$$

into our calculator and employ the rref program to obtain

$$\begin{bmatrix} 1 & 0 & 0 & 0 & 0 & 0.22 \\ 0 & 1 & 0 & 0 & 0 & 0.48 \\ 0 & 0 & 1 & 0 & 0 & -0.25 \\ 0 & 0 & 0 & 1 & 0 & 0.25 \\ 0 & 0 & 0 & 0 & 1 & -0.16 \end{bmatrix}.$$

Thus the solution is

$x = 0.22,\ y = 0.48,\ z = -0.25,\ w = 0.25,\ z = -0.16.$ ▲

Balancing a Chemical Equation

Conservation laws are fundamental to science. The Law of Conservation of Mass states that there is no detectable change in mass during the course of an ordinary (nonnuclear) chemical reaction. This means that there must be as many atoms of each element, combined or uncombined, after a chemical reaction as before the reaction. This law is illustrated by the following two examples.

Example 2.8.5

When copper is heated in air, a black copper oxide is produced. The chemical reaction is $Cu + O_2 \rightarrow CuO$. To balance this reaction, means to determine the number of atoms of each chemical (copper, oxygen) that is needed to satisfy the law of conservation of mass.

Solution:

We begin the solution process by using variables to denote the number of molecules in each chemical combination involved in the reaction. We let

x_1 = number of molecules of Cu (copper) each of which contain of 1 atom of Cu.
x_2 = number of molecules of O_2 (oxygen) each of which contain of 2 atoms of 0.
x_3 = number of molecules of CuO (copper oxide) each of which contain 1 atom of Cu and 1 atom of 0.

Applying the law of conservation of mass to copper (Cu) yields

$x_1 = x_3$ (This says that there are x_1 atoms of copper before the reaction and x_3 atoms of copper after the reaction.)

Applying the law of conservation of mass to oxygen (O) yields

$2x_2 = x_3$ (This says that there are $2x_2$ atoms of oxygen before the reaction and x_3 atoms of oxygen after the reaction.)

Thus in order to balance the equation, we need to find positive integer values for x_1, x_2, and x_3 that satisfy the following systems of equations (this is the mathematical model of the situation):

$$\begin{cases} x_1 = x_3 \\ 2x_2 = x_3 \end{cases}$$ or rewriting with all variables on the left-hand side

$$\begin{cases} x_1 \quad -x_3 = 0 \\ \quad 2x_2 - x_3 = 0. \end{cases}$$

By inspection $x_1 = x_3 = 2x_2$, thus we can probably guess that $x_1 = 2$, $x_2 = 1$, $x_3 = 2$ is a solution. However, our purpose is to illustrate a general method for balancing chemical reactions, and thus we continue.

We form the augmented matrix for this system and then apply the rref method.

Augmented matrix: $\begin{bmatrix} 1 & 0 & -1 & 0 \\ 0 & 2 & 1 & 0 \end{bmatrix}$

rref form: $\begin{bmatrix} 1 & 0 & -1 & 0 \\ 0 & 1 & -.5 & 0 \end{bmatrix}$. This form indicates infinitely many solutions (because there are fewer rows than unknowns).

The equivalent system of equations is $\begin{cases} x_1 - x_3 = 0 \\ x_2 - .5x_3 = 0 \end{cases}$ or $\begin{cases} x_1 = x_3 \\ x_2 = .5x_3 \end{cases}$.

Because x_1 and x_2 are expressed in terms of x_3, any value for x_3 yields corresponding values for x_1 and x_2. This means that there are infinitely many solutions as the form of the rref matrix indicated. Because the unknowns represent numbers of atoms, they must have positive integer values. By letting $x_3 = 2$, we obtain the solution that we had guessed earlier: $x_1 = 2, x_2 = 1, x_3 = 2$. Thus the balanced equation is $2Cu + O_2 = 2CuO$. ▲

We now illustrate the process of balancing chemical reactions with a more involved example. Everyone has seen rust: a rusty nail, rust on a car fender, rusty shovel, and so on. Rust results from combining iron and water. The chemical reaction is: iron plus water produces iron oxide (rust) plus hydrogen gas. The chemists express this reaction by writing

$$Fe + H_2O \rightarrow Fe_3O_4 + H_2.$$

Example 2.8.6

Balance the chemical reaction: $Fe + H_2O \rightarrow Fe_3O_4 + H_2$. That is, determine the number of atoms of each chemical (iron, hydrogen, oxygen) in order to satisfy the law of conservation of mass.

Solution:

We begin the solution process by using variables to denote the number of atoms in each chemical combination involved in the reaction. We let

x_1 = number of molecules of Fe (iron) each of which contain 1 atom of Fe.
x_2 = number of molecules of H_2O (water) each of which contain 2 atoms of H and 1 atom of 0.
x_3 = number of molecules of Fe_3O_4 (iron oxide) each of which contain 3 atoms of Fe and 4 atoms of 0.
x_4 = number of molecules of H_2 (hydrogen) each of which contains 2 atoms of 0.

Applying the law of conservation of mass to iron (Fe) yields

$x_1 = 3x_3$. (This says that there are x_1 atoms of iron before the reaction and $3x_3$ atoms of iron after the reaction.)

Applying the law of conservation of mass to hydrogen (H) yields

$2x_2 = 2x_4$.

Applying the law of conservation of mass to oxygen (O) yields

$x_2 = 4x_3$.

Thus in order to balance the equation, we need to find positive integer values for x_1, x_2, x_3, and x_4 that satisfy the following systems of equations (this is the mathematical model of the situation):

$$\begin{cases} x_1 \quad -3x_3 \quad = 0 \\ 2x_2 \quad -2x_4 = 0 \\ x_2 \quad -4x_3 \quad = 0 \end{cases}$$

We solve this system by applying the rref method to the augmented matrix

$$\begin{bmatrix} 1 & 0 & -3 & 0 & 0 \\ 0 & 2 & 0 & -2 & 0 \\ 0 & 1 & -4 & 0 & 0 \end{bmatrix}.$$

The result is $\begin{bmatrix} 1 & 0 & 0 & -.75 & 0 \\ 0 & 1 & 0 & -1 & 0 \\ 0 & 0 & 1 & -.25 & 0 \end{bmatrix}$. This form indicates infinitely many solutions. (How?)

The equivalent system is $\begin{cases} x_1 - .75x_4 = 0 \\ x_2 - x_4 = 0 \\ x_3 - .25x_4 = 0. \end{cases}$

Thus the mathematical solution of this system: $x_1 = .75x_4$, $x_2 = x_4$, $x_3 = .25x_4$ defines x_1, x_2, and x_3 in terms of x_4. This means assigning a value to x_4, determines the values for x_1, x_2, and x_3. Because the number of atoms must be positive integers, we choose $x_4 = 4$. Then $x_1 = 3, x_2 = 4, x_3 = 1$ and the balanced equation is

$$3Fe + 4H_2O = 1Fe_3O_4 + 4H_2. \quad \blacktriangle$$

Exercises 2.8

1. Solve the system $\begin{cases} x + 2y = 4 \\ 2x + 3y = 6 \end{cases}$ graphically and then check your answer by solving the system using the substitution method and the matrix method.

2. Solve the system $\begin{cases} 5x + 1y = 4 \\ -3x + 4y = 6 \end{cases}$ using the matrix method and then check your answer by solving the system using the substitution method and the graphical method.

3. Use a calculator to:

a. Solve for $\{x, y, z\}$ in $\begin{cases} x + 3y - z = 2 \\ 2x + y + 2z = 5 \\ 3x - 2y + z = 4 \end{cases}$

b. Solve for $\{x, y, z\}$ in terms of w in $\begin{cases} x + 3y - z = 2w \\ 2x + y + 2z = 5w \\ 3x - 2y + z = 4w \end{cases}$

4. Make up a system of two equations in two unknowns that does not have a solution. Graphically show that your system has no solution.

5. Determine the equation of the straight line, $y = mx + b$, that passes through the points $(1,2)$ and $(5,3)$ by solving for m and b using the following method. Each data point determines an equation in two unknowns m and b, obtained by substituting the coordinates of the data point into the equation of the straight line. (For example, substituting the coordinates of $(1,2)$ into the equation of the line gives the equation $m + b = 2$.) The two data points thus determine a system of two equations in the two unknowns, m and b. Solve this system for m and b and then substitute these values into the equation of the straight line.

6. Use the method of Exercise 5 to determine the equation of the straight line, $y = mx + b$, that passes through the points $(2, 3)$ and $(5, -2)$.

7. Apply the "what-if" technique to Example 2.8.1 by including a third study group whose representative, Justin, spends \$3.30 buying two cans of soda, one bottle of water, and three bags of chips. Model this new situation and then determine the price of a can of soda, a bottle of water, and a bag of chips.

8. The local deli had a mix-and-match sale on apples, oranges, and grapes. They offered the following three options, each for \$3.00.

   Option #1 24 apples, 12 oranges, 1 bunch of grapes

   Option #2 12 apples, 18 oranges, 1 bunch of grapes

   Option #3 8 apples, 8 oranges, 3 bunches of grapes.

   Based on this sale, determine the price for a single apple, a single orange, and one bunch of grapes.

9. Atlanta's Best Pizza shop filled 3 take-out orders. The first was for three regular, one large, and two extra large pizzas, and the charge was \$51. The second order was for four regular and two large pizzas, and the charge was \$48. The third order was for two regular, two large, and two extra large pizzas, and the charge was \$56. Determine the price for each of the three sizes of pizza.

10. An oil company's three refineries are designed to process different amounts of heating oil, diesel oil, and gasoline. Their production capabilities from one barrel of crude oil are given by the following table.

Product	Refinery 1	Refinery 2	Refinery 3
Heating oil	9 gal.	4 gal.	2 gal.
Diesel oil	2 gal.	6 gal.	4 gal.
Gasoline	3 gal.	4 gal.	7 gal.

Table 2.8.1

   If the demand is for 5 million gallons of heating oil, 8 million gallons of diesel oil, and 8 million gallons of gasoline, determine the number of barrels of crude oil that should be processed by each of the refineries.

11. How would you advise the oil company in Exercise 10 to distribute its crude oil if the demand for diesel oil drops to 5 million gallons?

12. Balance these chemical reactions:

    a. Ethone plus oxygen produces carbon dioxide plus water: $C_2H_6 + O_2 \to CO_2 + H_2O$.

    b. Chlorine plus potassium hydroxide produces potassium chloride plus potassium chlorate plus water:
    $Cl_2 + KOH \to KCl + KClO_3 + H_2O$.

13. Balance these chemical reactions:

a. Nitropropane produces water plus carbon monoxide plus nitrogen gas plus oxygen gas: $C_3H_5(NO_3)_3 \rightarrow H_2O + CO + N_2 + O_2$.

b. Sodium carbonate (Na_2CO_3) plus bromine gas (Br_2) produces sodium bromide ($NaBr$) plus sodium bromate ($NaBrO_3$) plus carbon dioxide (CO_2): $Na_2CO_3 + Br_2 \rightarrow NaBr + NaBr0_3 + C0_2$.

14. Balance these chemical reactions:

a. $HCL + MnO_2 \rightarrow MnCL_2 + H_2O + CL_2$

b. $Fe + MnO_4 + H \rightarrow Mn + Fe + H_2O$

15. (Small-group activity) The graph of a linear equation in three variables is a plane. Illustrate, by holding two pieces of cardboard, that geometrically two planes have the same three possibilities as do two lines. That is, they can be parallel, coincide, or intersect. (Think of the relationships between the planes represented by the walls, floor, and ceiling of your room.) Generalize the small-group activity in this section to planes, That is,

A. Write the equations for three different planes that intersect. Form a system of these three equations and then transform the corresponding augmented matrix. Explain how can you tell from the transformed matrix that there is a unique solution?

B. Write two equations for the same plane. Form a system of these two equations and then transform the corresponding augmented matrix. Explain how can you tell from the transformed matrix that there are infinitely many solutions.

C. Write the equations for two parallel planes. Form a system of these two equations and then transform the corresponding augmented matrix. Explain how can you tell from the transformed matrix that there is no solution.

16. Determine the equation of the line that is the intersection of the two planes $x + y + z = 2$ and $x - z = 3$.

17. (Small-group activity) A 3x3 magic square is a square array of integers 1, 2, 3, . . . 9 such that the sum of the elements in each row, each column, and along both diagonals is the same. Determine values for the unknowns $x, y, z, u, v,$ and w to make the following a magic square. $\begin{matrix} 8 & 1 & 6 \\ x & y & z \\ u & v & w \end{matrix}$. Hint: Form a system of equations representing the row, column, and diagonal sums.

18. (small-group activity) A 3x3 magic square is a square array of integers 1, 2, 3, . . . 9 such that the sum of the elements in each row, each column, and along both diagonals is the same. Determine values for the unknowns $x, y, z, u, v,$ and w to make the following a magic square. $\begin{matrix} 4 & 1 & 6 \\ x & 5 & z \\ u & v & 6 \end{matrix}$. Hint: Form a system of equations representing the row, column, and diagonal sums.

19. Does there exist a magic square whose first row has elements 8, 6, and 7. Explain why there does exist such a square or explain why such a square is impossible.

2.9 Linear Inequalities

Given any two real numbers, either they are equal or one is smaller than the other. This is called the Trichotomy Law of Real Numbers. (Does the Trichotomy Law apply to points in the plane?) Symbolically the law is stated as follows.

If c and d are any two real numbers, then one and only one of the following relations is true:

$$c = d$$
$$c < d$$
$$d < c.$$

The $<$ is called an inequality sign and $c < d$ means that c is smaller than d, or equivalently, d is larger than c. (Note that $c < d$ may also be written as $d > c$.)

If c and d are numbers, then $c < d$ means there is a positive number b such that $c + b = d$ as shown in the following picture.

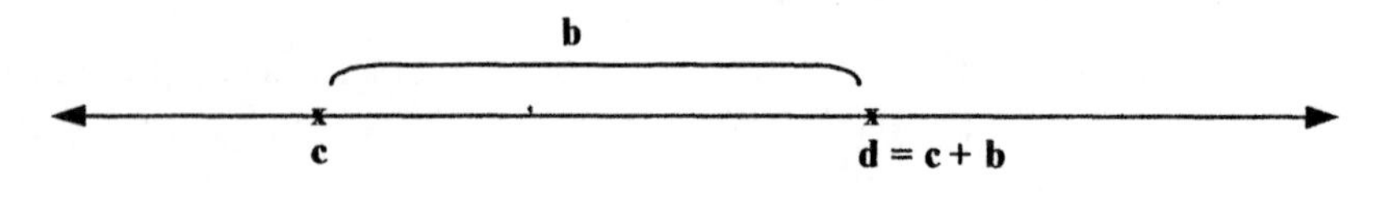

Figure 2.9.1

The inequality symbol is used in comparing two quantities. The quantities may be numbers such as

$$4 < \sqrt{17}$$

or, expressions such as

$$3 + \sqrt{21} < 10 - \sqrt{2} \quad \text{(check this out)}$$

or, sets of numbers such as

$$x < 5 \quad \text{(that is, the set of all numbers less than 5).}$$

Example 2.9.1

Determine the real numbers x for which $4x + 7 < 2x + 11$ is a true statement.

Solution (graphical): We follow the same procedure that was used in graphically solving for the roots of an equation. Namely, subtract the right-hand side from the left-hand side and compare the result to zero. Doing this transforms the inequality to

$$(4x - 7) - (2x + 11) < 0 \quad \text{or} \quad 2x - 18 < 0.$$

Our problem is now to determine the real numbers x for which $2x - 18 < 0$ is a true statement.

Using a graphing calculator, plot $y = 2x - 4$. The answer is the set of values of x for which the plot is below the x-axis. The answer appears to be all real numbers < 2.

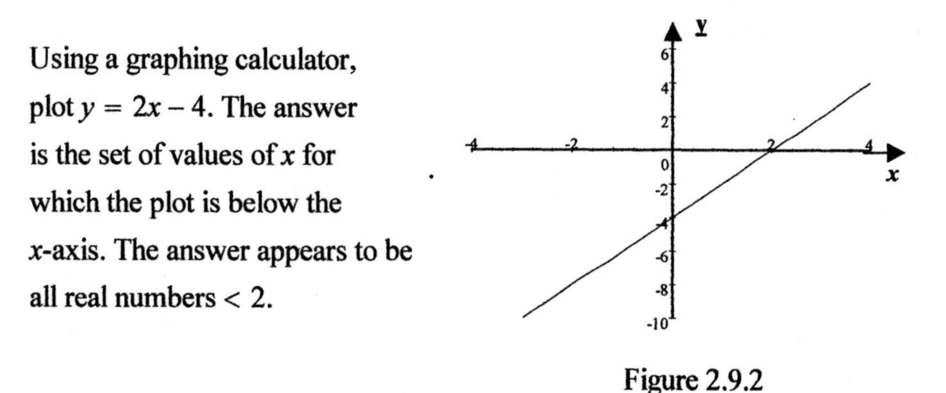

Figure 2.9.2 ▲

We now confirm the graphic result by analytically solving the inequality.

	$4x + 7$	$<$	$2x + 11$	stated problem
implies	$4x$	$<$	$2x + 4$	subtract 7 from both sides
implies	$2x$	$<$	4	subtract $2x$ from both sides
implies	x	$<$	2	divide both sides by 2

Thus the answer is the set of real numbers less than 2. ▲

We apply the "what-if" technique to the previous exercise by replacing the $2x$ with $5x$ to obtain $4x + 7 < 5x + 11$. In order to solve the inequality problem graphically, we begin by subtracting the right-hand side from the left-hand side. This gives us the equivalent problem $-x - 4 < 0$

The plot $y = -x - 4$ suggests the solution is the set of all real numbers $x > -4$.

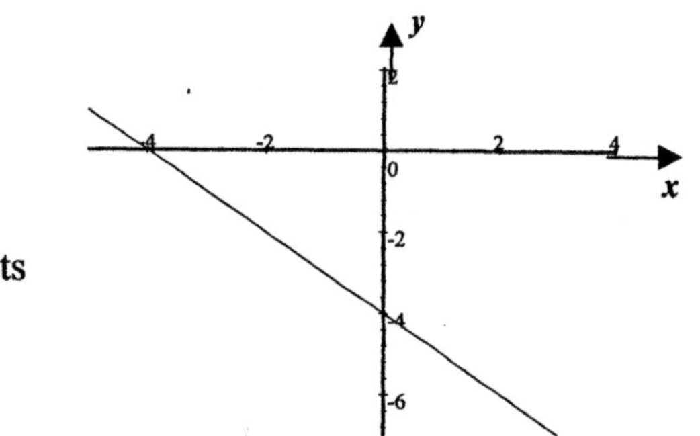

Figure 2.9.3

We now attempt to confirm this analytically as we did in Problem 1.

	$4x + 7$	$<$	$5x + 11$	stated problem
implies	$4x$	$<$	$5x + 4$	subtract 7 from both sides
implies	$-x$	$<$	4	subtract $5x$ from both sides
???	x	$<$	-4	divide both sides by -1

Does the last step make sense? That is, is $x < -4$ equivalent to $-x < 4$? (Is it legitimate to divide both sides of an inequality by a negative number?) Let us check by selecting a number less than negative 4, say –6, and substituting it for x in the inequality $-x < 4$. This gives

$$-(-6) < 4 \quad \text{or} \quad 6 < 4.$$

Certainly 6 is not less than 4. Thus $x < -4$ is not equivalent to $-x < 4$. What is wrong with our reasoning?

We now solve the problem in a different way that does not require dividing by a negative number.

Solution:

	$4x + 7$	$<$	$5x + 11$	stated problem
implies	$4x - 4$	$<$	$5x$	subtract 11 from both sides
implies	-4	$<$	x	subtract $4x$ from both sides

Thus the answer is the set of all real numbers greater than –4 as was suggested by the graphical approach. (Check this result by substituting numbers greater than –4 for x in the problem and then simplifying the arithmetic.)

This example and its variations illustrate the need to establish arithmetic rules for working with inequalities.

Arithmetic Rules for Inequalities

1. A number may be added or subtracted from both sides of an inequality without changing the sense (the direction) of the inequality symbol. (For example, $2x < 5 \Rightarrow 2x + 4 < 5 + 4$.)
2. A positive number may multiply or divide both sides of an inequality without changing the sense (the direction) of the inequality. (For example, $2x < 8 \Rightarrow x < 4$.)
3. A negative number multiplying or dividing an inequality changes the sense (the direction) of the inequality. (For example, $-3x < 6 \Rightarrow x > -2$.) The incorrect answer obtained in the first attempt to solve the previous problem resulted from violating this rule.

Inequalities involving one variable are used to divide the number line. We will see in the next section how inequalities involving two variables divide the plane.

If you want to express the set of real numbers that are less than or equal to 3, write $x \leqslant 3$. This is called a **conditional inequality**. Strict and conditional inequalities are used, along with braces and parentheses, to denote open and closed intervals.

Closed Interval	$[2,5]$	$2 \leqslant x \leqslant 5$
Half Closed Interval	$[2,5)$	$2 \leqslant x < 5$
Half Closed Interval	$(2,5]$	$2 < x \leqslant 5$
Open Interval	$(2,5)$	$2 < x < 5$

Table 2.9.1

Absolute Value

Another way to describe a closed or open interval is to use the absolute value function with an inequality. The absolute value function measures the distance between two objects. Because the distance between two points is always zero or positive, the absolute value of an expression must always be zero or positive. For example, the distance between 7 and 2 is the same as the distance between 2 and 7. In both cases, the distance is 5. Therefore we define the absolute value $|7-2| = |5| = 5$ and $|2-7| = |-5| = -(-5) = 5$. That is, the absolute value of a positive number is just that number and the absolute value of a negative number is the opposite of that number.

The distance between numbers a and b is defined by the **absolute value** as follows.

$$|a-b| = \begin{cases} a-b & \text{for } a-b \geqslant 0 \\ -(a-b) & \text{for } a-b < 0 \end{cases}$$

In particular the size of x or the distance between x and 0 is

$$|x-0| = |x| = \begin{cases} x & \text{for } x \geqslant 0 \\ -x & \text{for } x < 0 \end{cases}.$$

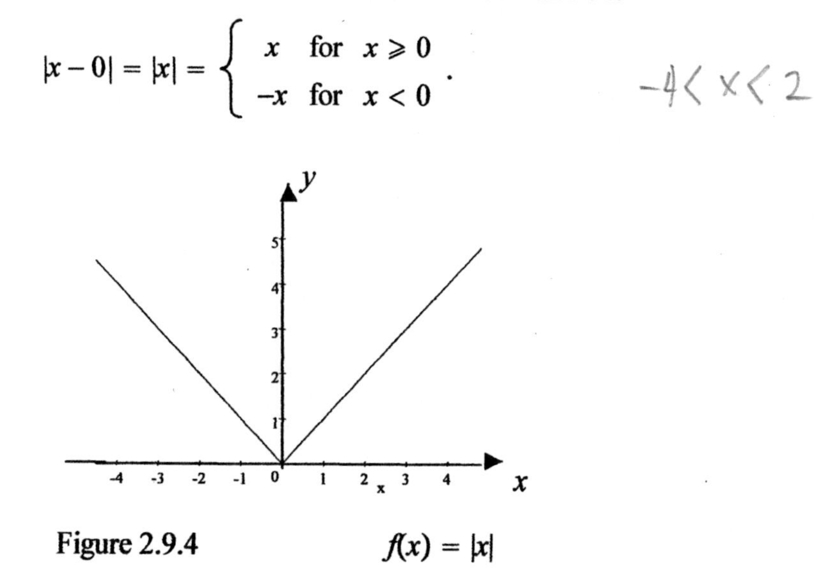

Figure 2.9.4 $f(x) = |x|$

Note that the graph of the absolute value function consists of two half lines formed into a V shape.

Query 1.

What does the graph of $f(x) = |x-2| + 3$ look like?

It is very important to understand the definition of the absolute value of an expression when the expression is other than just a number. We illustrate with the following:

The set of all points that are within 3 units of 2 can be expressed by $|x-2| < 3$.

Because $|x-2| < 3$ and $|x-2| = \begin{cases} x-2 & \text{for } x-2 \geqslant 0 \\ -(x-2) & \text{for } x-2 < 0 \end{cases}$,

we have

for $x-2 \geqslant 0$, $x-2 < 3$ or $x < 5$, and

for $x-2 < 0$, $-(x-2) < 3$ or $-x+2 < 3$ or $-x < 1$ or $x > -1$.

Thus $|x-2| < 3$ means the same as $-1 < x < 5$.

This result makes sense, because the set of all points that are within 3 units of 2 is an open interval centered at 2 with radius 3. Therefore the endpoints of the interval are $2-3 = -1$ and $2+3 = 5$, thus $-1 < x < 5$.

This result is illustrated in the following plot of a portion of the real line.

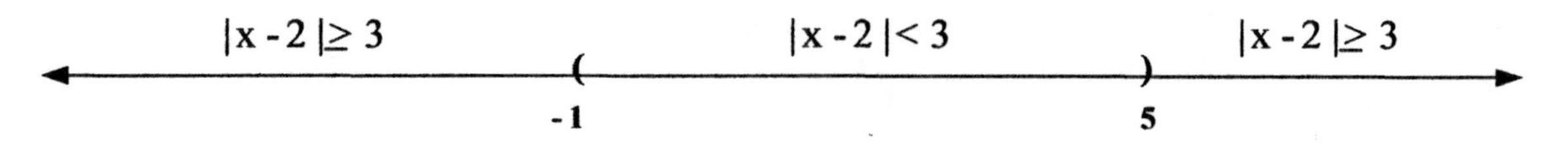

Figure 2.9.5

The left parenthesis at -1 opening to the right indicates that the endpoint -1 is not included. Similarly, the right parenthesis at 5 indicates that the endpoint 5 is not included.

Query 2.

1. Solve the absolute value equation, $|x-2| = 3$, using the definition of absolute value.
2. Solve the absolute value equation, $|x-2| = 3$, using the interpretation that this equation represents the set of all points 3 units from 2.
3. Use a calculator to solve the absolute value equation, $|x-2| = 3$. Hint: subtract 3 from both sides of the equality and then plot the left-hand side, $y = |x-2| - 3$. The solution(s) are the point(s) where the curve meets the x-axis.

Example 2.9.2

Use a calculator to graphically approximate the values of x which satisfy the inequality equation

$$|3x+4| - x < |x-5|.$$

Solution:

Subtract the right-hand side of the inequality from both sides of the inequality

$$|3x+4|-x-|x-5|<0$$

and then plot the left-hand side of the inequality.

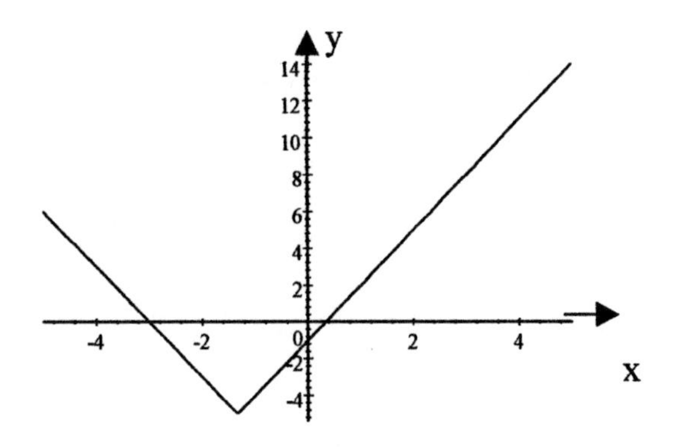

Figure 2.9.6

The values of x that will make the left-hand side of the inequality negative are precisely the values of x for which the plot is below the x-axis. So approximating where the plot intersects the x-axis, we obtain an answer $-3 < x < 0.33$. (Zooming in on the points of intersection could give a more accurate answer.) ▲

Small-Group Activity: Setting Traffic Light Signals

Traffic signals on high-volume highways allow for cross traffic and pedestrian crossing to occur. Particularly in large urban regions, traffic engineers time the traffic signals on high-volume highways to control traffic speed and to allow platoons of vehicles traveling in the same direction to clear all the signals without interruption. A traffic signal has three phases: green, amber, and red. The cycle time is the time (seconds) required for a signal to complete its green-amber-red sequence. Assume a cycle time of 85 seconds with 40 seconds for green, 5 seconds for amber, and 40 seconds for red. *Signal offset* is a technical term used in transportation engineering and planning that measures the lag time in the green phases of two successive traffic signals. For example, let A and B be successive traffic signals. The offset for B is the time between A turning green and B turning green. The offset for a traffic signal is defined to be the smallest time (seconds) difference between the preceding traffic signal turning green and the beginning of its green phase. Consider a sequence of five traffic signals, denoted by $A, B, C, D,$ and E on a high-volume highway. The distances of the traffic signals from A are B (2,000 ft), C (5,000 ft), D (7,000 ft), and E (9,000 ft).

1. Determine a common offset for the traffic signals $B, C, D,$ and E that allows a vehicle traveling 30 mph to clear all of the signals without interruption.
2. If the offset determined in part a were shortened, would a driver have to speed up or slow down in order to clear all of the signals without interruption? Explain.

Hint: let the offset be a variable. Locate each of the traffic signals on a time line. Determine a green interval for each of the traffic signal points that represents when the signal is green. Describe each of these intervals by a (double) inequality statement that gives upper and lower bounds for the offset.

Exercises 2.9

1. (Computational Skill) Determine if the following inequalities are correct.
   a. $7 - 3 < \frac{5+10}{3}$
   b. $\frac{2}{5} + \frac{3}{4} - 1 < 1$
   c. $(\frac{1}{2} + \frac{2}{5})(\frac{3}{4} - \frac{7}{8}) < (\frac{3}{2} - \frac{2}{3})$
2. (Computational Skill) Solve, graphically and analytically, for all values of x that satisfy the following inequality statements.
   a. $3x + 4 < -2(x - 5)$
   b. $2(3 - x) > 4x - 6$
   c. $3 < \frac{4}{x+2}$
3. (Computational Skill) Make up and solve three exercises similar to Exercises 1a–c, 2a–c.
4. (Calculator Skill) Use a calculator to graphically approximate the values of x that satisfy the following inequality statements.
   a. $3x + 4 < -2(x - 5)$
   b. $2(3 - x) > 4x - 6$
   c. $3 < \frac{4}{x+2}$
5. Express each of the following intervals, given in absolute value notation, in terms of a double inequality or two single inequality expressions (e.g., $|x - 2| < 3$ means $-1 < x < 5$).
   a. $|x + 2| < 3$
   b. $|x - 5| < 4$
   c. $|x - 1| \leqslant 5$
   d. $|x - 3| > 5$
   e. $|x - 3| > 2x$
6. Express each of the following intervals, given in inequality notation, in terms of absolute values.
   a. $-2 < x < 5$
   b. $2 < x < 6$
   c. $-6 < x < -2$
   d. $-3 \leqslant x \leqslant 6$
   e. $x < -2$ or $x > 6$

7. Sketch (by hand), on the same axes, the graphs of $f(x) = |x - 2|$ and $g(x) = |x| - 2$. Label your graphs and then check your work by plotting the two functions on your calculator.

8. Sketch (by hand), on the same axes, the graphs of $f(x) = |x + 2|$ and $g(x) = |x| + 2$. Label your graphs and then check your work by plotting the two functions on your calculator.

9. Sketch (by hand), on the same axes, the graphs of $f(x) = |2x|$ and $g(x) = 2|x|$. Label your graphs and then check your work by plotting the two functions on your calculator.

10. Select four different numerical values for c. Sketch, on the same axes, the graphs of $f(x) = |x - c|$ and then use your calculator to check your work.

11. Select four different numerical values for c. Sketch, on the same axes, the graphs of $f(x) = |x| - c$ and then use your calculator to check your work.

12. Select four different numerical values for c. Sketch, on the same axes, the graphs of $f(x) = |cx|$ and then use your calculator to check your work.

13. Select four different numerical values for c. Sketch, on the same axes, the graphs of $f(x) = c|x|$ and then use your calculator to check your work.

14. Write a paragraph explaining the following:
    a. How the graph of $f(x) = |x - c|$ is transformed for different values of c
    b. How the graph of $f(x) = |x| - c$ is transformed for different values of c
    c. How the graph of $f(x) = |cx|$ is transformed for different values of c
    d. How the graph of $f(x) = c|x|$ is transformed for different values of c

2.10 Linear Programming

Suppose a farmer has 200 acres of land on which she can plant any combination of the two crops, corn and wheat. Corn requires four worker-days of labor for each acre planted, while wheat requires one worker-day for each acre planted. Suppose also that corn produces $60 of revenue per acre and wheat produces $40 per acre. If the farmer has 320 worker-days of labor available for the year, what is her most profitable planting strategy?

In order to transform the statement of this problem into mathematical notation, we need to first identify and define the variables. The number of acres planted in corn and the number of acres planted in wheat are the variables. So let us define C and W as follows:

$$C = \text{number of acres of corn}$$
$$W = \text{number of acres of wheat.}$$

The constraints on land and labor are given by the following system of four linear inequalities:

Land	$C + W$	$\leqslant 200$
Labor	$4C + W$	$\leqslant 320$
Corn	C	$\geqslant 0$
Wheat	W	$\geqslant 0.$

The farmer's objective is to maximize her total revenue subject to the given constraints. (She receives $60 per acre for corn and $40 per acre for wheat.) Thus she wants to determine C and W in order to

$$\text{Maximize } 60C + 40W.$$

This linear expression $60C + 40W$ is called the **objective function**. That is, a function to be maximized subject to a set of constraints. The problem of maximizing or minimizing a linear expression subject to a set of linear equations or inequalities is called **linear programming**. Linear programming is one of the most important tools in management science.

Three geometric observations hold the key to solving this problem.

1. The first observation is that a straight line divides the plane into two half-planes. For example, the line given by the equation $3x + 2y = 6$ divides the plane into half-planes A and B as shown in the following plot:

Two Half Planes

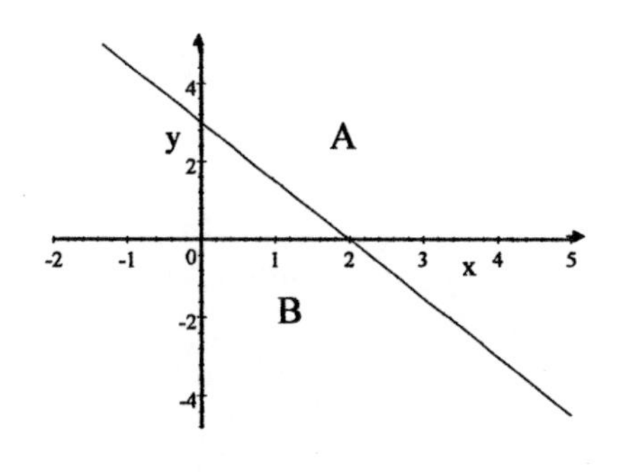

Figure 2.10.1

The linear inequality, $2x + 3y \leqslant 6$, then defines a half-plane, but which half-plane? Is this half-plane A or B? Because the origin point $(0, 0)$ satisfies the linear inequality $2x + 3y \leqslant 6$, B, the half-plane containing the origin, is defined by $2x + 3y \leqslant 6$. Thus the other half-plane, A, is defined by $2x + 3y > 6$. (The procedure for identifying a half-plane determined by a linear inequality is to select a point and check to see if it satisfies the linear inequality.) Geometrically

a linear inequality defines a half-plane.

Because the farmer's problem has four constraints given by four linear inequalities, the solution of the problem must lie in the region that is common to all four half-planes determined by the constraints. This common region is called the **feasible region.** The following shows the multiplot of the four constraints and the feasible region.

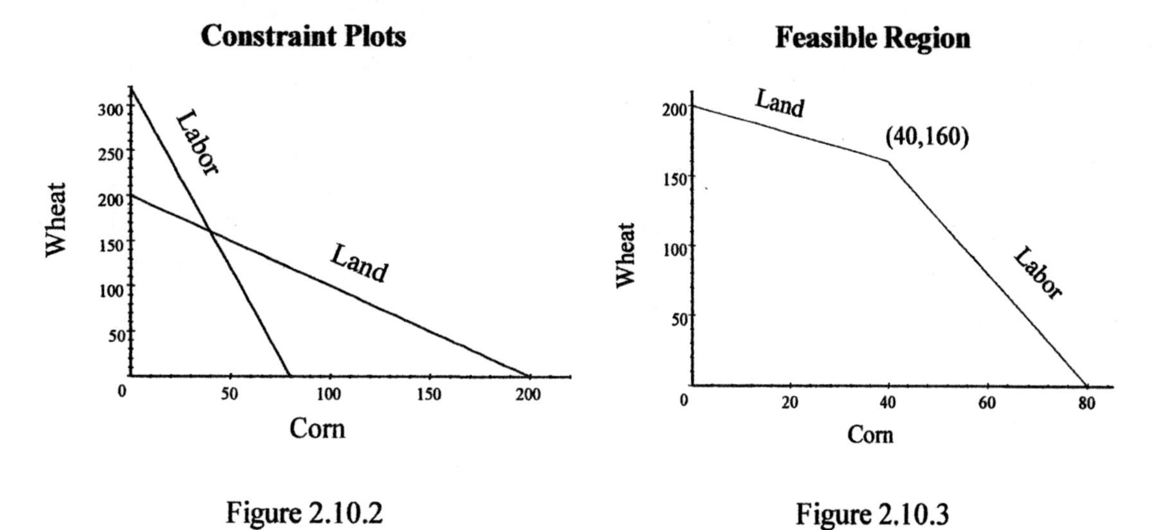

Figure 2.10.2

Figure 2.10.3

2. The second observation is that increasing either C or W will increase the value of the objective function: $60C + 40W$. Thus if you select any point inside the feasible region and then move to the right (increasing C) or up (increasing W), you will increase the value of the objective function $60C + 40W$. Therefore the maximum revenue must occur at some point on one of the boundary

line segments of the feasible region.

3. The objective function, $60C + 40W,$ is linear and thus when evaluated over a straight-line segment will assume its maximum value at one end of the line segment. (Why?) This means that the maximum value of the objective function occurs at a corner point of the feasible region. (Why?)

We give a graphical argument to illustrate Observation 3 as well as again illustrating the exploratory approach to mathematics. The farmer wants to maximize the value of $60C + 40W$. We make a guess for the value for the objective function, $60C + 40W$: say it is $4,000. That is, $60C + 40W = 4,000$. We superimpose the graph of this equation on the plot of the feasible region to see what we can observe.

Objective Function

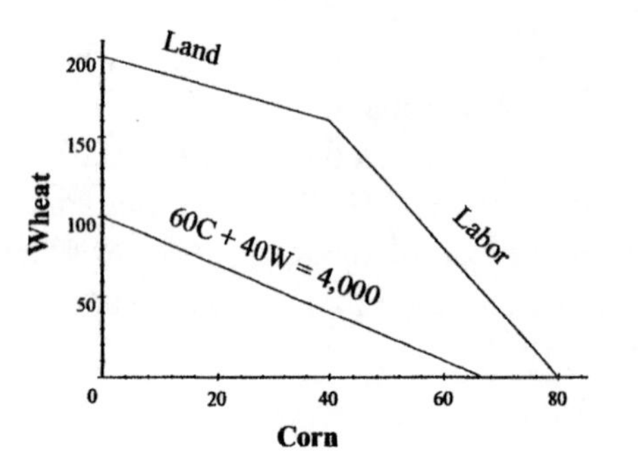

Figure 2.10.4

We observe that the line segment $60C + 40W = 4,000$ lies within the feasible region. This means that every point on the line segment satisfies all four constraints, and for each point the farmer would receive $4,000 in revenue. For example the point $(20, 70)$ lies on the line segment. When the coordinates of the point $C = 20$ and $W = 70$ are substituted into the constraint inequalities, each inequality is satisfied and when the coordinates are substituted into the objective function the result is 4,000.

We also observe that the farmer can do better because each point on the line segment can be moved either upward or to the right, both of which would increase the value of the objective function (that is, the farmer's revenue). Thus let us make a second guess for the value of the objective function, say $10,000. That is, $60C + 40W = 10,000$. We superimpose the graph of this line on the previous plot.

Objective Function

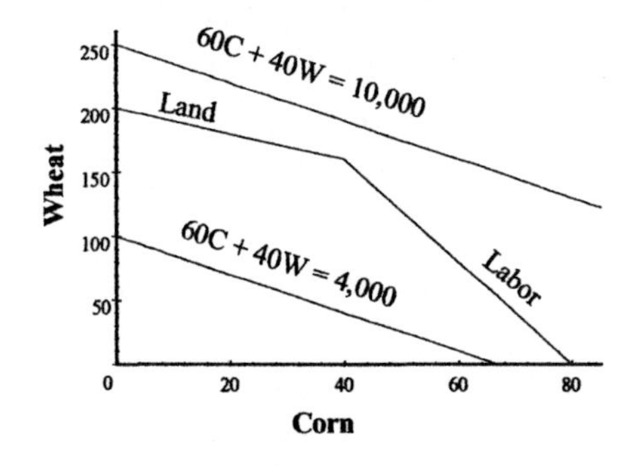

Figure 2.10.5

The second line, $60C + 40W = 10,000$, lies outside the feasible region. This means that no point on the line satisfies all four constraint inequalities and therefore no matter what division the farmer makes of her land between corn and wheat, she can never receive $10,000 for her crops. Thus the maximum revenue possible must be between $4,000 and $10,000.

We observe that the two lines are parallel, because changing the value of the objective function does not change the slope, which is determined by the coefficients of the variables C and D. Rewriting this last equation in slope-intercept form, we have

$60C + 40W = 10,000$ or $40W = -60C + 10,000$ or $W = -\frac{3}{2}C + 250.$

Thus, in this problem, the slope is $-\frac{3}{2}$, regardless of the value of the objective function.

We also observe that increasing the value of the objective function shifts the line upward. This observation suggests that if we gradually increase the value of the objective function starting at $4,000, we will eventually obtain a value where the plot of the resulting equation just touches one corner of the feasible region. This is what happens when the objective function has the value $8,000 as illustrated in the following plot.

Objective Function

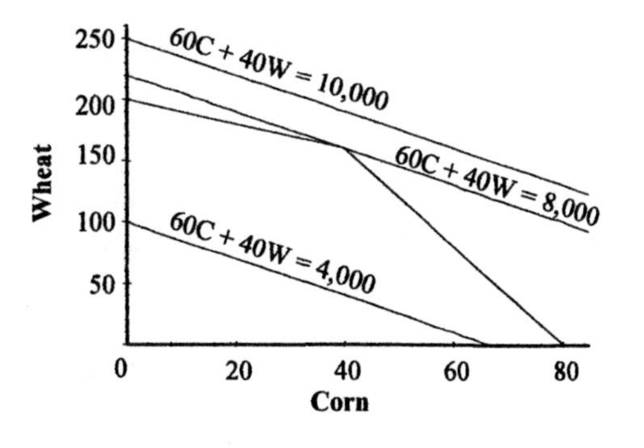

Figure 2.10.6

Therefore, there is a solution when the objective function is equal to 8,000, but for no larger value. This is exactly what Observation 3 claimed.

Thus a method for solving a linear programming problem is to:

First plot the constraint equations to determine the feasible region and then evaluate the objective function at each of the corner points of the feasible region.

The coordinates of a corner point of two intersecting constraint lines are determined by solving the corresponding two linear constraint equations. For example, to find the coordinates of the corner point of the land and labor lines, we do the following

$$\begin{cases} \text{Land} & C+W = 200 \\ \text{Labor} & 4C+W = 320 \end{cases}$$

W	$=$	$200-C$	Solve for W in land equation
$4C+200-C$	$=$	320	Substitute for W in labor equation
$3C+200$	$=$	320	Simplify
$3C$	$=$	120	Subtract 200
C	$=$	40	Divide by 3
W	$=$	$200-40$	Substitute for C in the land equation
W	$=$	160	Simplify.

Thus $(40,160)$ is the corner point in the feasible region where the labor and capital constraint lines intersect. The remaining corner points are found in a similar manner. The results are

(C,W) : **Corner Coordinates**	**Intersecting Boundary Segments**	**Objective Function**
$(0,0)$	$C \geqslant 0$ and $W \geqslant 0$	0
$(0,200)$	$C \geqslant 0$ and Land	8,000
(40,160)	**Land and Labor**	**8,800**
$(80,0)$	Labor and $W > 0$	4,800

Thus the farmer's optimal strategy is to plant 40 acres of corn and 160 acres of wheat. The income for this strategy is $8,800.

We now revisit the previous problem and make it a bit more realistic by including a capital constraint. We will show that the solution of this revised problem illustrates a portion of the federal program of price supports for farmers.

Revised Problem

Suppose a farmer has 200 acres of land on which she can plant any combination of the two crops, corn and wheat. Corn requires four worker-days of labor and $20 of capital for each acre planted, whereas wheat requires 1 worker-day of labor and $10 capital for each acre planted. Suppose also that corn produces $60 of revenue per acre and wheat produces $40 per acre. If the farmer has $2200 of capital and 320 worker-days of labor available for the year, what is her most profitable planting strategy?

We define the variables, C and W, as before.

$$C = \text{number of acres of corn}$$
$$W = \text{number of acres of wheat}$$

Including the new constraint on capital, we now have the constraints given by the following system of five linear inequalities.

Land	$C + W$	$\leqslant 200$
Labor	$4C + W$	$\leqslant 320$
Capital	$20C + 10W$	$\leqslant 2200$
Corn	C	$\geqslant 0$
Wheat	W	$\geqslant 0$

The farmer's objective is to maximize her total revenue subject to the given constraints. (She receives \$60 per acre for corn and \$40 per acre for wheat.) That is, she wants to determine C and W in order to maximize her objective function $60C + 40W$.

The three observations stated in the previous problem also apply in this revised problem. The plots of the constraints and the feasible region are

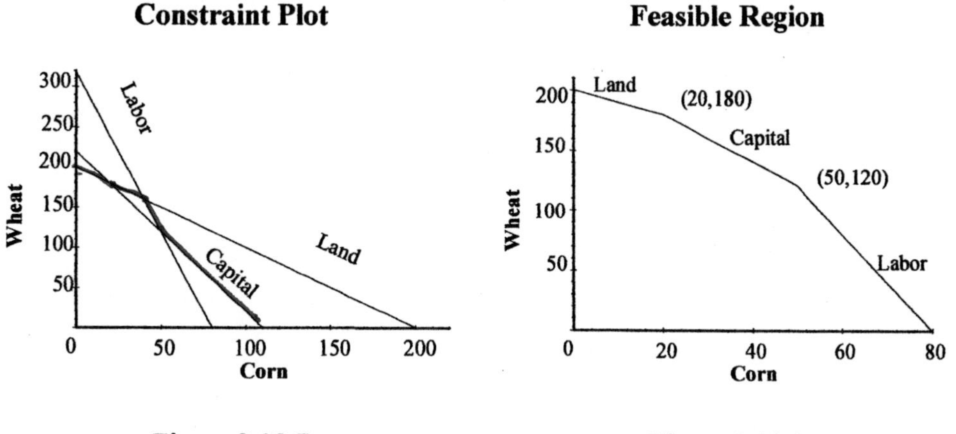

Figure 2.10.7

Figure 2.10.8

As in the previous problem, we graphically show the results of guessing values for the objective function. We guess 4,000 and 10,000 as before.

Objective Function

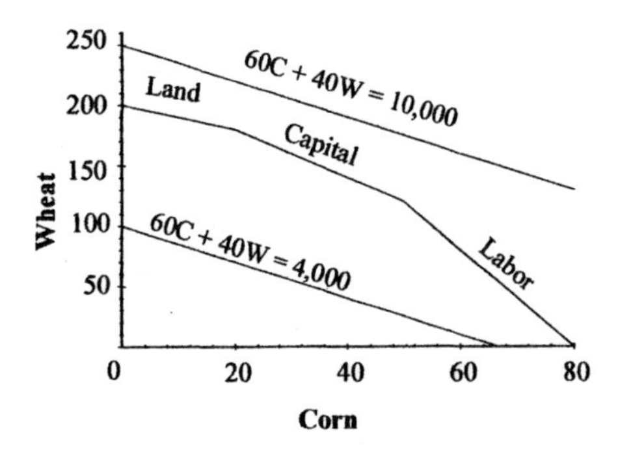

Figure 2.10.9

We observe that the line segment $60C + 40W = 4,000$ lies within the feasible region. This means that the farmer can do better. That is, she can increase the value of her objective function. However, as the plot shows, she cannot increase the value to $\$10,000$ as the line $60C + 40W = 10,000$ lies outside the feasible region. Thus the maximum revenue possible must be between $4,000 and $10,000.

As in the previous problem the lines for different values of the objective function are parallel, because changing the value of the objective function does not change the slope, which is determined by the coefficients of the variables C and D, which in this problem is $-\frac{3}{2}$.

The following plot shows the graph of the objective function meets the feasible region in just one point when the objective function has value $8,400.

Objective Function

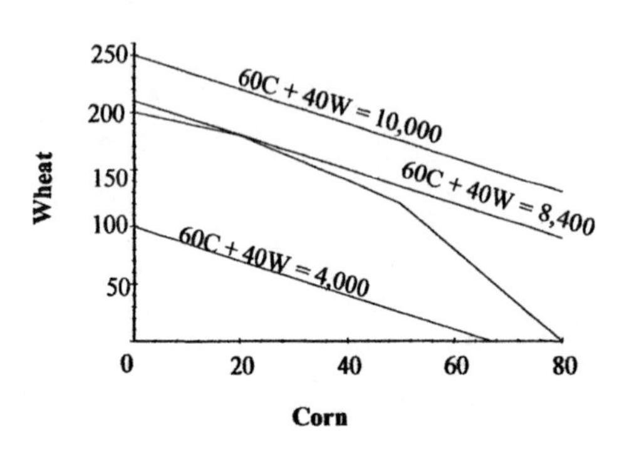

Figure 2.10.10

Therefore, there is a solution when the objective function is equal to 8,400, but for no larger value.

We obtain the coordinates of the corner points as before by solving the equations of pairs of intersecting constraint equations. For example, to find the coordinates of the corner point of the labor and capital lines, we do the following

$$\begin{cases} \text{Labor} & 4C + W = 320 \\ \text{Capital} & 20C + 10W = 2200 \end{cases}$$

W	$=$	$320 - 4C$	Solve for W in labor equation
$20C + 10(320 - 4C)$	$=$	2200	Substitute for W in capital equation
$20C + 3200 - 40C$	$=$	2200	Simplify
$-20C$	$=$	-1000	Simplify, subtract
C	$=$	50	Divide by -20
W	$=$	$320 - 200$	Substitute into labor equation
W	$=$	120	Simplify

Thus $(50, 120)$ is the corner point in the feasible region between the labor and capital constraints. The remaining corner points are found in a similar manner. The results are as follows.

Corner Coordinates	Intersecting Boundary Segments	Objective Function
$(0,0)$	$C \geqslant 0$ and $W \geqslant 0$	0
$(0,200)$	$C \geqslant 0$ and Land	8,000
$(20,180)$	**Land and Capital**	**8,400**
$(50,120)$	Capital and Labor	7,800
$(80,0)$	Labor and $W > 0$	4,800

Thus the farmer's optimal strategy is to plant 20 acres of corn and 180 acres of wheat. The income using this strategy is \$8,400.

Query 1.

Verify that the coordinates used for the corner points in the preceding problem are correct. Hint: To find the coordinates of a corner point, solve the system of the two equations for the two intersecting boundary line segments.

Before leaving this problem, we note that the farmer's constraints (land, labor, and capital) determined the feasible region, and the marketplace (prices for corn and wheat) determined the objective function. If the market prices for corn and wheat change, or equivalently, the farmer receives a subsidy for one of the crops, the optimal solution may change.

For many years, the federal farm program has offered crop subsidies in order to influence both the types of crops grown and the total number of acres planted. To illustrate how a crop subsidy can cause land to be taken out of production, let us suppose that in the preceding problem the farmer receives a subsidy for corn that increases her revenue from \$60 to \$90 per acre. The objective function is then $90C + 40W$. The values of this new revenue function at the corner points of the feasible region are shown in the following table.

Corner Coordinates	Intersecting Boundary Segments	Objective Function
$(0,0)$	$C \geqslant 0$ and $W \geqslant 0$	0
$(0,200)$	$C \geqslant 0$ and Land	8,000
$(20,180)$	Land and Capital	9,000
$(50,120)$	**Capital and Labor**	**9,300**
$(80,0)$	Labor and $W > 0$	7,200

The new optimal strategy is to plant 50 acres of corn and 120 acres of wheat, for a total of 170 acres. The subsidy has resulted in the farmer removing 30 acres from production.

The four observations from the preceding problem provide a graphical method for solving a linear programming problem.

1. Plot the lines corresponding to the constraints in order to determine the feasible region.
2. Determine the coordinates of each of the corner points on the boundary of the feasible region.
3. Evaluate the objective function at each of the corner points.
4. The largest value is the maximum value of the objective function subject to the constraints.

Query 2.

Is it possible for the solution of a linear programming problem to occur at every point along a boundary segment as well as at a corner point? Explain.

People in business as well as in their daily lives are often faced with optimization problems that involve a number of constraints such as capital, time, volume, equipment, labor and so on. When the constraints can be expressed as linear inequalities and the objective function is linear, the optimization problem is a linear programming problem. The petroleum industry is the largest user of linear programming. Problems involving 400 to 600 or more activities and 200 or more constraints are common in the industry. Before the 1950s, when linear programming was developed, refinery managers would approximate solutions to these complex problems based on trial-and-error methods and experience. There was no way to determine how optimal these approximate solutions were. An important aspect of linear programming is that it provides an optimal solution. In an industry such as the petroleum industry that refines thousands or millions of barrels per day, a small change in the profit per barrel can cause a major change in the total profit.

Exercises 2.10

1. (Computation Skill) In the following, determine if the given point satisfies the inequality.
   a. Point $(x,y) = (1,3)$, inequality $2x + 3 < y + 3$
   b. Point $(x,y) = (-1,-2)$, inequality $2x + 3 < y + 2$
   c. Point $(x,y) = (0,0)$, inequality $2x + 3y < 5$

2. (Computational Skill) In each of the following, sketch a picture showing how the given inequality divides the plane into two half planes and then shade the half plane defined by the inequality.
   a. $2x + 3 < y + 3$
   b. $2x + y < 4$
   c. $3(x - y) < 2x + 5$
3. (Computational Skill) Make up and solve three exercises similar to Exercises 1a-c, 2a-c.
4. Answer the Queries in this section.
5. In reference to the Revised Problem in this section, determine a government subsidy for wheat that would encourage farmers to only plant wheat. (Assume that corn produces $60 of revenue per acre.)
6. A carpenter makes straight chairs and rocking chairs. He sells a straight chair for $15 and a rocking chair for $20. Cutting and assembling takes 2 hours for a straight chair and 4 hours for a rocking chair. Sanding, painting, and polishing takes 2 hours for each type of chair. If he spends 5 hours per day cutting and assembling and 3 hours per day sanding, painting, and polishing, how many chairs of each kind should he make in a 6-day week, working 8 hours per day?
7. Rework Exercise 6 with the assumption that the carpenter can sell a rocking chair for $30 and a straight chair for $15.
8. The new Pizza-Pizza shop is open Tuesday through Saturday from 11:00 A.M. to midnight and closed on Sundays and Mondays. The manager runs two overlapping shifts: the first is from 11:00 A.M. to 7:00 P.M. and the second is from 4:00 P.M. to midnight. She estimates that she needs at least 4 people during the 11 A.M. to 4 P.M. period, 12 people during the 4:00 P.M. to 7:00 P.M. period, and 6 people from 7:00 P.M. to midnight. She wants you to answer the following questions.
   a. If she pays workers on the first shift $7 per hour and workers on the second shift $8 per hour, how many people should be hired on each shift in order to minimize the 40-hour-per-week payroll?
   b. If she pays workers on the first shift $7.50 per hour and workers on the second shift $8 per hour, how many people should be hired on each shift in order to minimize the 40-hour-per-week payroll?
9. Suppose that a meal must contain at least 500 units of vitamin A, 1000 units of vitamin C, 200 units of iron, and 50 units of protein. A dietician provides the following information.

   Meat: one serving of meat has 20 units of vitamin A, 30 units of vitamin C, 10 units of iron, and 15 units of protein.

   Fruit: one serving of fruit has 50 units of vitamin A, 100 units of vitamin C, 1 unit of iron, and 2 units of protein.

   A serving of meat costs 50 cents and a serving of fruit costs 40 cents.

   a. State the linear programming problem of minimizing the cost of a meal of meat and fruit that meets all the minimum nutritional requirements.
   b. Plot the feasible region for your linear programming problem.
   c. Solve the linear programming problem

10. A dairy farmer wants to supplement four vitamins in the feed that he buys. He has found two products that contain the four vitamins that he wants to supplement. His question is, should he buy just one or the other of the two products or should he mix them in order to meet (or exceed) the minimum vitamin requirements and do so at minimum cost? The pertinent facts for the situation are summarized in the following table.

	Product 1	Product 2
Cost per ounce	3 cents	4 cents
Vitamin 1 per ounce	5 units	25 units
Vitamin 2 per ounce	25 units	10 units
Vitamin 3 per ounce	10 units	10 units
Vitamin 4 per ounce	20 units	20 units

The farmer needs to provide, for each hundred pounds of feed, at least

50 units of vitamin 1
100 units of vitamin 2
60 units of vitamin 3
100 units of vitamin 4.

Determine the objective function to be minimized, express each of the constraints in the form of a linear inequality, plot the feasible region, and then answer the farmer's question, including the minimum cost.

11. The manager of a large curtain and drapery shop agrees to hire at least three college students during the summer as apprentices. She pays her experienced seamstresses $12 per hour and will pay the apprentices $7 per hour. In order to maintain the shop's reputation for high quality, she wants to have at least three times as many experienced seamstresses as apprentices. She considers that an apprentice accomplishes about 70% as much as an experienced person. The manager needs to hire the equivalent of at least 15 experienced people and also wants to minimize her 40-hour-per-week payroll. Determine how many apprentices and how many experienced seamstress she should hire. Only whole numbers are acceptable answers.

2.11 Chapter Summary

Algebra is the study of variables, operations on them, and dependency relations between them.

Variables are mathematical objects that may assume different numerical or qualitative values. Information about a variable is called **data.** Section 2.1 describes five ways in which data is commonly displayed: table, bar chart, pie chart, scatter plot, and line graph. Data on pairs of related variables is usually represented by either a graph or a table. When represented by a graph, one variable is denoted as the **independent variable**, and its values are measured along the horizontal axis. The other variable is denoted as the **dependent variable**, and its values are measured along the vertical axis. The context of the situation will determine which variable is independent and which is dependent. (Review the four questions at the beginning of Section 2.4.) A point on the graph is an **ordered pair**. The first entry is the value of the independent variable, and the second entry is the corresponding value of the dependent variable.

In order to communicate efficiently, we will usually label variables by letters that are suggestive of the name of the variable. For example, a cost variable may be labeled by the letter *c*. Although this procedure makes a lot of sense, it is not always followed. In fact, there is a long mathematical tradition for denoting a variable by the letter *x*. In some texts the convention is to use letters near the beginning of the alphabet for labeling constants and letters near the end of the alphabet for labeling variables. Within the description of a problem and its solution, each variable will have its own label. The same label will never be applied to different variables within the same problem and solution. However, a particular label may be applied to different variables in different problems. (The label "*x*" has been applied to millions of variables.) Thus it is very important to begin every solution procedure by defining the labels for the variables that will be used in that problem and solution.

Large amounts of data and sometimes even small amounts of data contain so much information that we can be overwhelmed and have a great deal of difficulty extracting useful information. For example, suppose we had a listing of all of the foul shots that Michael Jordan has taken as a professional basketball player with an indication of which ones he made. How could we extract useful information from such a list? Statisticians summarize data in several different ways. In this text, we use the three most common methods: **average, median,** and **mode**. These methods are described in section 2.3. We need to be very careful to understand the differences between these three methods of summarizing data, and to realize that any summary will focus on a particular aspect of the data and possibly hide other aspects.

Two variables are linearly related when the points on their graph lie on a straight line. Section 2.5 describes how to compute the equation of the straight line determined by two linearly related variables. In particular, we have the following characterization of a straight line: A straight line is completely determined by its slope and a point on the line.

The **slope** of a line represents the average **rate of change** of the dependent variable with respect to the independent variable. Because we can compute the rate of change (slope) from any two points on a straight line, we can determine the equation of the line from any two points on the line. The ability to compute the equation of a straight line and to be able to determine the slope (rate of change) from the equation of a straight line is fundamentally important in the study of mathematics. This is the emphasis of Section 2.6.

Applications of linear equations and linear inequalities (Sections 2.7–2.9) are at the heart of applied mathematics. In many applied situations the first major step is to "linearize the data," that is, to assume that the relations between variables are linear. This leads to systems of linear equations or systems of linear inequalities. The latter are involved in solving linear programming problems using methods we developed in Section 2.10. Whether you plan to farm, operate a dressmaking shop, be a rocket scientist, or any other profession, you need to develop skill in solving linear equations and linear inequalities.

Linear programming, the subject of Section 2.10, was developed by George Dantzig and others in the 1950s. (The first problem he solved using his new technique of linear programming was a minimum cost diet problem.) The applications of linear programming to problems with hundreds of constraints expanded as computers became more and more powerful. Petroleum and chemical industries are major users of linear programming for minimizing costs of blending fuels. The telephone industry is another major user of linear programming for developing schedules and routing telephone calls.

Fun Projects

Fun Projects are small-group (three to five students), out-of-class projects. They are designed for six to ten hours of work. Instructors are strongly urged to provide a full class period for student groups to meet and work on their project as well as class time for student groups to present and discuss their work. Instructors are also encouraged to formulate and present an interesting story line to introduce a project.

Projects culminate in a written report consisting of:

- Cover Page (creative design by students)
- Title Page (project name, date, instructor name, students' names)
- Table of Contents
- Executive Summary (one-page abstract of the problem, approach used, results obtained)
- Supporting Data (computations, labeled drawings, labeled computer plots, and/or printouts)
- Group Log (time, date, location, and brief description of each meeting)
- Evaluation Summary of the group's learning experience in working on the project
- List of References Consulted

All group members should be involved in answering each of the questions. In addition, each member of the group should be assigned a particular responsibility in connection with the project such as the following:.

Leader: Responsible for developing the group. Responsible for seeing that the projecat is completed in a satisfactory manner and on time.
Recorder: Arranges group meetings and records group activities.
Checker: Checks accuracy of all computations. Checks to see that all questions are answered.
Typist: Types Executive and Evaluation Summaries.
Reader: Responsible for proofreading and final assembly of the report.

These real-world projects are aimed at developing the whole student, not just their mathematics ability. For example, Fun Projects may require students to do comparison shopping, interview officials, write letters to businesses, conduct surveys, research a topic, as well as compiling their work into a formal report and making a formal presentation. Learning how to work effectively as a team member is a valued skill today in business and industry where working in teams is the standard practice rather than an exception. Working on Fun Projects addresses the primary course goal of developing students to become exploratory learners in the quest to develop them into confident and competent problem solvers as well as the six other goals described in the Preface.

Instructors are strongly encouraged to modify the Fun Projects in order to individualize them to their classes. (The pictures accompanying the following Fun Projects as well as the Fun Projects following Chapters 3 and 4 were scanned from student reports.)

1. Doubling Leads to Large Numbers

(Purposes: Provide an introduction to using a calculator; provide experience in converting from one system to another; raise questions of how to think about large numbers that are seen in newspapers, magazines, on TV, and so on; provide a writing exercise; provide a small-group experience.)

Part I: The Power of Doubling

Consider an 8 x 8 checkerboard. Number the blocks 1 through 64. Place two pennies on Block 1, four on Block 2, eight on Block 3, and so on, doubling the number of pennies as you move from one block to the next. Answer the following questions.

1. Determine the number of pennies on:
   Block 8________Block 16______Block 32_______Block 64_______.
2. How much money (in dollars) is on Block 64?
   How much money (in dollars) is on Block 63?
3. What is the average amount of money on all blocks?
   What is the block number that contains the closest amount to the average?
4. How much money (in dollars) would be on Block 64 if the first block had only one penny instead of two? Explain your reasoning.
5. Estimate and then compute the height of the pennies on Block 64 if all of the pennies were stacked one on top of another. (Assume Block 1 holds two pennies.)
   a. Record each group member's estimate of the height (in feet).
   b. Compute:
      i. Average thickness of a penny (inches)_________ (centimeters)_________.
      ii. Height of pennies on Block 64 in feet, miles, meters, kilometers.
6. Compare the height of the stack of pennies on Block 64 to the distance from
   a. Earth to the moon
   b. Earth to the sun
   c. Earth to Proxima Centuri, the closest star to earth (~4 light years, light travels ~186,000 miles/sec)

Part II: Visualizing the Number of Pennies on Block 64

1. Count the number of beans in a pound. (Identify the type of bean.) If the pennies were beans, compute how many pounds of beans would be on Block 64.
2. Determine the dimensions of a cubical box that would just hold the beans on Block 64. Write a short paragraph explaining how you determined your answer.
3. If cubical box A just holds all the beans on Block 32 and cubical box B just holds all of the beans on Block 64, what is the ratio of the side dimension of box A to the side dimension of box B?

Part III: How to Understand the Relative Sizes of Thousand, Million, Billion, Trillion

1. Consider the following: the national debt is measured in trillions, the Brookings Institute estimated the United States spending in the nuclear arms race from 1940 through 1996 was $5.5 trillion; a B-2 bomber costs 2.5 billion; Fox, ABC/ESPN, and CBS paid $17.6 billion to the NFL network for TV rights for eight years; the price tag on the Cleveland Browns franchise is reportedly $500 million and their proposed new stadium is expected to cost $250 million; superstars are demanding and receiving salary contracts for more than $100 million, as are some corporate executives. How do we comprehend these large figures?
2. Draw a line, then mark one end zero and the other end one billion. On your line, have each member of the group draw (without computing) a hash mark representing one thousand and another for one million. Then measure the length of the line and compute the correct location of the hash marks.
3. Draw a line, then mark one end zero and the other end one light year. On your line, have each member of the group draw (without computing) a hash mark representing the distance to the moon and another the distance to the sun. Then measure the length of the line and compute the correct location of the hash marks.
4. Write a paragraph describing a creative way to help your parents and friends develop a sense of the comparative sizes of large magnitudes. For example, what is a meaningful way to comprehend the size of a thousand, million, billion, trillion, or light year in terms of a smaller quantity, say one hundred?

2. City and Country Populations

(Purposes: Research on the Internet; raise a a real-life issue and challenge the students to provide a creative solution; provide a creative writing exercise; provide a small-group experience.)

Introduction: Concerns over population growth invade all aspects of life. Urban planners are challenged to plan for the growth of urban centers, environmentalists are concerned about the over-or-underpopulation of certain species, transportation engineers and planners are tasked with responsibilities to provide facilities for moving people and goods, and the agriculture sector has the daunting task of providing food for an expanding world population.

In 1798, the British economist Thomas Malthus predicted that the world's population would eventually starve to death. His reasoning was that the growth of the food supply was linear whereas the population was growing at an exponential rate. Thus the growth of the population would eventually exhaust the world's food supply and a worldwide famine would ensue.

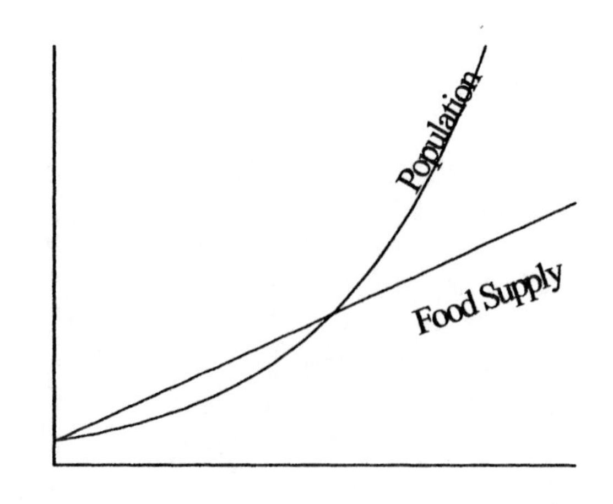

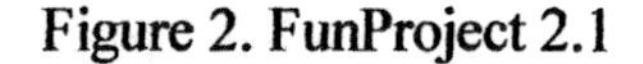
Figure 2. FunProject 2.1

In the mid-1800s a Belgian mathematician, P. J. Verhulst, modified the Malthus theory by predicting that there was a maximum population carrying capacity, M, that the world could sustain. Furthermore as the population approached M, the rate of the population growth would decrease. The resulting S-shaped curve is called a **logistic curve**.

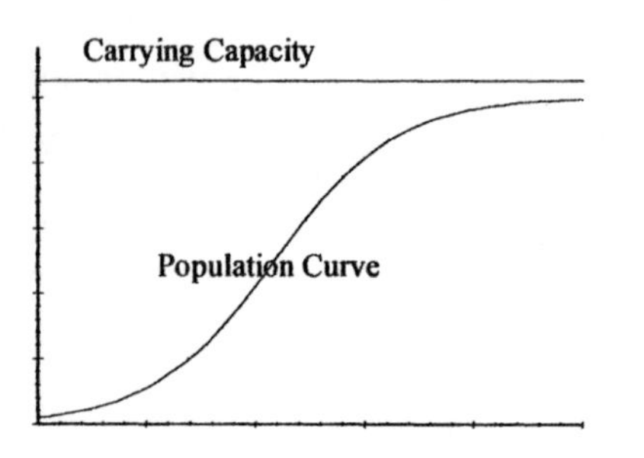

Figure 2. FunProject 2.2

Anyone who has experienced traffic jams in large cities such as Houston, Dallas, Atlanta, or New York has a sense of overpopulation. In contrast, there are areas in every state where the population density is so low as to raise questions about the seriousness of overpopulation concerns. Complete the following tasks.

1. List the names and the populations of the ten largest cities in the United States. Explain how you obtained the data. (Hint: U.S. census information can be found on the Internet at http://www.census.gov and http://www.census.gov/population/censusdata)
2. Compute the percentage of the U.S. population that lives in the ten largest cities. Explain how you

obtained the data.

3. Select a state and then determine:
   a. The percentage of the state's population that lives in the state's largest city.
   b. Assume the shape of the largest city is a disk. Estimate the radius of the disk and then compute the corresponding approximate area of the city.
   c. Compute the percentage of the state's area contained within that city.
4. Write a two-page analysis and recommendation for creating a new city in a sparsely settled region outside a major city in order to relieve the population growth of the city. For example, consider a location thirty to fifty miles outside of Atlanta or midway between Houston and Dallas or fifty miles north or west of New York City. Your analysis should address, but not be restricted to:
   a. Transportation
   b. How to attract business and industry to the new site
   c. Quality of life
   d. Responsibility to plan and implement (should it be a group of citizens, the city, the state, or the national government)

3. Pepsi, Coke, or...?

(Purposes: Provide experience collecting, displaying, and analyzing data; provide an inquiry and writing exercise; provide a small-group experience.)

Introduction: A new vending company called *Drink Up* is planning to submit a bid for the vending services on your campus. They have asked your group to help them determine what varieties of soda they should offer in their canned drink machines and also what portion of the capacity of a machine should be devoted to each of the varieties. The machines can accommodate no more than six varieties. Also the maximum capacity of a machine is 480 cans. Finally, they ask you to recommend, with reasons, how often their machines should be serviced (refilled).

In particular, the *Drink Up* company asks your group to do the following:

1. Survey 50 or more students in order to determine the six most favorite varieties of soda and to determine a prediction for the number of cans of each variety that will be bought in a typical Monday through Friday week. Provide a written description of your survey. (For example, what questions did you ask? Where did you survey?) You also need to provide a written explanation for how you determined your predictions.
2. List, in order with the most popular listed first, the six most favorite choices of soda.
3. Predict the average number of cans and the corresponding percentages of the total sales for each of the top six varieties that will be bought each day on your campus. Explain the reasoning involved in your prediction.
4. For a validity check of your choices of soda and your predictions, you should do the following:
   a. Interview the manager of the food service company on your campus to determine the choices of soda offered in the dining halls, the volume consumed, and percentage of total volume for each variety for a typical Monday through Friday week. Compare these results against those your group developed.
   b. Interview the manager of a fast-food restaurant near campus to determine the choices of soda offered, the volume consumed, and the percentage of the total volume for each variety for a typical Monday through Friday week. Compare these results against those your group developed.
5. Display the data on soda variety, volume, and percentage determined from your survey, interview with the dining hall manager, and interview with the manager of the fast-food restaurant. If there are major discrepancies in the data, offer suggestions to explain the discrepancies.
6. If the total number of purchased drinks were spread approximately equally over ten machines on campus, how often should the *Drink Up* company plan to refill their machines?

4. The Chain Letter

(Purposes: Recognize exponential growth in a simple context; experience converting from one system to another; gaining a sense for large quantities; research on the Internet; provide a writing exercise; provide a small-group experience.)

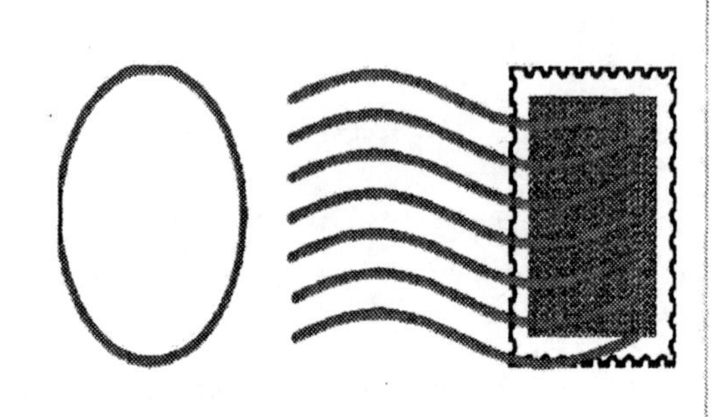

Introduction: The fall of communism in Albania in the early 1990s left a vacuum in the banking and regulatory system. Corrupt pyramid savings type schemes, called Ponzi schemes, spread across Albania. The schemes became accepted and legitimized by the government, which led many of the common people to become involved. As more people became involved, it became harder and harder for the schemes to raise additional money by finding new people to join or convincing those already involved to invest more funds. In 1997, the Albanian pyramid savings funds collapsed, touching off a total breakdown of law and order as furious citizens went on rampages sacking their own state in attempts to retrieve their money. Foreign diplomats were evacuated for their safety as the government fell.

A recipe chain letter is similar to a pyramid scheme, except there is no money involved. In such a scheme each person invites a fixed number, say n, new people to join. Each of these people invite n others to join and these n people invite n others to join, and so on. The number of persons involved can be pictured as a pyramid with one person at the top (the instigator), n people on the second line (first iteration of invitations), n^2 people on the third line (second iteration of invitations), n^3 people on the fourth line (third iteration of invitations), and so forth. Thus the name *pyramid* for the name of the scheme. Each person on the bottom level has a path to the person at the top of the pyramid. This path is represented by a ordered list of names. The top or first name is that of the person at the top of the pyramid, the second name is that of the person on the second level, and so on. To invite a person to join, you send them your genealogical path list with instructions to send a recipe to the top person on the list, cross off this name, add their name to the bottom of the list, and then send the revised list to n potential joiners. Of course, there are variations of this scheme. For example, if the chain letter asked the receiver to also send $5 with the recipe, the chain letter would be a pyramid scheme.

There are substantial costs involved in a chain letter such as time and cost of paper and postage. Questions 1-6 are designed to provide a "sense" of the total cost of a an innocent chain letter. Question 7 asks you to do some research and then write a two-page essay.

Dear Friend,

This is a recipe tree. There is no money involved so it is fun and legal. Please send a copy of your favorite recipe to person A at the bottom of this letter. Cross off person A's name, move my name from position B to position A and insert your name in position B. Copy the revised letter and send it to five different people. Do not send a recipe to them. It will be fun to receive the recipes and interesting to note their origin. You should receive 25 recipes. If you cannot do this within one week, send this back to me or it will spoil the fun for the rest of us. Thanks and happy eating.

Sincerely,
Joe

A Jane Dish, Serve Street, Anywhere, U.S.A
B Joe Pot, Bake Street, Somewhere, U.S.A

Answer the following questions assuming that:

a. Each person who receives a copy of the recipe letter forwards it to five people and this continues for fifteen iterations.
b. No person receives more than one of these letters.

1. Compute the number of persons who received this letter and compare that number to the population of the United States and to the population of the world. (Describe how you determined the populations of the United States and the world.)
2. Assuming that a 34 cent postage stamp is used to send each letter, compute the cost of the postage to send all of these letters. Compare the cost of postage to the gross domestic product of Grenada

(approximately $300 million).

3. Paper is usually sold in reams. Assume that a ream of paper contains 480 sheets. How many reams of paper are required for the letters, assuming one sheet per letter?

4. If all the reams of paper were stacked on top of each other, how tall (kilometers, km) is the resulting stack, assuming a ream of paper is 4 centimeters thick. Would the stack extend into the earth's troposphere (approximately 17 km), into the earth's stratosphere (approximately 50 km), into the earth's mesosphere (approximately 90 km), into the earth's ionosphere?

5. Assuming each sender takes an average of five minutes to produce, address, stamp, and mail each letter, how many years will have been spent on this chain letter?

6. Assuming an average person begins working at age 21 retires at age 65, and works 48 forty-hour weeks each year, compute the number of lifetimes of work involved in this chain letter.

7. Write a two-page paper discussing one of the following (Hint: Check out pyramid and Ponzi schemes on the Internet):

   a. The difference between chain letters, pyramid schemes, and Ponzi schemes
   b. Why all pyramid schemes must collapse
   c. Why pyramid and Ponzi schemes are fraudulent

5. Packaging

(Purposes: Explore relations between area and circumference and between volume and surface area; research dimensions of rectangles, cubes, spheres, and cones; provide a writing exercise; provide a small-group experience.)

Introduction: A trip to any supermarket reveals examples of different forms of packaging. In general, the shapes of two-dimensional packages are usually rectangular, circular, or triangular whereas three-dimensional packages are usually box, cylindrical, or conical.

1. Ruby has 100 feet of fencing to fence in five rectangular runs for a small kennel. The runs are to be identical in size, and each run is to be twice as long as wide. (See the following diagram.) Determine the dimensions of the runs.

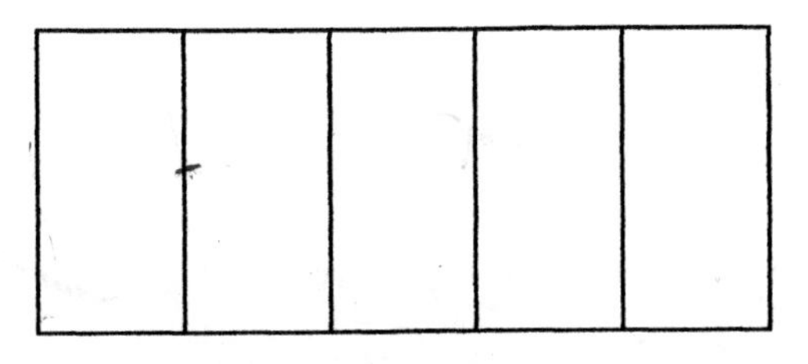

Figure 2. FunProject 5.1

2. Determine the dimensions of a cube that has the same surface area as a sphere of radius 6 inches. Compute the volume of the cube and the volume of the sphere. Which is larger?

3. Determine the maximum volume of a conical cup that can be constructed by cutting a wedge out of a circular disk and then joining the cut edges. (See the following diagrams.)

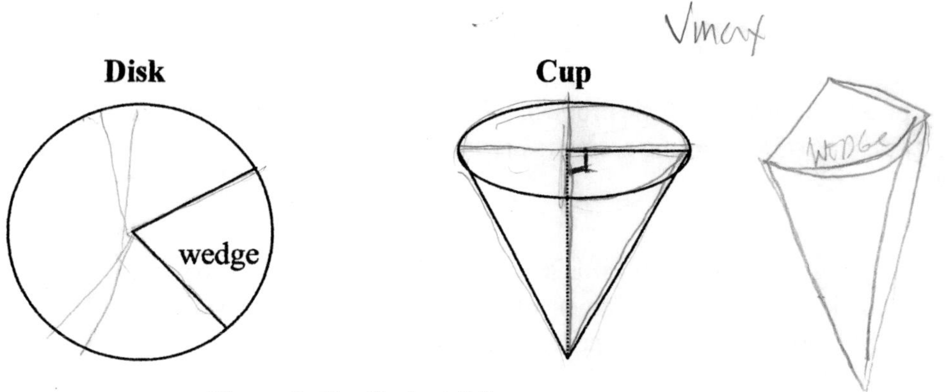

Figure 2. FunProject 5.2

Hint: how does changing the wedge angle affect the volume of the cup? What is the arc length of the wedge? Is the radius of the cup the same as the radius of the disk? How do you determine the slant height of the cup? What relationship exists between the radius of the cup, the depth of the cup, and the slant height of the cup?

4. Write an essay explaining your reasons why dry foods are usually packaged in boxes and wet foods are usually packaged in cylindrical containers. (Note that oatmeal and milk are exceptions.)

Abstract:

Packaging is the most important process in the production of foods, drinks and more. Packaging are the one that attract customers / reach the customers to buy the product. We [illegible] their design [illegible] about their containers. Why do wet foods comes in more dimensional packages like cones, cylinders, &

6. Kicking Field Goals

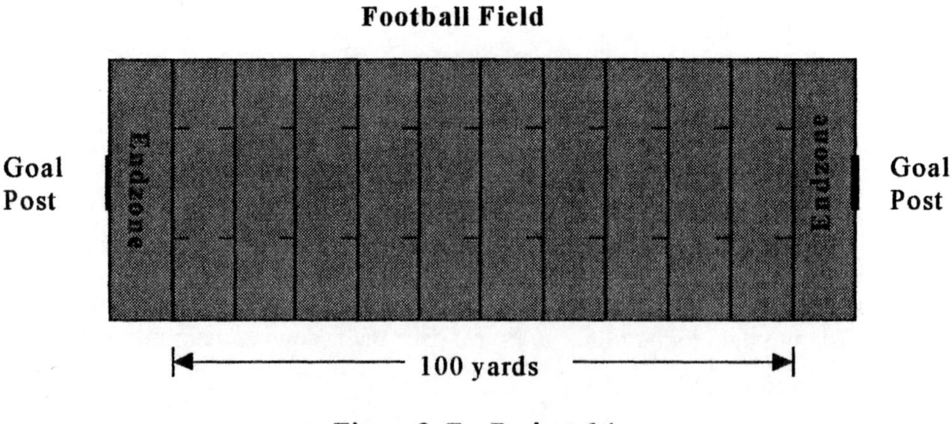

Figure 2. FunProject 6.1

(Purposes: Provide a realistic problem whose solution involves the cosine and inverse cosine functions; provide a writing assignment; provide a small-group problem-solving experience.)

A college football field is 120 yards long (including two 10-yard end zones) and is 160 feet wide. The field is marked off with lines drawn parallel to the end zones. The lines are 10 yards apart. There are two sets of hash marks running the length of the field and located 53 feet and 4 inches from the sidelines. Goalposts are centered at the back edge of the end zones (10 yards beyond the goal line). The distance between the uprights of the goalposts is 23 feet and 4 inches. The "kicking angle" is the angle formed by the lines drawn from the football to either side of the goalposts. The larger the kicking angle, the greater the success rate of the field goal kicker.

In the National Football League, the size of the football field is the same as in college. However the span between the uprights of the goalposts is narrower, 18 feet and 6 inches, and the hash marks run from the edge of one goalpost to the other goalpost. (Thus the hashmarks are 70 feet and 9 inches from the sidelines.)

1. College Football: Assume the football is kicked from a hash mark. Determine the distance of the ball from the back edge of the end zone that results in the largest kicking angle by implementing the following steps:

a. Draw a picture of a college football field. On your picture, indicate a hash mark where you want the football to be, and then draw a right triangle with one leg of the triangle being the perpendicular distance of the football to the back of the end zone and the hypotenuse of the triangle being the line from the football to the nearest edge of the goalpost. Label the leg to the back of the end zone *d* (for distance).

b. Develop a function for the distance of the football to the nearest edge of the goalpost (the hypotenuse of your triangle) in terms of *d*, the distance of the football to the back of the end zone.

c. Let *A* be the angle in your triangle at the football vertex. Write out the expression for the cos(*A*). Use the inverse cosine function in your calculator to solve for *A* as a function of *d*. (The cosine function maps an angle into a number and the inverse cosine function reverses this mapping, mapping a number into an angle.)

d. Superimpose a second right triangle on the picture drawn in Part a. The football is at the same hash mark, the distance to the back of the end zone is still *d*, but this time the hypotenuse is drawn from the football to the furthest edge of the goalposts. Let *B* denote the angle at the football. Repeat part c for your new triangle.

e. Determine a function for the kicking angle in terms of *d*, the distance of the football to the back edge of the end zone, plot this function, and determine from your plot the value of *d* that gives the largest kicking angle.

2. National Football: Assume the football is kicked from a hash mark. Determine the distance from the back edge of the end zone that gives the largest kicking angle.

3. Write a summary describing how the kicking angle changes as a function of the distance to the back edge of the end zone in college football compared to the National Football League.

7. The Cost of Driving

(Purposes: Provide an experience analyzing a real-world situation that affect students; provide experience in converting from one system to another; use ratios, averages, and percentage; inquiry aspect involving banks, insurance companies, and automobile companies; provide a writing exercise; provide a small-group experience.)

"Hey Dad, may I take the wheels tonight?" is a question on the lips of many teenagers. The sense of freedom, enjoyment of driving, convenience, and oftentimes the necessity of personal transportation are among many reasons for getting us into our vehicles without considering the expenses. What are the expenses?

For this project, you are tasked to estimate the cost per mile for operating a leased vehicle of your group's choice. You begin by deciding on a vehicle that is available for a 3-year lease. Then estimate the number of miles the vehicle will be driven over the 3-year lease period. Compile a list of anticipated expenses over the 3-year period. Your list should include, but not be limited to the following.

a. Leasing cost (This may depend on the estimated mileage.)
b. Insurance cost
c. Registration and license fees

d. Fuel cost

e. Maintenance

You should add at least three more categories to this list. Determine the estimated cost per mile by averaging the estimated total expense over 3-years by the estimated number of miles driven.

In your report, describe the vehicle, the type of insurance, and list the references (automobile dealers, insurance companies, State Vehicle Office) including names of individuals contacted. Clearly show how you transformed individual expenses to the 3-year period. For example, how do you determine the lease cost for the 3-year period when it involves some fixed fees for the lease period as well as monthly payments?

8. Daily Recommended Amount of Sodium in a 2,000 Calorie per day Diet

(Purposes: Understand round off and its effects on nutrition labels; understand percentages; extract information from data; provide an inquiry activity; provide a small-group experience.)

The Nutrition Labeling and Education Act of 1990 let to the establishment of recommended daily values for several food ingredients, including sodium. This law requires nuitrition labels for most food products that enables the public to observe and comprehend the information readily and to understand its relative significance in the context of a total daily diet.

On a plane ride to Montana, a stewardess gave Don a can of Welch's orange juice. The nutrition label on the can listed 15 milligrams (mg) of sodium which represented 1% of the recommended daily value (DV) based on a 2,000 calorie/day diet. Don figured that this implied that the DV is 1500 mg per day. He then noticed the person sitting next to him had a can of Canada Dry Ginger Ale whose nuitrition label listed 90 mg of sodium representing 4% of the DV, thus implying that the DV is 2500 mg per day. This apparent contradition raised several questions in Don's mind. Can both of these nutrition labels be correct? Can you tell the DV for a 2,000 calorie/day diet from the information on a nutrition label? How accurate is the information on nutrition labels? In particular, would it make a difference if the weight or the percentage figure were rounded off to a full integer?

Checking the nutrition labels on the other varieties of soda in the stewardess' cart (with the stewardess' permission), Don complied the following chart

Soda	Sodium (mg)	DV Percentage
Welch's Orange Juice	15	1%
Canadra Dry Ginger Ale	90	4%
Pepsi Cola	35	1%
Sprite	45	2%

For this Fun Project, your tasks are:

1. Expand Don's chart by including at least ten other varieties of soda along with their weight (mg) and percentage amounts of sodium as listed in their nutrition labels. (In your report, state the source of your data - for example, if you went to a store give the name of the store.)
2. For each variety, determine the allowable range of DV assuming that both the weight and percentage figures were rounded off to full integers.
3. Determine if there are any contradictions on the data that you collected.
4. Display your data, including the implied allowable ranges of DV, in an easily understood manner.
5. What information concerning the DV for sodium in a 2,000 calorie/day diet can you extract from your data?
6. What is the official Food and Drug Administration's DV for sodium in a 2,000 calorie/day diet? (Hint: Research the FDA website.)
7. Include in your written report, a one page essay on nutrition labels. Base your essay on the FDA article "Scouting for Sodium" that first appeared in the FDA *Consumer* in September 1994. The article was revised and reprinted in September 1995. The article can be found at http://www.fda.gov/fdac/foodlabel/sodium.html.

Evelyn Boyd Granville

Evelyn Boyd Granville was one of the first two African-American women to earn a Ph.D. in mathematics. She was born (1924) and raised in Washington, D.C. She attended Smith College with scholarship aid and money she earned while working summers at the National Bureau of Standards. Graduating summa cum laude with membership in Phi Beta Kappa, she received numerous fellowships to study at Yale University. She earned her Ph.D. (1949) from Yale under the direction of Einar Hille, the reknowned functional analyst and former president of the American Mathematical Society. The following year, she joined the faculty at Fisk University, where she inspired two women, Etta Falconer and Vivienne Malone Mays, to pursue Ph.D.s in mathematics.

Dr. Granville left teaching to work as an applied mathematician at the Diamon Fuze Laboratories and then later with the Project Vanguard and Project Mercury programs. She also worked with the Computation and Data Reducton Center of the U.S. Space Technology Laboratories. She completed her industrial career as a senior mathematician at IBM.

Dr. Granville returned to education as a faculty member at the California State University at Los Angles. In 1985, she and her husband moved to Tyler, Texas, where they purchased a farm. She accepted a position at Texas College in Tyler and in 1990 was appointed to the Sam A. Lindsey Chair of the University of Texas at Tyler.

Dr. Granville has served as a member on several boards and committees including the United States Service Panel of Examiners of the Department of Commerce, the examining committee of the Board of Medical Examiners, advisory committee of the National Defense Education Act Title IV Graduate Fellowship Program, and the Board of Trustees of the Center for the Improvement of Mathematical Education. In addition, she has been very active in the National Council of Teachers of Mathematics and the American Association of University Women. She was the first African-American women mathematician to receive an honorary doctorate degree from Smith College.

Her pioneering career and warm personality has inspired many. She says that her life has been rich—she has been blessed with a fine family, received an excellent education, made many wonderful friends, and last, but not least, has a wonderful marriage.

Chapter 3 Functions

Data is transformed into information by extracting relations from data sets. As indicated in Sections 2.6–2.9, linear relations play a fundamental role in the application of mathematics to real life. Part of the appeal of linear relations is that they are easy to recognize (constant rate of change) and easy to formulate (equation of a straight line). Nonlinear relations are more difficult to recognize and to formulate. In order to facilitate transforming data into information, we focus on dependency relations—both linear and nonlinear. In Section 2.4, we called dependency relations functions. Understanding the function concept, is critical to transforming data into information. In this chapter, we will rigorously define function, show how to shift and scale graphs of functions, and develop an algebra of functions that will enable us to generalize to a broad spectrum of functions from five basic function categories. We then focus on extracting function relations from data sets by fitting a curve to the data. We use the graphical approach of "conjecture and verify" and the analytical approach of regression analysis.

3.1 Displaying Functions

The function concept is one of the most important concepts in mathematics. Functions, as stated in Section 2.4, are special types of relations. They are dependency relations in which the dependent variable depends on the independent variable in a unique manner (see Section 2.4). We often visualize a function as an input–output process in which the independent variable is referred to as the input and the dependent variable is referred to as the output. We illustrate with a list of dependency relations.

- The letter grade you receive in college algebra depends on (is a function of) your average numerical score according to the following table:

Numerical Average	**Letter Grade**
90–100	*A*
80–89	*B*
70–79	*C*
60–69	*D*
< 60	*F*

Table 3.1.1

The table defines the grade function with the numerical average being the independent or input variable and the corresponding letter grade being the dependent or output variable.

- A person's shoe size depends on (is a function of) the size of the person's foot. The independent or input variable is the size of the person's foot.

- The area of a circle depends on (is a function of) the radius, r. We denote the area function by the letter a and indicate the value of the function for a circle of radius r by writing $a(r)$. Thus $a(r) = \pi r^2$. The independent or input variable is r.

- The volume of a sphere depends on (is a function of) the radius, r. We denote the volume function by the letter v and indicate the value of the function for a sphere of radius r by writing $v(r)$. Thus $v(r) = \frac{4}{3}\pi r^3$. The independent or input variable is r.

- The volume of a circular cylinder depends on (is a function of) both the radius, r, and the height, h. We denote the volume function by the letter v and indicate the value of the function for a cylinder of radius r and height h by writing $v(r,h)$. Thus $v(r,h) = \pi r^2 h$. The independent or input variables are r and h.

- The amount of a 6% sales tax depends on (is a function of) the cost, c, of the item purchased. We denote the tax function by the letter t and indicate the value of the tax on an item whose purchased price is c by writing $t(c)$. Thus $t(c) = .06c$. The independent or input variable is c.

- The recorded average monthly temperature in Houston (1993) depends on (is a function of) the month. The temperature function is given by the following bar chart graph:

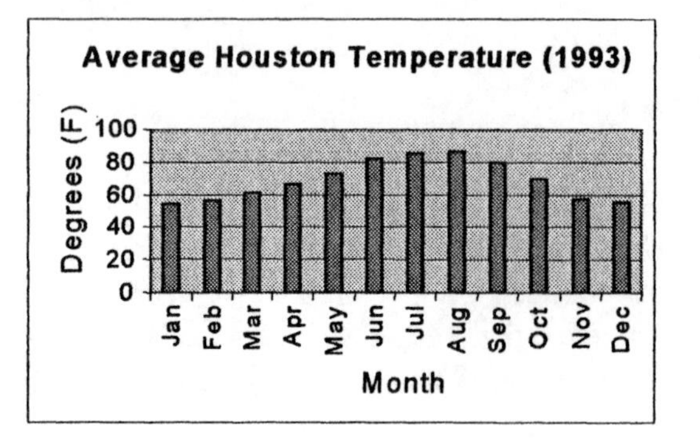

Figure 3.1.1

Note that functions, like data, are displayed in different ways. The most common ways are graphic, symbolic, and numeric. We use the following example to illustrate the three ways.

Example 3.1.1.

Don had a well drilled next to his camp. The well hole is 6 inches in diameter and is 140 feet deep. Model the volume of water (in gallons) in the well as a function of the depth of the water in the well. That is, define a function whose independent variable is the depth of the water and whose dependent variable is the volume of the water.

Graphic Solution

Because the volume of the water is equal to the depth times the constant cross-section area of the hole, the volume is a linear function of the depth. Thus the graphical model is a straight line measuring volume against depth. To determine the desired line we need 2 points (or a point and a slope). When the depth is 0, the volume is 0 (dry well) and so one point is $(0,0)$. Now for the second point, consider the depth of the water to be 140 feet (well is full). To determine the corresponding number of gallons, we first determine the volume of the well in cubic feet and then transform cubic feet into gallons. Because the diameter of the well hole is 6 inches, the radius is 3 inches or $\frac{1}{4}$ foot. Thus

$$\begin{aligned} \textbf{volume} &= \textbf{(depth) x (cross-section area)} \\ &= 140 \times (\pi r^2) \\ &= 140 \times \tfrac{\pi}{16} \\ &= \tfrac{140\pi}{16} \text{ ft}^3 \end{aligned}$$

The conversion factor from cubic feet to gallons is

$$0.134 \text{ ft}^3 = 1 \text{ gallon or } 1 \text{ ft}^3 = \tfrac{1}{0.134} \text{ gallon.}$$

Thus when the depth is 140 feet, the volume is $(\frac{1}{0.134})(\frac{140\pi}{16}) = 205.14$ gallons. Hence our second point is $(140, 205.14)$. The graph of the function showing gallons of water (vertical axis) versus depth of water (horizontal axis) is as shown.

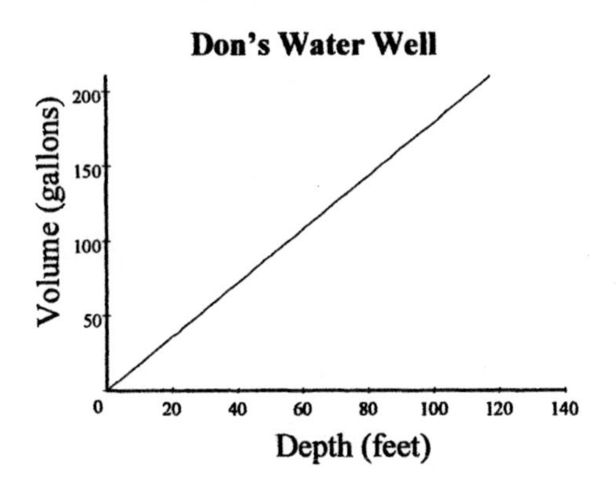

Figure 3.1.2

To determine the number of gallons of water in the well for a given depth, mark the depth point on the horizontal axis, move vertically from that point to the line, and then move horizontally from the line to the vertical axis. ▲

Note that the depth is restricted to the interval [0,140] because the well is 140 feet deep.

Symbolic Solution

We begin by defining our function symbol and our variables. Let

v be the name of our volume function

d = depth of the water

$v(d)$ = volume of the water in gallons when the depth of water = d.

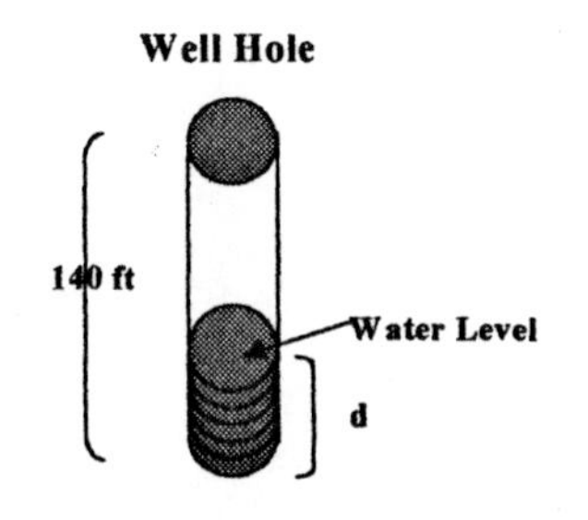

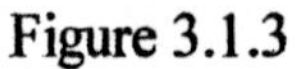
Figure 3.1.3

Because the well hole has a fixed diameter of 6 inches, the volume of water is equal to the depth times the constant cross section area of the hole. Recall that the area of a circle is πr^2. Because the depth is measured in feet, we express the radius in feet,

$$\text{radius} = 3\ in = \tfrac{1}{4}\text{ ft and so } \pi r^2 = \tfrac{\pi}{16}\text{ ft}^2.$$

Because volume equals (depth) x (cross-section area),

$$v(d) = (d)(\tfrac{\pi}{16})\text{ ft}^3.$$

We want to express volume in terms of gallons, and so we transform the volume expression from cubic feet into gallons. The conversion factor from cubic feet to gallons is 1 $\text{ft}^3 = \frac{1}{0.134}$ gallons. Thus the desired model is

$$v(d) = (d)(\tfrac{\pi}{16})(\tfrac{1}{0.134}) = 1.465d \quad \text{for } 0 \le d \le 140.$$

Note that $v(140) = 205.14$ gallons, which agrees with the graphical solution. ▲

Numeric or Tabular Solution

Using the function formula from the symbolic solution for the volume of water, we display depth and volume readings at 10 foot intervals.

Depth (ft)	Volume (ft^3)	Volume (gal)
0	0	0
10	1.964	14.653
20	3.927	29.306
30	5.891	43.959
40	7.854	58.612
50	9.818	73.265
60	11.781	87.918
70	13.744	102.57
80	15.708	117.220
90	17.671	131.880
100	19.635	146.530
110	21.598	161.180
120	23.562	175.840
130	25.525	190.49
140	27.489	205.140

Table 3.1.2

Because the table only gives readings at specified depths, we say that this numeric or tabular form is an *approximate* (or partial) solution. Although the data values presented are accurate, interpolating between data values requires an approximation process and thus the term *approximation* is associated with a tabular solution. ▲

In the previous example, we were able to determine an exact symbolic solution. This is not always possible, as is illustrated in this next example, where we use a data set as a numerical approximation of a function.

Exercise 3.1.2.

Margaret knows that the temperature of a can of soda taken from the refrigerator and left on the counter will gradually warm to the temperature of the room. Thus she reasons that the temperature of the can of soda is a function of the time since it was taken out of the refrigerator. What is this temperature function?

Numerical or Tabular Solution

Margaret does not know how to determine the temperature function in symbolic (that is, a formula) form. Thus she conducts the following experiment: she removes a can of soda from the refrigerator and leaves it on the counter for one hour. The temperature of the soda in the refrigerator is 35°F and the room temperature is 75°F. Margaret records the temperature of the soda every 10 minutes obtaining the following data

Time	Temperature (F)
0	35.00°
10	50.00°
20	59.38°
30	65.23°
40	69.90°
50	71.19°
60	72.62°

Table 3.1.3

This data provides Margaret with a numerical approximation to the temperature function. Because the changes in the temperature over 10-minute intervals are not constant, Margaret knows that the function is not linear (that is, its graph is not a straight line). To confirm this, she plots the data points to get a graphical approximation to the temperature function.

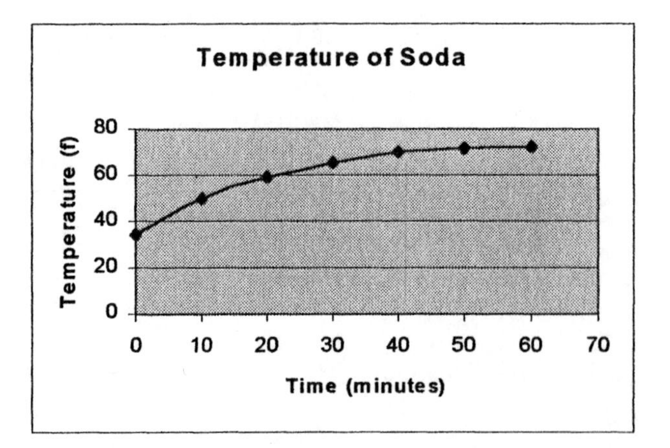

Figure 3.1.4

Connecting the points to obtain a line plot enables Margaret to approximate the temperature of the soda can at times other than the recorded times. ▲

How could the approximation be improved? (Answer: Record temperature readings over shorter periods of time.)

Later in the text, we shall show how to obtain a symbolic form for the temperature function by fitting a curve to these points.

Which of the three methods of displaying a function (graphic, symbolic, numeric) is used depends on the purpose for the display and the audience. For example, a multiplot would probably be the best choice to compare the general trend of two functions, whereas symbolic expressions would be preferred if the two functions were to be multiplied together. We now discuss the advantages and disadvantages of each of the three methods.

Graphic displays of functions provide a visual picture and a qualitative feeling for the function relation between the independent or input variable (measured along the horizontal axis) and the dependent or output variable (measured along the vertical axis). Several properties of the function can be approximated by inspection. Some examples are

- The maximum and minimum values (values of the dependent variable at the high and low points of the curve)
- Where the function is positive (values of the independent variable where the curve lies above the horizontal axis)
- Where the function is negative (values of the independent variable where the curve lies below the horizontal axis)
- Where the function is increasing (values of the independent variable where the curve is "rising to the right")
- Where the function is decreasing (values of the independent variable where the curve is "falling to the right")

- The zeros of the function (values of the independent variable where the curve touches the horizontal axis)
- The general trend of the function relation

The major disadvantages to the graphical method are

- Only approximate results are obtained. (A computer or calculator graph is drawn by plotting points and then connecting the points, that is, a line plot. Therefore the behavior of the function between points is approximated.)
- Performing algebraic operations (that is, add, subtract, multiply, divide, compose) on graphs is difficult. (Try multiplying two functions that are represented by graphs.)
- The shape of the curve obtained by a calculator or computer can be artificially distorted by changing the graphing window.
- Relations involving four or more variables cannot be plotted. (The graph of a relation involving two variables is in the plane, and the graph of a relation involving three variables is in 3–D space.)

Graphic displays of functions are commonly found in the popular press, that is, newspapers and magazines.

Symbolic (or formula) displays of functions provide for exact calculations. The main advantage of the symbolic form is that it provides a formula to which computations can be applied. Algorithms or procedures can be written for determining the function properties (increasing, maximum value, etc.), and these in turn can be programmed for calculators and computers. The development of software for calculators and computers requires that functions be displayed symbolically. The major disadvantage of symbolic displays is the amount of mathematics that one is required to understand to precisely formulate the dependency relation.

Symbolic displays of functions are commonly found in mathematics and science books where the emphasis is on (1) exactness, (2) developing computational algorithms, and (3) understanding the dependency relations. The modern paradigm is *graphics suggest and symbolics confirm.*

Numeric (or tabular) displays of functions are often used to approximate functions with a finite set of data points. The results of experiments, surveys, and digital recording devices are often presented in a table of values that represents an approximation to an unknown function, for example, the temperature function for the can of soda. The primary advantage of a numeric display is that it provides an easily described sampling of the function. In experimental sciences, the first step in trying to understand a phenomena (say, a dependency relationship) is usually to conduct an experiment and record the data in tabular and graphic forms. The role of the mathematician is often to then extract a function in symbolic form from the data. (This process will be illustrated repeatedly in the modeling problems in Chapter 4.) Two major disadvantages of displaying functions numerically are the approximate nature of the information and the difficulty in computing with tables.

Numeric displays of functions are commonly used in the popular and business press to exhibit data (for example, census reports).

Exercises 3.1

1. (Computational Skill) Evaluate the function $f(x) = \frac{x^2 - x}{x+3}$ and simplify the numerical result for
   - **a.** $x = 2$
   - **b.** $x = -2$
   - **c.** $x = \frac{1}{4}$

2. (Computational Skill) Evaluate the function $f(x) = (\frac{1}{x} + \frac{x}{3})^2$ and simplify the numerical result for
   - **a.** $x = 2$
   - **b.** $x = -2$
   - **c.** $x = \frac{1}{4}$

3. (Computational Skill) Make up and solve three exercises similar to Exercises 1a–c, 2a–c.

4. (Calculator Skill) Graphically solve for the zeros of the following functions. (A zero of a function is a point where the plot of the function meets the horizontal axis.)
   - **a.** $f(x) = x^2 - 3x + 2$
   - **b.** $g(x) = (x^2 - 4)(x + 2)$
   - **c.** $h(x) = (x - 3)^2(x + 4)$ (Explain why the plot crosses the x-axis at –4, but not at 3.)

For each of the relations in Exercises 5–14, determine if the relation is a dependency relation that satisfies a uniqueness condition (each element of the independent vaiable is paired with exactly one element of the dependent variable). For each dependency relation:

- **a.** Determine which variable is independent, which is dependent.
- **b.** Sketch a reasonable graph to illustrate the relationship.
- **c.** Write one or two sentences justifying the shape and behavior of your sketch.

5. The (outside) temperature at your school measured over a day
6. The cost of pizza and the size of pizza
7. Your course grade and your shoe size
8. The amount of a person's education and the person's salary
9. The amount of interest paid on a credit card account and the balance in the account
10. The day of the year and the number of hours of daylight
11. Postage required for a first-class letter and the weight of the letter
12. Your weight and your height
13. Collect five examples of functions displayed in a newspaper or magazine; identify the display as graphic, symbolic, or numeric; and write a short explanation of why you think the authors used the display method that they did.

3.2 Definitions

We treated relations and functions in an informal manner in Sections 2.4 and 3.1 in order to develop a sense and a feeling for these concepts. Because relations and functions are fundamental to this course and all future mathematics that you study, we need to understand their formal definitions.

A **relation** from one set to another set is a pairing of the elements in the first set with the elements in the second set.

Thus a relation is a set of **ordered pairs**. The term ordered refers to the requirement that the first component of each pair belongs to the first set and the second component of each pair belongs to the second set.

Query 1.
Does a two-column table define a relation? Why?

Example 3.2.1.

The following list was copied from the Orange County, New York Telephone Directory.

Phillips, Robert J 198C E Mtd Rd S Cld Sp 10516	265 – 9095
Phillips, Ross 10 New Cld Sp10516	265 – 3184
Phillips, Samuel B 60 N Elm Becb 12508	831 – 3326
Phillips, Scott 63 Main Port Jervis 12771	856 – 5825
Phillips, Sean & Ramsi 128 Murrey Ave Goshn 10924	294 – 0773 294 – 5096
Phillips, Stanley 1 State Midtown 10940	343 – 2256 ▲

Table 3.2.1

Does this partial telephone directory define a relation? Why?

Note that there are two phone numbers listed for Phillips, Sean & Ramsi. Thus there are two ordered pairs that have the same first element, but have different second elements.

Example 3.2.2.

A "Speeding" relation is defined by the following schedule of fines for speeding, published by the Harris County Justice Court.

Violation	Fine
1–10 mph over limit	\$20
11–15 mph over limit	\$70
16–20 mph over limit	\$90
21–25 mph over limit	\$110
26–30 mph over limit	\$130
More than 30 mph over limit	\$190
Racing, Contest of Speed	\$190

Table 3.2.2

The ordered pairs (8 mph over limit, \$20), (23 mph over limit, \$110) are elements of this speeding relation. Note that (\$70, 12 mph over limit) is not an ordered pair of this speeding relation. Why? ▲

Query 2.

Can a graph define a relation? Why?

The following graph of a circle of radius 2 defines a "circle" relation. The point or points where any vertical line intersects the circle are ordered pairs in the circle relation. For example, $(1, \sqrt{3})$ and $(1, -\sqrt{3})$ are two points in the relation. The point (3,4) is not in the relation. Why?

Circle Plot

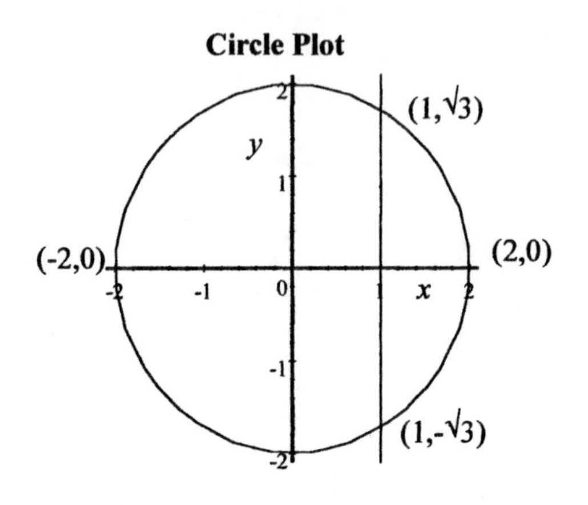

Figure 3.2.1 Circle: $x^2 + y^2 = 4$

Example 3.2.3.

The "squaring" relation relates a number to the square of that number. The following are a few of the ordered pairs of this relation.

$(2,4), (5,25), (1,1), (-2,4), (-13,169), (10,100), (1.5,2.25)$

If we let x denote an arbitrary number, then x^2 denotes the square of the number. Thus (x,x^2) represents an arbitrary ordered pair of the squaring relation, and the squaring relation can be described symbolically using set notation by writing

$$\{(x,x^2) : x \text{ is a number}\}.$$

This is read, "the set of all ordered pairs (x,x^2) such that x is a number." ▲

Query 3. How do you read $\{(x,x^3) : x \text{ is a number}\}$? Give three ordered pairs in this relation.

Note: The curly brackets, { }, are called **set brackets**, the entry before the colon represents an arbitrary element of the set, the colon is read "such that," and the entry after the colon describes the restrictions on the variable.

Query 4.
How can you express the relation $y = 1/x$ for $x > 0$ in set bracket notation?

Example 3.2.4.

The squaring relation can be partially described numerically by the following table.

First Component	Second Component
2	4
5	25
1	1
–2	4
–13	169
10	100
1.5	2.25

Table 3.2.3

We say that the table only partially describes the squaring relation because it only shows seven of the infinitely many ordered pairs that make up the squaring relation. ▲

Example 3.2.5.

The squaring relation can be partially described graphically by the following graph.

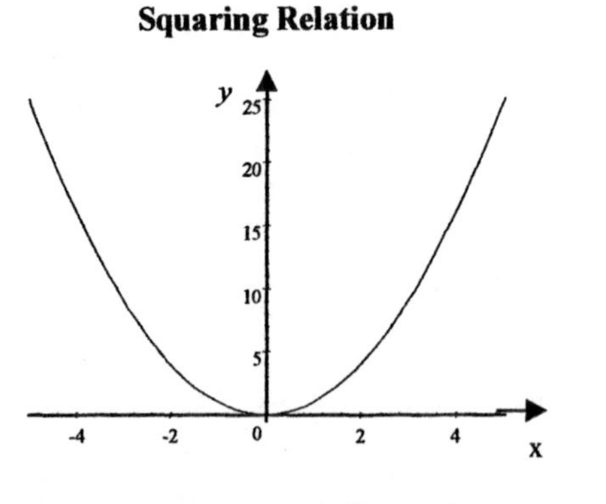

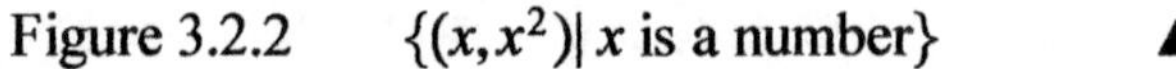
Figure 3.2.2 $\{(x,x^2)|\, x \text{ is a number}\}$ ▲

We have spoken informally of function as a special type of relation, namely a dependency relation that satisfies a uniqueness condition. Before we begin the development of a formal mathematical definition, let us consult a dictionary for the definition of the word function. The *Funk & Wagnalls Standard Desk Dictionary* (1977) gives the following five definitions of function.

1. The specific, natural, or proper action or activity of anything.
2. The special duties of action required of anyone in an occupation, office, or role.
3. Any more or less formal or elaborate social gathering or ceremony.
4. Any fact, quality, or thing depending upon or varying with another.
5. A quantity whose value is dependent on the value of some other quantity.

Query 5.
Which of these definitions correspond to our informal definition of a function as a dependency relation satisfying a uniqueness condition?

Now for the mathematical definition of function.

Definition

A **function** is a **relation** linking two sets called **domain** and **range** in which each element of the domain is paired with **exactly one** element of the range. In the ordered pairs, the first component is called the **independent or input variable** and the second component is called the **dependent or output variable.**

We comment on the boldfaced terms.

Relation

A relation between two sets is a set of ordered pairs in which the first elements are in one set and the second elements are in the other set. For example, $\{(a,2),(b,4),(c,1),(d,6)\}$ is one relation between the sets $A = \{a,b,c,d,e\}$ and $B = \{1,2,3,4,5,6,7,8\}$. There are several relations between the sets A and B.

Function

A function is a special type of relation and thus a function is a set of ordered pairs.

Domain and Range

Because a function is a relation, a function relates two sets. The first set is the set of first components in the ordered pairs—this is called the domain of the function. The second set is the set of second components in the ordered pairs—this is called the range of the function.

Exactly One

This is the crucial uniqueness condition that distinguishes a function as a special type of relation. This condition requires each domain element to be paired with a *unique* (one and only one) range element. That is, there cannot exist two different ordered pairs with the same first component. The circle relation in the previous example does not satisfy this *exactly one* condition (why?) and therefore the circle relation is *not* a function.

Independent (input) Variable and Dependent (output) Variable

The *exactly one* condition requires that the first component, the independent variable or input variable, in an ordered pair uniquely determines the second component, the dependent variable or output variable. Thus the dependent variable is dependent on the independent variable (the output is dependent on the input). This is the reason for saying that a function is a dependency relation.

The word **map** is often used when wanting to portray a sense of motion in the matching of a domain element of a function with its unique range element. For example, we would say "the function $f(x) = 3x - 2$ maps 2 into 4" because $f(2) = 3(2) - 2 = 4$.

Query 5.

Explain why the squaring relation is a function and the circle relation is not a function.

The exactly one condition has a graphical interpretation that is called the **vertical line test.** Under this test, a graph defines a function provided any vertical line intersects the graph in at most one point. Note that a vertical line may not intersect the graph at all. Figure 3.2.1 clearly shows by the vertical line test that the circle does not define a function.

Example 3.2.6.

Use the vertical line test to show that the graph of $y = \sqrt{x}$ defines y as a function of x whereas the graph of $y^2 = x$ does not define y as a function of x.

Solution

Vertical lines A and B in Figures 3.2.3 and 3.2.4 represent the only possibilities for vertical lines with respect to the plots.

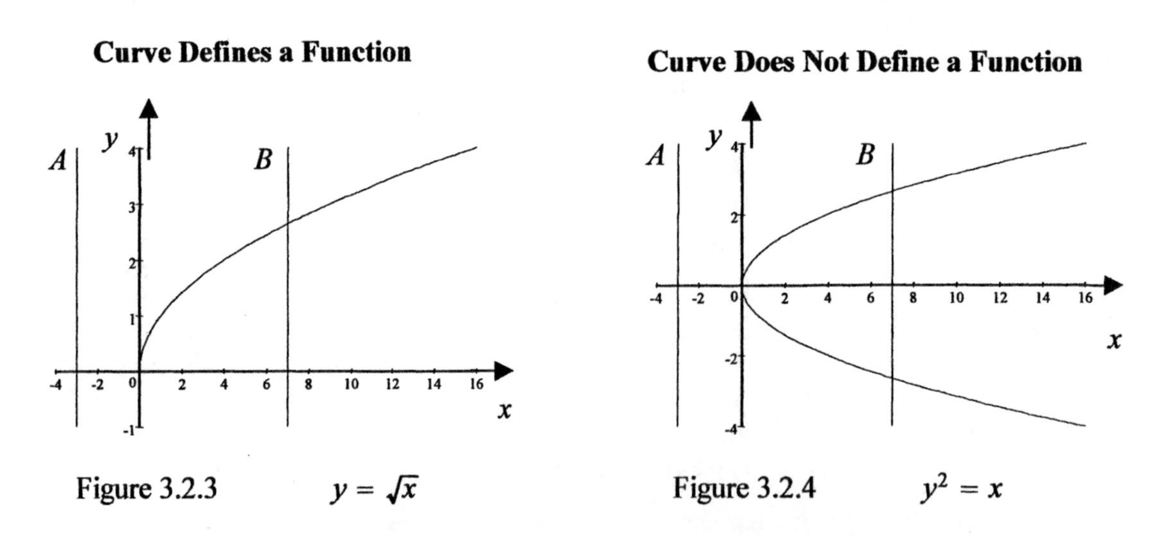

Figure 3.2.3 $y = \sqrt{x}$

Figure 3.2.4 $y^2 = x$

In Figure 3.2.3, the curve satisfies the vertical line test, whereas in Figure 3.2.4, the curve does not satisfy the vertical line test. ▲

In Section 3.1, we illustrated different methods of displaying functions.

What does it mean to display a function? Answer: To display a function is to show how to pair an arbitrary element from the domain with its corresponding element from the range. Or given a value for the independent variable, show how to determine the value of the corresponding dependent variable. (That is, given an input value, how do you determine the corresponding output value?)

Example 3.2.7.

$f(x) = 2x - 3$ is a linear function displayed using the symbolic or formula method. The function is called linear because its graph is a straight line. ▲

Note that

f is the name of the function,
x is the input variable or the independent variable,
$2x - 3$ is the output expression or dependent variable or range expression, and
$f(x)$ denotes the output expression or dependent variable or range expression.

Thus $(x, f(x))$ is an ordered pair in the function f, and the function is expressed in set notation as

$$f = \{(x, 2x - 3) : x \text{ is a real number}\}.$$

Now consider evaluating the function f at $x = 4$:

$$f(4) = 2(4) - 3 = 5.$$

Thus f maps 4 onto 5. That is, the ordered pair (4,5) is an element of the function f.

Example 3.2.8.

$f(x) = x^2 + \sqrt{x}$ is a function displayed using the symbolic or formula method. ▲

Note that

f is the name of the function,
x is the input variable or the independent variable,
$x^2 + \sqrt{x}$ is the output expression or dependent variable or range expression, and
$f(x)$ denotes the output expression or dependent variable or range expression.

Thus $(x, f(x))$ is an ordered pair in the function f and the function is expressed in set notation as

$$f = \{(x, x^2 + \sqrt{x}) : x \text{ is a nonnegative real number}\}.$$

Now consider evaluating the function f at $x = 4$

$$f(4) = 4^2 + \sqrt{4} = 16 + 2 = 18$$

Thus f "maps" 4 onto 18. That is, the ordered pair $(4, 18)$ is an element of the function f.

The following plot and table display this same function $f(x) = x^2 + \sqrt{x}$ graphically and numerically.

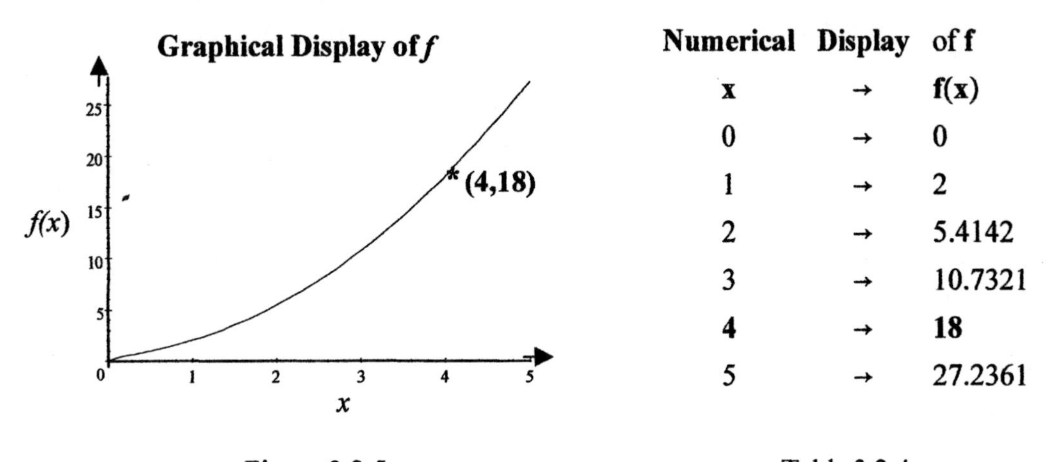

Numerical Display of f

x	→	f(x)
0	→	0
1	→	2
2	→	5.4142
3	→	10.7321
4	→	**18**
5	→	27.2361

Figure 3.2.5

Table 3.2.4

Because the range expression is only defined for nonnegative real numbers, the domain of f is the set of nonnegative real numbers ($\sqrt{x}$ is not defined for negative numbers).

Because the range expression is 0 when $x = 0$, positive when x is positive, and there is no largest value, the range of f is the set of nonnegative numbers.

The **natural domain** of a function is the set of input values (independent variable) for which the range expression is defined. For example the domain of $f(x) = \sqrt{x}$ is the set of non-negative real numbers because $\sqrt{x}$ is not defined for real numbers when x is negative. We will always mean the natural domain when we say domain unless an explicit domain set is specified.

Notation and Convention

We read "$f(x)$" as "f of x" or as "f at x" and mean that the function f is evaluated at the input value x and the expression $f(x)$ represents the corresponding output value. We stress that $f(x)$ does not mean f times x.

Example 3.2.3 illustrates a common way to symbolically denote a function. Because both $f(x)$ and $x^2 + \sqrt{x}$ denote the output, they must be equal to each other. Thus, we write the equation $f(x) = x^2 + \sqrt{x}$. Another convention is to use the letter y to represent the output or dependent variable. This gives us three different symbols—y, $f(x)$, $x^2 + \sqrt{x}$—representing the output or dependent variable. Therefore the following three equations are all equivalent and are all used to express a function.

$$y = f(x) \qquad y = x^2 + \sqrt{x} \qquad f(x) = x^2 + \sqrt{x}$$

However, a function is not an equation. The point we are making is that even though a function is presented as an equation, a function is much more than an equation. A function is a special type of relation linking two sets, whereas an equation is merely the statement that the expressions on the left- and right-hand sides are equal.

Small-Group Activity

Make up two relations, each consisting of six ordered pairs, such that the first is a function and the second is not. Display each of your relations graphically, symbolically, and numerically. Present and explain your examples to the class.

Each point on a graph represents an ordered pair in which the first component is the horizontal value of the point and the second component is the vertical value of the point. Thus a graph is a display of a set of ordered pairs; thus, each graph is the display of a relation. Conversely, each relation can be graphed. Therefore, because some relations are functions and some are not, some graphs are function displays and some are not.

Query 6.

How do you tell if a given graph is the graph of a function? For example, does the following graph represent a function?

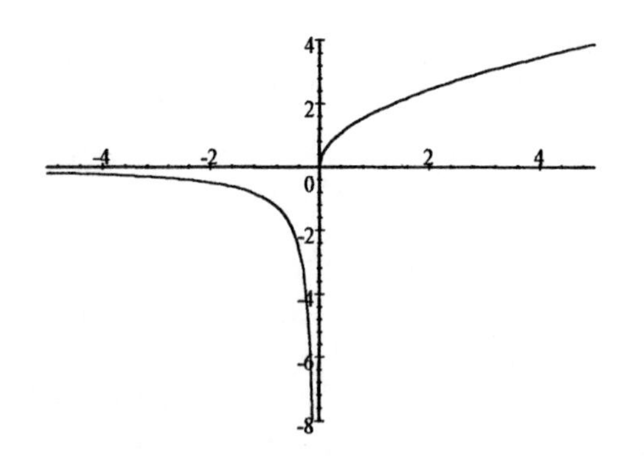

Figure 3.2.6

Small-Group Activity

Sketch five different graphs, two that display functions and three that display relations that are not functions. Explain to your class how the vertical line test can be used to determine which of your five graphs represent functions.

Query 7.

Consider a Wal-Mart store in which the checkout clerks scan the bar code of each purchased item into the memory of their cash register. Does the cash register have a built-in (cash) function? If so, describe the function. What are the inputs? What are the outputs? Is the function displayed within the cash register in a graphical, symbolic, or numerical form?

The need to make predictions and to study rates of change are two major reasons for the central importance of functions in the study and application of mathematics.

1. Prediction. An important purpose of collecting and analyzing data is to be able to make predictions. For example, the analysis of the race questions on the U.S. Census provides the basis for several policy decisions. Another example involves the minimum wage as described in Example 3.2.9. Because two-variable data can be plotted (scatter plot) or displayed in a two-column table, the ordered pairs of data can be considered to be part of a function. If the underlying function can be determined or at least approximated, it can be used for predictive purposes. For example, consider only the table of values in the squaring relation. If by analyzing this table you realized that the underlying function is the squaring function, then you could easily predict that 11 would be paired with 121.
2. Rates of change. Understanding the rate of change of the dependent variable with respect to the independent variable is possibly the most important aspect in applying mathematics to cause–effect situations. For example determining maximum and minimum values, depends on understanding rates of change. (For linear functions, the rate of change is called the **slope.**) Because a function provides a unique relationship between the dependent and independent variables, the function concept is well suited to this type of analysis.

Because of the interest in studying rates of change, we will think of functions in their dynamic form of mapping an independent variable onto a dependent variable and speak of a function as a *mapping*. In order to emphasize the dynamic sense, we will also refer to the independent variable as the **input** variable and the dependent variable as the **output** variable.

Example 3.2.9.

During 1996, emotional political debates arose over raising the minimum wage. If the minimum wage is to be raised in 1996, how could you predict a fair wage based on the following historical record?

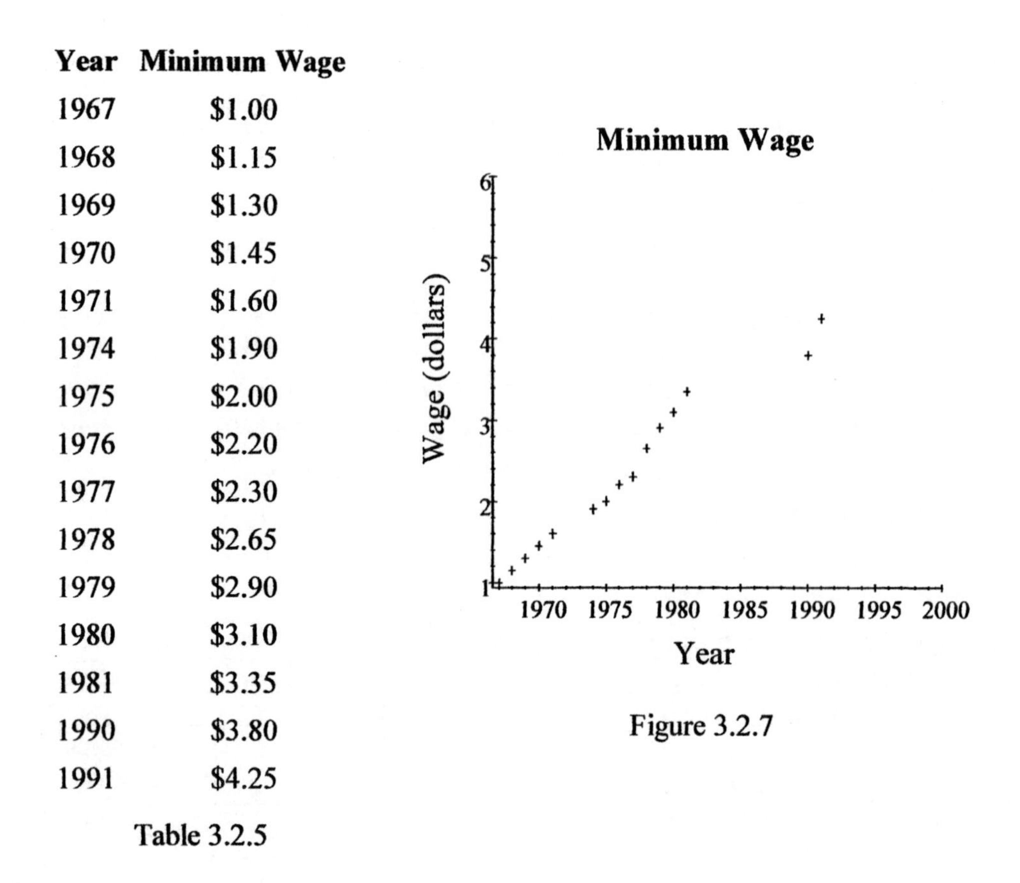

Year	Minimum Wage
1967	\$1.00
1968	\$1.15
1969	\$1.30
1970	\$1.45
1971	\$1.60
1974	\$1.90
1975	\$2.00
1976	\$2.20
1977	\$2.30
1978	\$2.65
1979	\$2.90
1980	\$3.10
1981	\$3.35
1990	\$3.80
1991	\$4.25

Table 3.2.5

Figure 3.2.7

Solution

Consider the data points in the table to approximate a linear function (a function whose graph is a straight line). Determine a line that gives a good approximation to the data, determine the function for this line, and then evaluate the function for 1996. At this point in our development we estimate the approximation line by selecting two data points to determine the line. (In Section 3.7, we will discuss how to determine the line that *best fits* the data.) We select the first data point $(1967, 1)$ and the last data point $(1991, 4.25)$. We now use the two-point method for determining the equation of a line (see Section 2.6).

Let (x,y) be an arbitrary point on the desired line with x representing time in years and y the corresponding minimum wage. Then

$$\text{slope} = \frac{y-1}{x-1967} \quad \text{and} \quad \text{slope} = \frac{4.25-1}{1991-1967} = 0.135.$$

Setting these two slope expressions equal to each other gives the equation of the line

$$\frac{y-1}{x-1967} = 0.135 \text{ and thus } y-1 = 0.135(x-1967)$$

or

$$y = 0.135x - 264.545.$$

The desired function is $f(x) = 0.135x - 264.545$. Evaluating this function for $x = 1996$, gives an approximation of \$4.92 for the minimum wage in 1996. ▲

We need to point out that \$4.92 is a very robust prediction in the sense that it resulted from computing a slope using the first and last data points in Table 3.2.6. If other points had been chosen, a different value would have been computed for the slope and thus a different prediction obtained. In Sections 3.6 and 3.7, we will develop methods for fitting a curve to data that minimizes the arbitrariness in the solution of Example 3.2.9.

The following example illustrates a way to determine the rate of change represented by data.

Example 3.2.10.

Determine the rate of growth in the monthly average of the number of Internet hosts for the period January 1999 through July 2001 based on the following data. (Source: http://www.netsizer.com) The months are numbered, with January 1999 being month 1 and July 2001 being month 31.

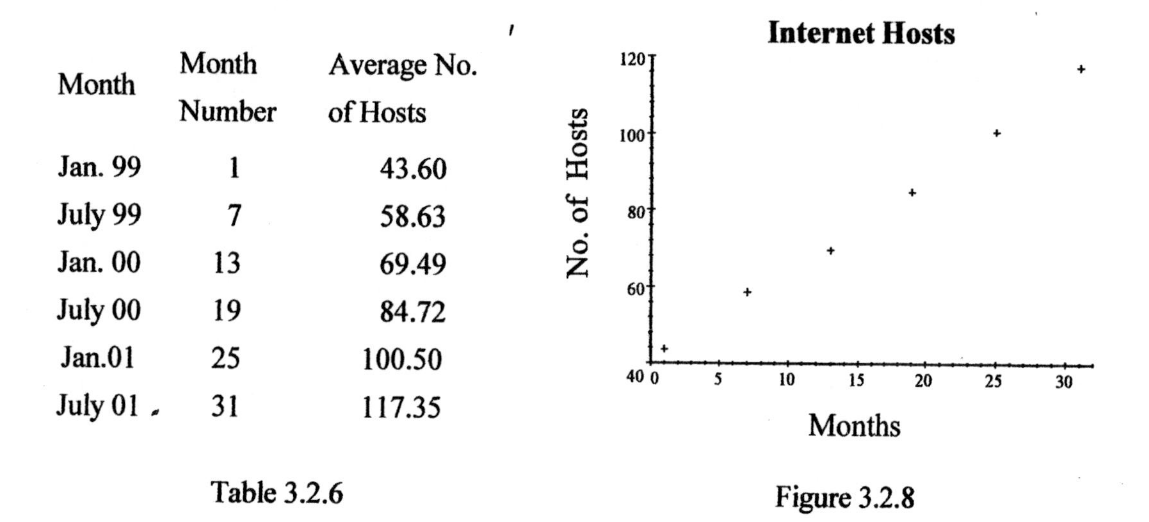

Month	Month Number	Average No. of Hosts
Jan. 99	1	43.60
July 99	7	58.63
Jan. 00	13	69.49
July 00	19	84.72
Jan.01	25	100.50
July 01	31	117.35

Table 3.2.6

Figure 3.2.8

Solution

The plot of the data points appear to be linear. Let us use the two-point method (see Section 2.6) for determining the equation of the line that "fits" these data points. We select two of the data points, say (1,43.6) and (31,117.35).
Let (x,y) be an arbitrary point on the desired line, with x representing month number and y the corresponding number of Internet hosts. Then

$$\text{slope} = \frac{y-43.6}{x-1} \quad \text{and} \quad \text{slope} = \frac{117.35-43.6}{31-1} = 2.46.$$

Setting these two slope expressions equal to each other gives the equation of the line

$$\frac{y-43.6}{x-1} = 2.46 \text{ and thus } y-43.6 = 2.46(x-1)$$

or

$$y = 2.46x + 41.14.$$

The slope of this function, 2.46, represents the rate of change in the number of Internet hosts per month for the period January 1999 to July 2001. ▲

Exercises 3.2

1. (Computational Skill) Evaluate the function $f(x) = 2^x + 3x^2$ and simplify the numerical result for the following:

   a. $x = 1$

   b. $x = 4$

   c. $x = \frac{1}{2}$

2. (Computational Skill) Evaluate the function $f(x) = \sqrt{2x} - \frac{x}{2}$ for the following:

   a. $x = t$

   b. $x = (t+2)^2$

   c. $x = \frac{1}{t}$

3. (Computational Skill) Make up and solve three exercises similar to Exercises 1a–c, 2a–c.

4. (Calculator Skill) Graphically solve for all of the zeros of the following functions. (A zero of a function is a point where the plot of the function meets the horizontal axis.) How do you know when you have found all of the zeros of a function?

   a. $f(x) = x - \sqrt{x}$

   b. $g(x) = x^2 - x^3$

   c. $h(x) = |x^2 - 3x| - \sqrt{x^2 + 4}$

5. Consider the function $f(x) = \sqrt{2x} - \frac{x}{2}$.

   a. Determine the domain of f.

   b. Graphically determine the input values that yield positive output values.

6. What is the domain of the circle relation?
   What is the range of the circle relation?

7. What is the domain of the squaring function?
   What is the range of the squaring function?

8. What is the domain of the square root function: $f(x) = \sqrt{x}$? Explain.
   What is the range of the square root function? Explain.

9. What is the domain of the cube function $f(x) = x^3$? Explain.
   What is the range of the cube function?

10. What is the domain of the cube root function: $f(x) = \sqrt[3]{x}$? Explain.
    What is the range of the cube root function?

11. Is the partial telephone directory relation reproduced in this section a function? Explain.

12. Write a paragraph describing the relationship (if any) between the squaring and square root functions and between the cube and cube root functions. (Is there a relationship between the range of the squaring function and the domain of the square root function? Is there a relationship between the range of the cube function and the domain of the cube root function?)

13. Does the listing of names and telephone numbers in a telephone book constitute a function? Explain.

14. Is the relation between individuals and their social security numbers a function? Explain.

15. Is the relation of mother to child a function? Explain.
Is the relation of child to mother a function? Explain.

16. Is the relation of driver to driver license number a function? Explain.

17. Is the relation of temperature to time a function? Explain.
Is the relation of time to temperature a function? Explain.

18. Is the relation of hours of daylight to days of the year a function? Explain.
Is the relation of days of the year to hours of daylight a function? Explain.

19. A function f is defined by the table

x	1	−2	4	2	5
$f(x)$	3	0	0	−1	2

Table 3.2.7

List the domain and range of f, and then plot the function.

20. A relation g is defined by the table

x	1	3	0	−1	2
$g(x)$	2	3	1	5	4

Table 3.2.8

Is g a function? Explain. List the domain and range of g, and then plot g.

21. A relation h is defined by the table

x	1	3	−2	1	5
$h(x)$	2	4	5	4	2

Table 3.2.9

Is h a function? Explain. List the domain and range of h, and then plot h.

22. Does the following graph define a function? Explain.
List the domain and range of the graph.

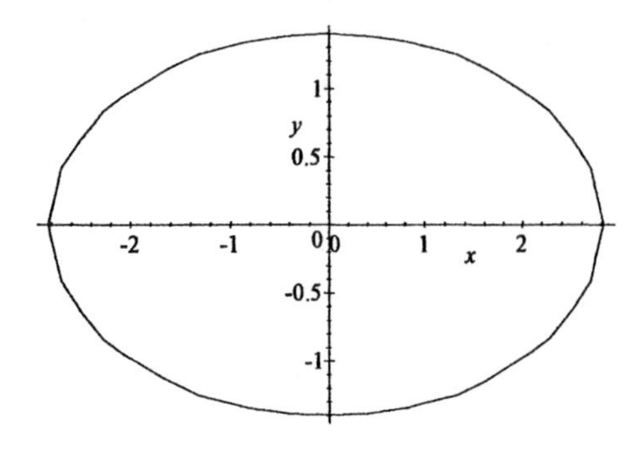

Figure 3.2.9

23. Does the following graph define a function? Explain.

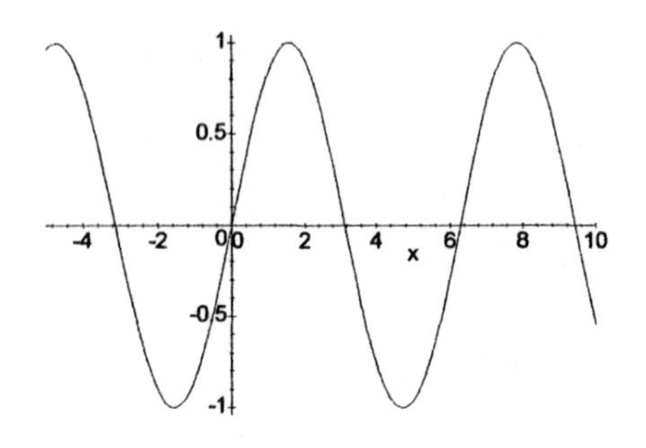

Figure 3.2.10

24. Does the following mapping chart define a function? If so describe the function by listing its ordered pairs. If the mapping chart does not define a function, explain why it does not.

1 → 3

4 → 2

3 → 0

5 → 1

6 → 4

Table 3.2.10

25. Does the following mapping chart define a function? If so, describe the function by listing its ordered pairs. If the mapping chart does not define a function, explain why it does not.

$2 \rightarrow 1$

$5 \rightarrow 3$

$3 \rightarrow 2$

$2 \rightarrow 4$

$0 \rightarrow 5$

Table 3.2.11

26. For each of the following conditions, make up an example or explain why no example can exist.

a. A function whose domain is $\{1,2,3,4\}$ and whose range is $\{2,3\}$

b. A relation that is not a function and whose domain is $\{1,2,3,4\}$ and whose range is $\{2,3\}$

c. An increasing function (i.e., a function whose graph "rises to the right")

d. A decreasing function (i.e., a function whose graph "falls to the right")

e. A function that is neither increasing nor decreasing

f. A function whose graph looks like a horseshoe opening to the right

27. (*Small Group*) Develop and carry out an experiment to approximate an unknown function (for example, the speed of a ball rolling down a hill). The record of your experiment should involve at least ten ordered pairs (independent variable, dependent variable). Display your results in both a numeric and graphic form. Explain how you would interpolate between data points or extrapolate beyond the last data point.

3.3 Predictions Based on Data

How can you make predictions based on data? That is, how can you interpolate between data points or extrapolate beyond data points? The following three worked problems illustrate a graphical approach to this question. Examples 3.3.2 and 3.3.3 show the need to extend our knowledge of functions beyond just linear functions. We conclude this section with a graphical presentation of the five most important categories of functions (power, radical, exponential, logarithmic, and periodic).

Example 3.3.1.

A partial cost schedule for mailing packages to Western Europe by the Global Priority Mail service is given by the following table. Based on this schedule, what would be a reasonable fee to mail a 4.5-lb package to Western Europe?

Weight (lbs)	Fee (dollars)
1	10.50
2	15.00
3	19.95
4	24.75

Table 3.3.1

Solution

A solution to this question can be obtained by plotting the data, determining a function whose graph contains (approximately) the data points, and then evaluating the function at the input value of 4.5. Determining such a function is called **fitting a curve to** data.

Postage Fee Schedule

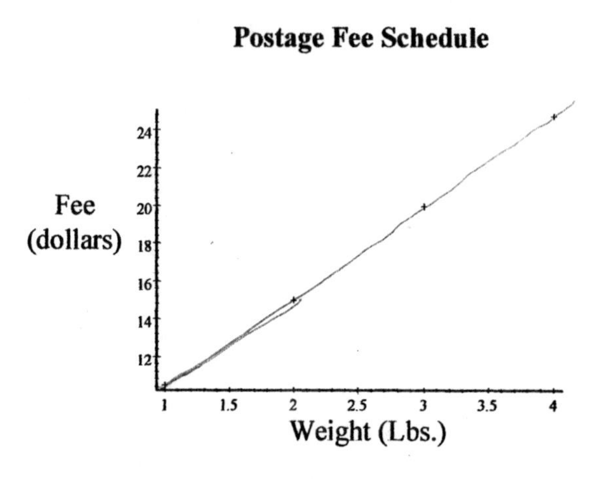

Figure 3.3.1

Because the four points appear to lie on a straight line, we will define a linear function whose graph contains (approximately) the data points. (Recall the graph of a linear function is a straight line.) In order to do this, we need to know a point on the desired line and the slope of the desired line.

We can choose any one of the four points for our point. Let us pick a middle point, say the point $(2, 15)$. One way to estimate the desired slope is to average the slopes of the line segments joining adjacent points.

The slope of the line connecting points $(1, 10.50)$ and $(2, 15.00)$ is $\frac{15.00-10.50}{2-1} = 4.5$.

The slope of the line connecting points $(2, 15.00)$ and $(3, 19.95)$ is $\frac{19.95-15.00}{3-2} = 4.95$.

The slope of the line connecting points $(3, 19.95)$ and $(4, 24.75)$ is $\frac{24.75-19.95}{4-3} = 4.8$.

The average of these three slopes is $\frac{4.5+4.95+4.8}{3} = 4.75$.

We define our variables by letting w denote weight in pounds and c denote the postage fee in dollars.

The equation of the line passing through the point $(2, 15)$ with slope 4.75 is $c = 4.75w + 5.50$. (Why?) Thus we define our cost function as

$$c(w) = 4.75w + 5.50$$

where $c(w)$ represents the cost of shipping a package that weights w pounds.

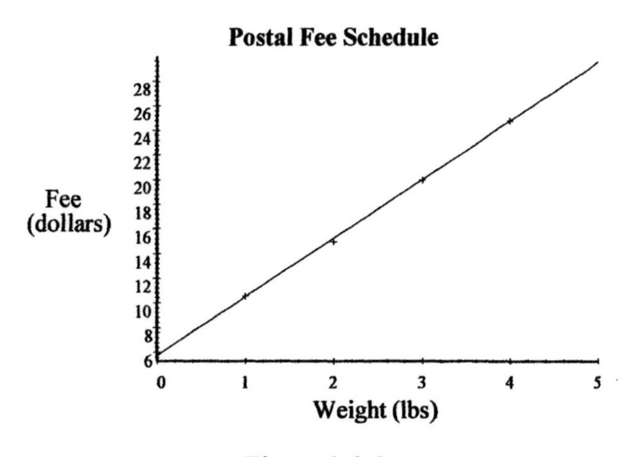

Figure 3.3.2

We predict the cost for shipping a 4.5-lb package by evaluating our function at $w = 4.5$,

$c(4.5) = (4.75)(4.5) + 5.50 = \$26.88$. (Cost is rounded up to next full cent.) ▲

Example 3.3.2.

Based on the following data taken from the Texas Drivers Handbook (1995), what is a reasonable stopping distance when traveling at 55 miles per hour?

Speed (mph)	Stopping Distance (ft)
20	45
30	78
40	125
50	188
60	272
70	381

Table 3.3.2

Solution

We employ the method used in the preceding example. That is, plot the data, determine a function whose graph contains (approximately) the data points, and then evaluate the function at the input value of 55.

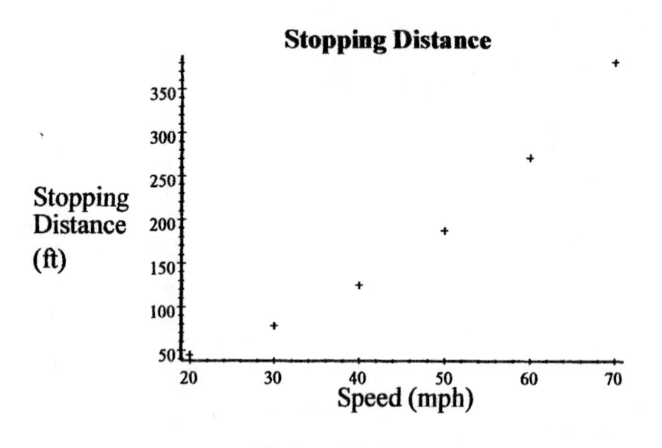

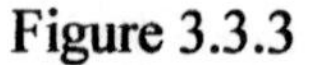
Figure 3.3.3

The data points appear to lie on a curve that passes through the origin and is curving upward. A parabolic curve (the graph of -a second degree polynomial) has these properties, and thus we try to fit a second degree polynomial to these data points. A reasonable approximation is given by the function $d(s) = 0.076s^2 + 0.15s$, where s represents the input speed and d represent the stopping distance. We will show how this function is determined in Sections 3.6 and 3.7. The following is the plot of this function superimposed on the data.

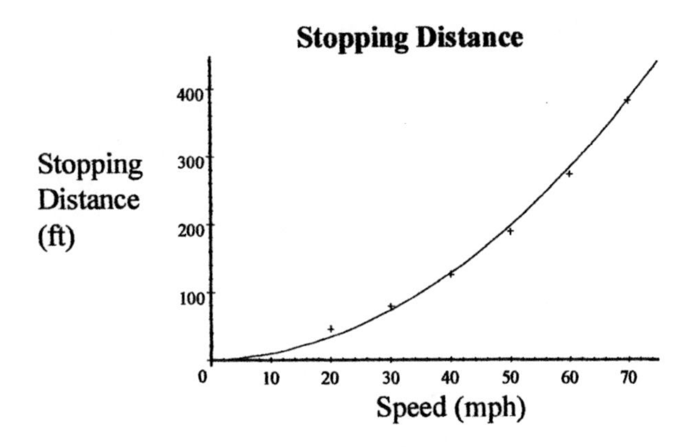

Figure 3.3.4

The function does not fit the data exactly, but is a good approximation. (How do you define "good approximation"?) We now predict a reasonable stopping distance when traveling at 55 miles per hour by evaluating our function at $s = 55$. This gives

$$d(55) = (0.076)(55^2) + (0.15)(55) = 238.15 \text{ feet.} \quad \blacktriangle$$

Example 3.3.3.

Predicting the size of the senior population (ages 65 and older) is important to several segments of our society: the medical profession, the recreation businesses, and the federal government, to name just three. (The cost of Medicaid is over 100 billion dollars and is increasing.) One prediction of the size of the senior population is given in the following table.

Year	Population (millions)
1900	3.1
1992	32.3
2000	35.3
2010	40.1
2020	53.3
2030	70.2

Table 3.3.3

Two reasonable questions to ask concerning this data are

(a) How accurate are the predictions?
(b) How can the predictions be extended beyond 2030?

Solution

One way to answer both of these questions is to follow the process illustrated in Examples 3.3.1 and 3.3.2. That is, plot the data and then determine a function whose graph contains the data points. Evaluating this function for years in which the population is known provides a check on the accuracy of the function. Evaluating this function for years beyond 2030 gives predictions beyond 2030. The following two plots indicate the data points do not lie on a line and also the points appear to rise more steeply than in the previous example. The second plot, which is a line plot, shows the basic shape of the data. A line plot is not a true representation of the actual situation because the population does not necessarily increase in a linear manner between data points.

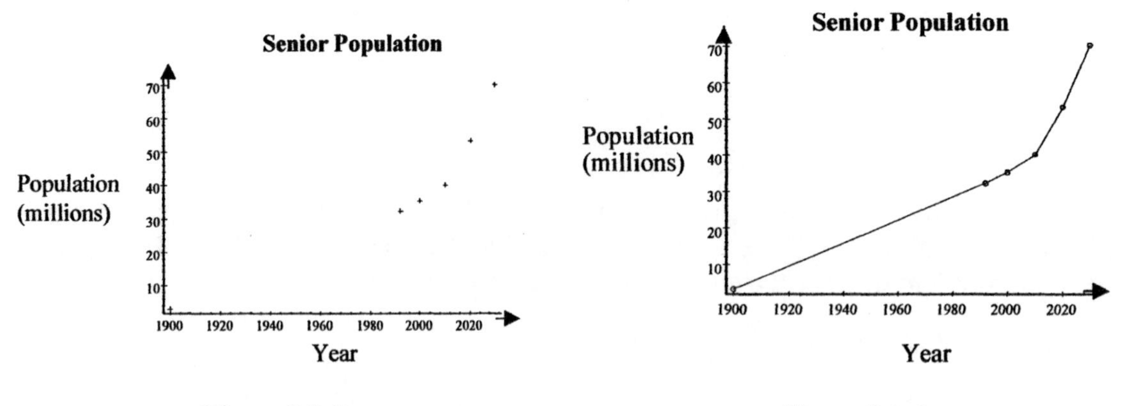

Figure 3.3.5

Figure 3.3.6

A more refined model for this example is obtained by using an exponential function, $p(x) = (7.26476 * 10^{-20}) * 1.02408^x$. (The asterisk is used to indicate multiplication.) We will show how this function is determined in Sections 3.6 and 3.7.

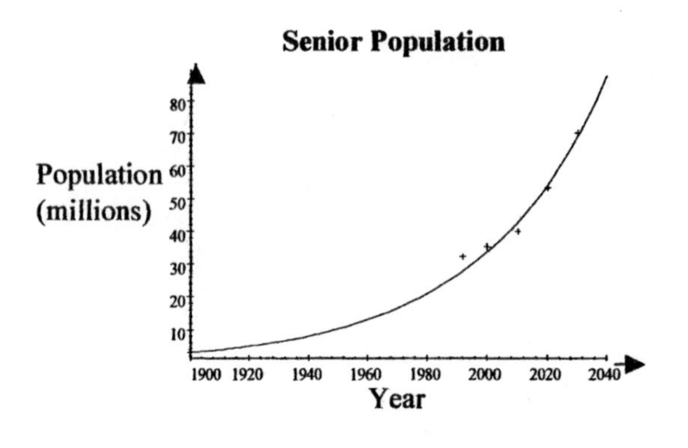

Figure 3.3.7

Based on the given data and this exponential function, we predict the population of seniors in 2040 to be

$$p(2040) = (7.26476 * 10^{-20}) * 1.02408^{2040} = 87,570,000. \quad \blacktriangle$$

A function whose graph fits a data set is called a **model** of the data. The model (that is, the function) is used to analyze the data and infer information about situations that are not explicitly given in the data such as making predictions based on the data.

Remark:

A fundamental reason for studying functions (and algebra) is to transform data into information and information into action.

The process of extracting a function from a data set by fitting a curve to the corresponding scatter plot is called **regression analysis.** This process often presents a challenge, which we will approach first using graphical methods (Section 2.6) and then secondly using analytical methods (Section 2.7). A key step in the process is recognizing the basic shape of the scatter plot. We now focus our attention on the shapes of the graphs of the five most important categories of functions.

1. **Power functions** are often used in modeling motion problems (for example, projectile, stopping times, falling body).

A power function is a variable raised to a positive integer power. For example $f(x) = x^2$ is a power function. Power functions are closely associated with **polynomial functions,** which are linear combinations of power functions. That is, a sum of power functions, each multiplied by a coefficient, and a constant. For instance, $p(x) = 3x^4 - 5x^2 + 2x - 7$ is a polynomial function. Explain why $h(x) = 2x^{3.5} - 4x + 3$ is not a polynomial function.

A polynomial is a linear combination of power terms, each multiplied by a coefficient, and a constant. For example, $2x^3 + 5x^2 - 3x + 4$ is a polynomial, but $2x^3 + 5x^{-2} - 3x + 4$ is not a polynomial. Why?

The largest exponent denotes the **degree** or **order** of the polynomial. Thus this polynomial has degree three (is third order). A linear polynomial is a polynomial of degree one. The last term, 4, is really $4x^0$. However $\mathbf{x}^0$ is defined to be 1, and thus $4x^0 = 4$. The custom is to omit the x^0 because it is defined to be one. A polynomial function is a function defined by a polynomial equation,

$$f(x) = 2x^3 + 5x^2 - 3x + 4.$$

2. **Radical function** is the inverse of a power function. For example,

$$g(x) = \sqrt{x} = x^{1/2}$$

is a radical function. Note that squaring and taking the square root are inverse operations of one another, and thus $f(x) = x^2$ and $g(x) = \sqrt{x}$, for $x \geq 0$, are said to be inverse functions of one another.

We show a plot of a polynomial, plots of two power functions, and two radical functions. In the exercises, you are asked to explore the plots of variations of these functions as well as the plots of other power and radical functions.

First degree polynomial function $f(x) = 3x + 2$ (The graph is a straight line.)

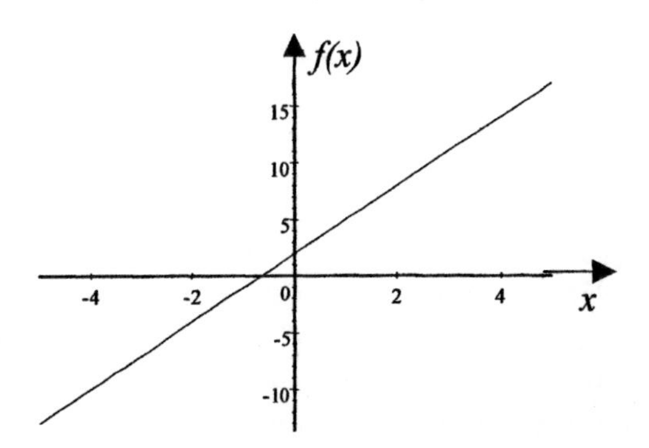

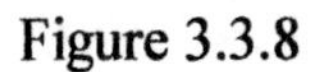
Figure 3.3.8

2nd degree power function $f(x) = x^2$

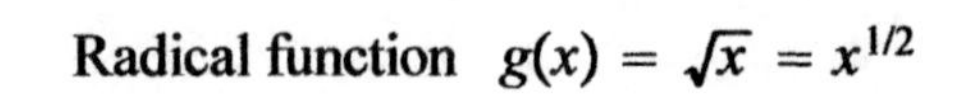
Radical function $g(x) = \sqrt{x} = x^{1/2}$

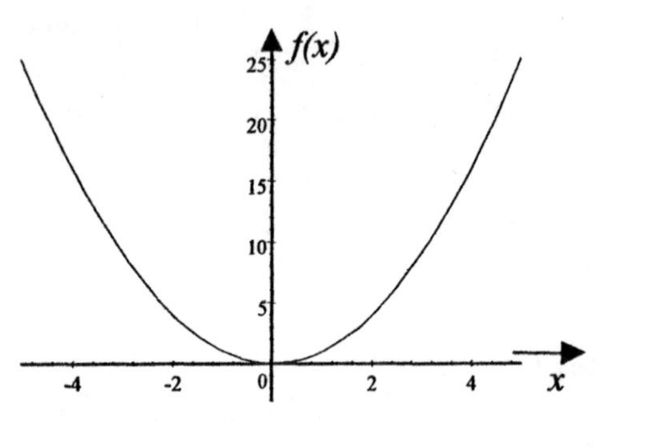

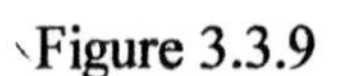
Figure 3.3.9

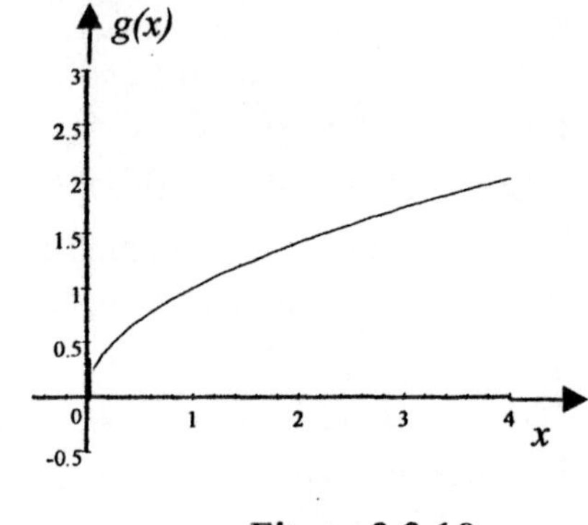

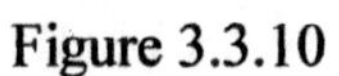
Figure 3.3.10

3rd degree power function $f(x) = x^3$

Radical function $f(x) = x^{1/3}$

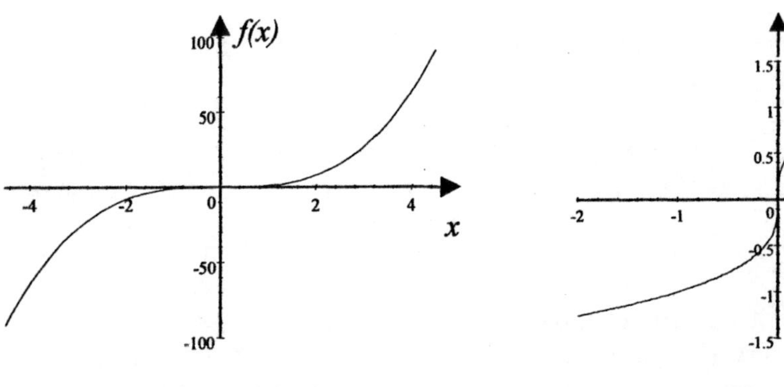

Figure 3.3.11

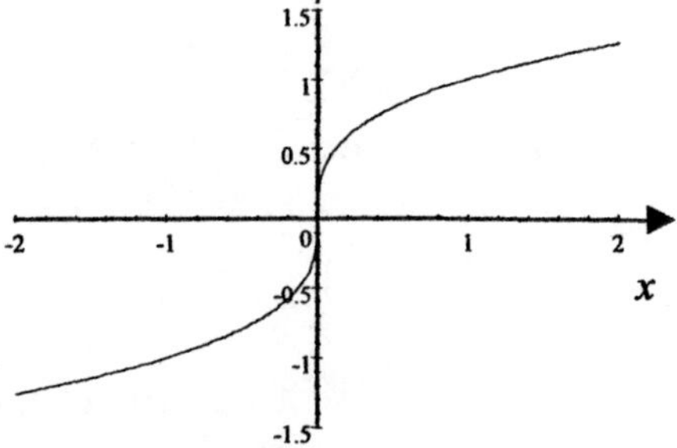

Figure 3.3.12

3. **Exponential functions** are used to model growth or decay situations, such as population growth or decay. Heating and cooling problems are also modeled with exponential functions as are savings accounts and mortgage loans.

Exponential functions contain terms in which the independent variable is the exponent of a number, say $f(x) = 10^x$.

Query.

What is the difference between a power term and an exponential term?

The population data given at the start of this section was modeled with the exponential function

$$p(x) = (7.26476 * 10^{-20}) * 1.02408^x.$$

Solving for the exponent x, given a value of $p(x)$, involves applying the inverse operation which is called the logarithmic operation.

4. **Logarithmic function** is the inverse of an exponential function, thus $\log(10^x) = x$. That is, the logarithmic operation is the inverse operation of raising to a power. The defining relation between an exponential function and a **logarithmic function** is given by

$$y = b^x \text{ if and only if } \log_b(y) = x$$

where b is the *base*. Any positive number can be used as the base. The two most commonly used bases are 10 and the irrational number $e = 2.718...$. Logarithms to the base 10 are called **common logarithms** and are denoted on the calculator by the LOG key. Logarithms to the base e are called **natural logarithms** and are denoted on the calculator by the LN key.

We illustrate plots of the common exponential function and its inverse, the common logarithm function. In the exercises, you are asked to explore the plots of variations of these functions as well as the plots of the natural exponential and natural logarithm functions.

Exponential Function $f(x) = 10^x$

Logarithmic Function $g(x) = \log_{10}(x)$

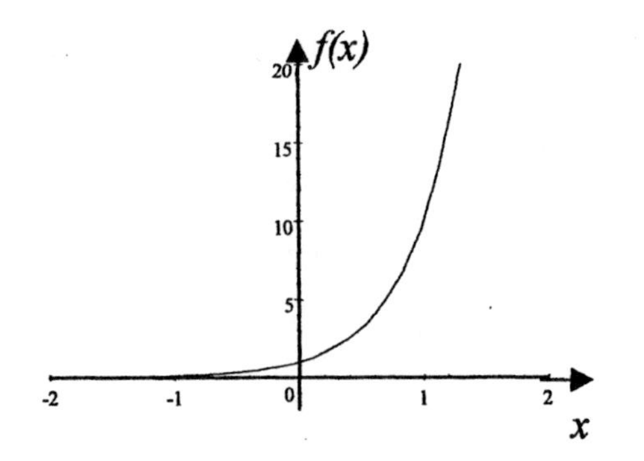

Figure 3.3.13

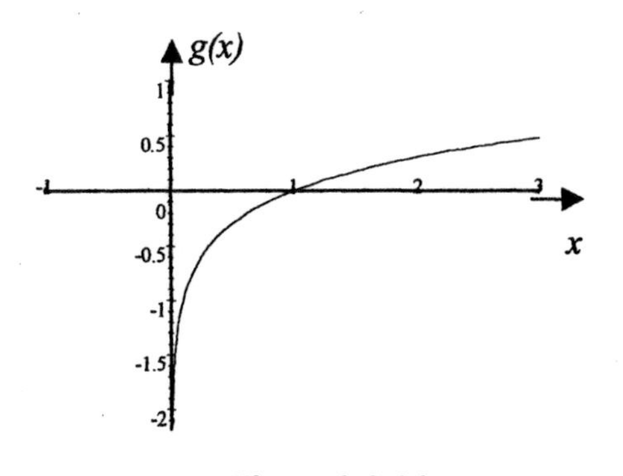

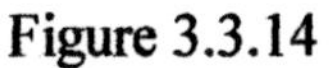
Figure 3.3.14

5. **Periodic functions** are used to model periodic behavior such as hours of daylight, the pendulum motion of a clock, sound waves, and light waves. Periodic functions are functions whose output values repeat in a systematic pattern. The sine and cosine functions are the primary periodic functions and are the ones most frequently used in applications.

We illustrate plots of sine and cosine functions. In the exercises, you are asked to explore the plots of variations of these functions.

Sine function $f(x) = \sin(x)$ Cosine function $f(x) = \cos(x)$

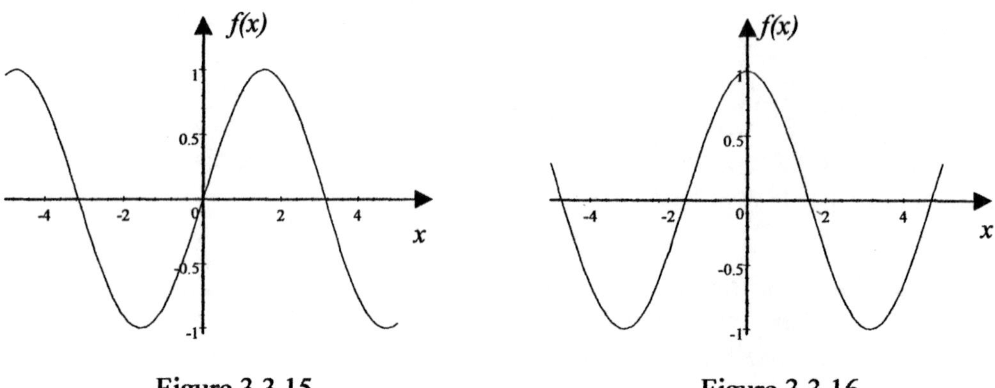

Figure 3.3.15 Figure 3.3.16

The importance of being able to recognize basic functions given their plots and, conversely, to recognize the plots of basic functions cannot be overemphasized. The purpose of the following exercises is to help us develop skill in relating basic shapes to basic functions. In these exploratory exercises, we investigate how changing the values of coefficients in the basic functions changes the basic shapes.

Exercises 3.3 (Use of a graphing calculator is highly recommended.)

1. (Computational Skill) Simplify:

a. $\dfrac{(4)(2^5)}{16}$

b. $\dfrac{(2+3)^3}{5}$

c. $\dfrac{2^3+4^2}{8}$

2. (Graphical Skill) Match each of the following functions with the appropriate plot:

a. $f(x) = x^3$ b. $g(x) = \log(x)$ c. $h(x) = -x^2 + 4$ d. $k(x) = e^{-x}$

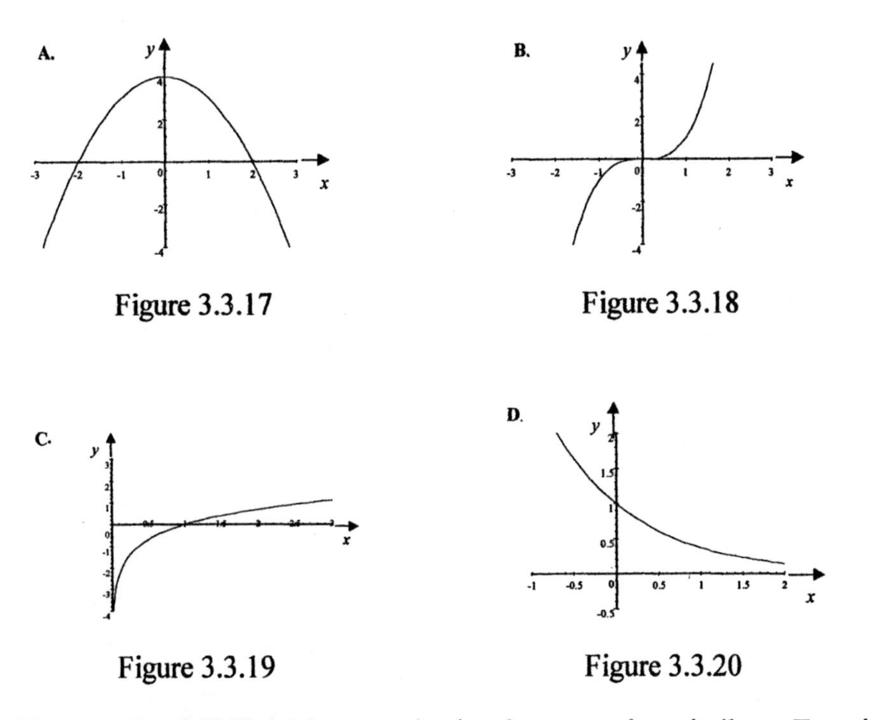

Figure 3.3.17

Figure 3.3.18

Figure 3.3.19

Figure 3.3.20

3. (Computational Skill) Make up and solve three exercises similar to Exercises 1a–c.

4. For each of the following plots, give a function whose graph approximates the plot.

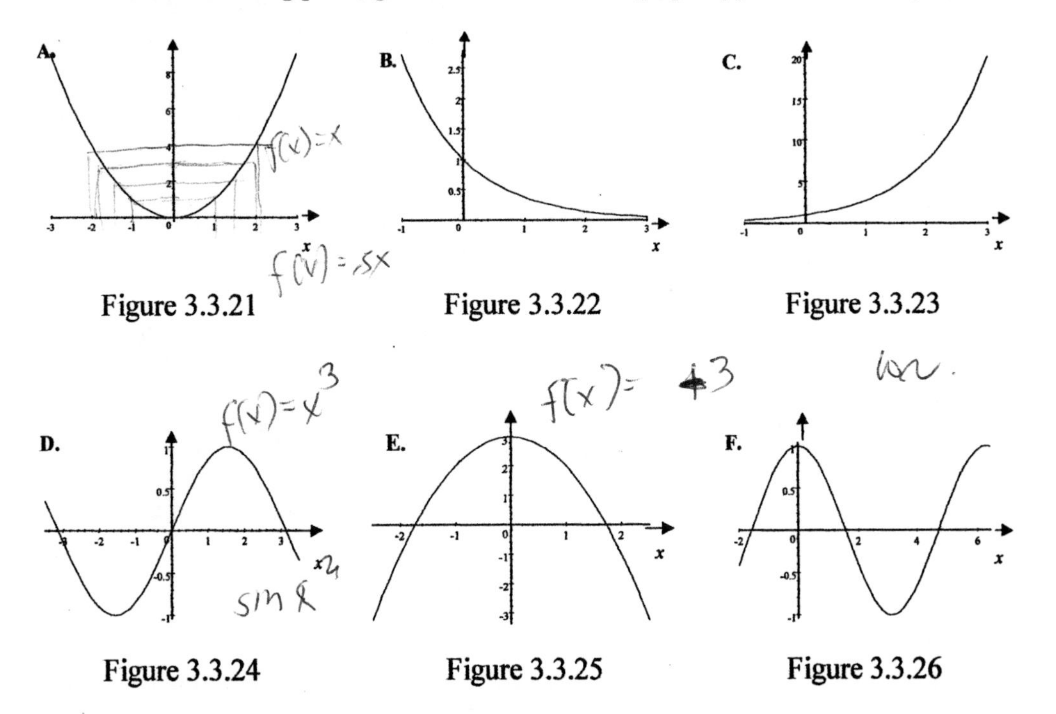

Figure 3.3.21

Figure 3.3.22

Figure 3.3.23

Figure 3.3.24

Figure 3.3.25

Figure 3.3.26

5. When investigating an automobile accident, police often approximate the speed of a vehicle, in mph, by the length of the skid mark. They use the function $s(d) = \sqrt{30cd}$ where d is the length of the skid mark and c is a variable representing the road condition. For example, $c = 0.8$ for dry concrete, $c = 0.4$ for wet concrete, $c = 1$ for dry tar, and $c = 0.5$ for wet tar.

   a. Approximate how fast a car was traveling if it leaves a skid mark of 100 ft. on dry concrete.

b. Approximate how fast a car was traveling if it leaves a skid mark of 100 ft. on wet tar.

c. What is the expected length of a skid mark on wet concrete that a car would leave if it had been traveling at 60 mph?

d. Does a doubling of the length of a skid mark indicate a doubling of the speed? Explain.

6. Investigate the graphical effects of changing the coefficient of x^2 in the polynomial function $f(x) = cx^2$.

a. Plot and label the functions $f(x) = x^2, g(x) = 2x^2, h(x) = 3x^2$ on the same pair of axes.

b. Plot and label the functions $f(x) = x^2, g(x) = -x^2, h(x) = -2x^2$ on the same pair of axes.

c. Plot and label the functions $f(x) = x^2, g(x) = \frac{1}{2}x^2, h(x) = \frac{1}{3}x^2$ on the same pair of axes.

d. Sketch (by hand) and label the graphs of $f(x) = x^2, g(x) = 5x^2, h(x) = \frac{1}{5}x^2$ and then check your graphs with your calculator.

e. Write a paragraph explaining how changing the coefficient of x^2 in the function $f(x) = cx^2$ changes the plot of f.

7. Repeat Exercise 6 with x^2 replaced with x^3.

8. Investigate the graphical effects of changing the exponent in the function $f(x) = x^n$.

a. Plot and label the functions $f(x) = x^n$ for $n = 1,2,3,4,5$ on the same axes.

b. Sketch (by hand) and label the graphs of $f(x) = x^2, g(x) = x^4, h(x) = x^6, k(x) = x^8$ and then check your graphs with your calculator.

c. Explain how the graphs of the even powers of x change as the powers increase.

d. Sketch (by hand) and label the graphs of $f(x) = x, g(x) = x^3, h(x) = x^5$ and then check your graphs with your calculator.

e. Explain how the graphs of the odd powers of x change as the powers increase.

9. Investigate the graphical effects of changing the coefficient of x in the function $f(x) = \sqrt{cx}$.

a. Plot and label the functions $f(x) = \sqrt{x}, g(x) = \sqrt{2x}, h(x) = \sqrt{3x}$ on the same pair of axes.

b. Sketch (by hand) and label the graphs of $f(x) = \sqrt{0.5x}, g(x) = \sqrt{4x}, h(x) = \sqrt{5x}$ and then check your graphs with your calculator.

c. Write a paragraph explaining how changing the coefficient of x in the function $f(x) = \sqrt{cx}$ changes the plot of f.

10. Investigate the graphical effects of changing the power of x in the function $f(x) = x^{\frac{1}{n}}$.

a. Plot and label the functions $f(x) = x^{\frac{1}{n}}$ for $n = 1,2,3,4$ on the same axes.

b. Sketch (by hand) and label the graphs of $f(x) = x^{\frac{1}{n}}$ for $n = 2,5,6$ and then check your graphs with your calculator.

c. Write a paragraph explaining how changing the power of x in the function $f(x) = x^{\frac{1}{n}}$ changes the plot of f.

11. Investigate the graphical effects of changing the coefficient c in the (common) exponential function $f(x) = c10^x$

a. Plot and label the functions $f(x) = 10^x, g(x) = 3 * 10^x, h(x) = 4 * 10^x$ on the same pair of axes.

b. Plot and label the functions $f(x) = -10^x, g(x) = -3 * 10^x, h(x) = -4 * 10^x$ on the same pair of axes.

c. Sketch (by hand) and label the graphs of $f(x) = 10^x, g(x) = 0.5 * 10^x, h(x) = 5 * 10^x$ and then check your graphs with your calculator.

d. Write a paragraph explaining how changing the coefficient of c in the function $f(x) = c10^x$, changes the plot of f.

12. Investigate the graphical effects of changing the coefficient c in the (common) logarithmic function to the base 10, $f(x) = \log(cx)$.

a. Plot and label the functions $f(x) = \log(x), g(x) = \log(2x), h(x) = \log(3x)$ on the same pair of axes.

b. Sketch (by hand) and label the graphs of $f(x) = \log(x), g(x) = \log(4x), h(x) = \log(5x)$ and then check your graphs with your calculator.

c. Write a paragraph explaining how changing the coefficient of c in the function $f(x) = \log(cx)$, changes the plot of f.

13. Graphically investigate the comparison between the natural exponential function, $f(x) = ce^x$ and the common exponential function, $g(x) = 10^x$.

a. Plot and label the functions $f(x) = e^x$ and $g(x) = 10^x$ on the same pair of axes.

b. Sketch (by hand) and label the graphs of $f(x) = e^{-x}$ and $g(x) = 10^{-x}$. Check your graphs with your calculator.

c. Write a paragraph explaining how changing the numerical sign of the exponent in the exponential function changes the plot of the function.

14. Investigate the graphical effects of changing the coefficient c in the trigonometric function $f(x) = c\sin(x)$.

a. Plot and label the functions $f(x) = \sin(x), g(x) = 2\sin(x), h(x) = 3\sin(x)$ on the same pair of axes.

b. Sketch (by hand) and label the graphs of $f(x) = \sin(x), g(x) = -\sin(x), h(x) = 4\sin(x)$ and then check your graphs with your calculator.

c. Write a paragraph explaining how changing the coefficient of c in the function $f(x) = c\sin(x)$ changes the plot of f.

15. Investigate the graphical effects of changing the coefficient c in the trigonometric function $f(x) = c\cos(x)$

a. Plot and label the functions $f(x) = \cos(x), g(x) = 2\cos(x), h(x) = 3\cos(x)$ on the same pair of axes.

b. Sketch (by hand) and label the graphs of $f(x) = \cos(x), g(x) = -\cos(x), h(x) = 4\cos(x)$ and then check your graphs with your calculator.

c. Write a paragraph explaining how changing the coefficient of c in the function $f(x) = c\cos(x)$, changes the plot of f.

16. Plot the graph of $f(x) = \sin(x)$ over each of the following intervals: $[-2,2], [-10,10], [-20,20]$. Based on your plots, determine the period of $\sin(x)$.

17. Plot the graph of $f(x) = \cos(x)$ over each of the following intervals: $[-2,2], [-10,10], [-20,20]$. Based on your plots, determine the period of $\cos(x)$.

18. Identify each of the following plots.

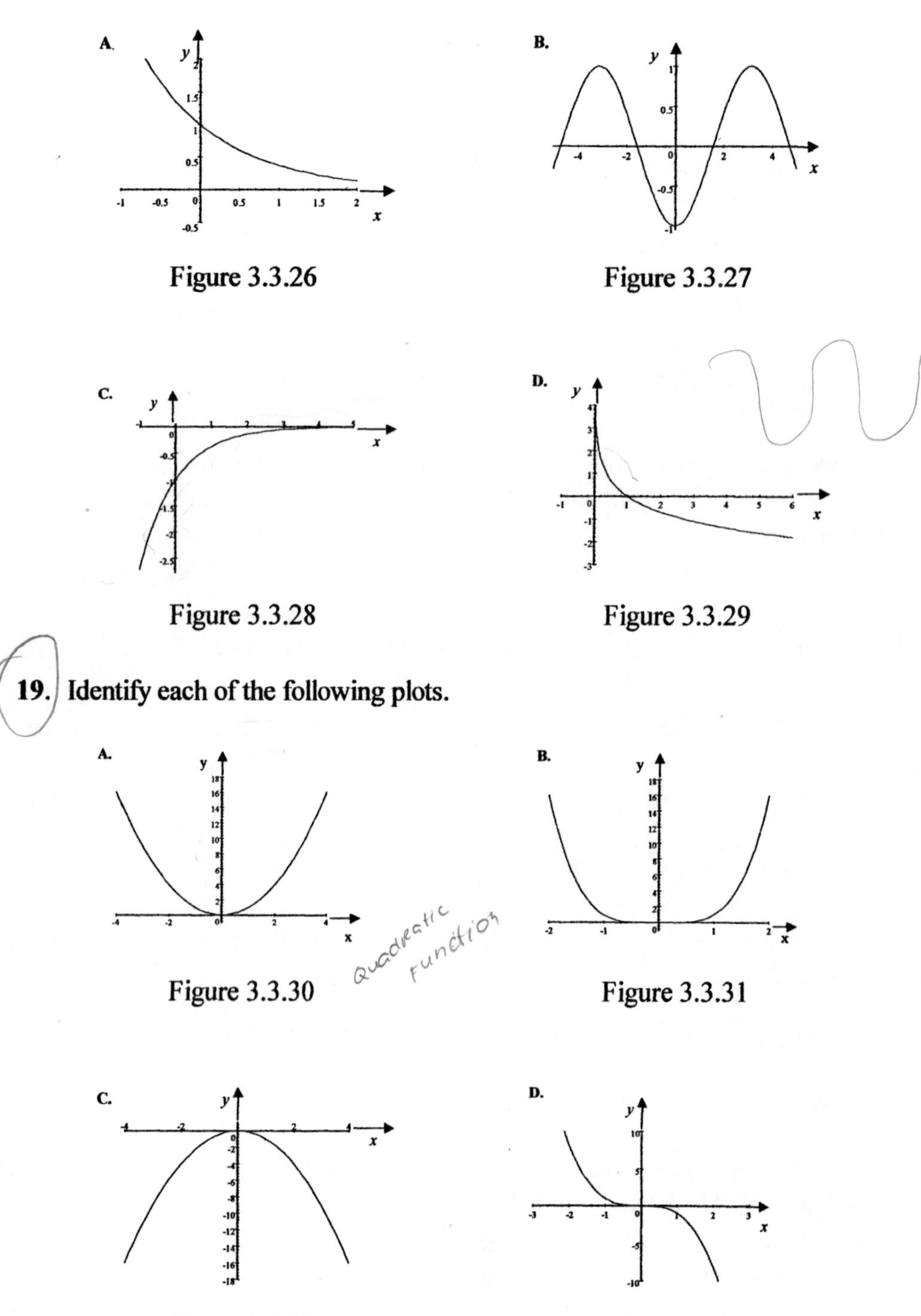

Figure 3.3.26

Figure 3.3.27

Figure 3.3.28

Figure 3.3.29

19. Identify each of the following plots.

Figure 3.3.30

Figure 3.3.31

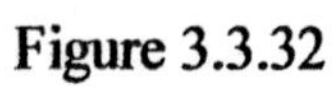

Figure 3.3.32

Figure 3.3.33

3.4 Shifting and Scaling Graphs

In this section, we continue to explore the shapes of the graphs of basic functions by focusing on moving graphs, without changing their shapes, and by resizing graphs. These operations are called ***shifting*** and ***scaling*** graphs.

Several computer screen savers are programs that move a fixed shape around the screen. How does the function expression for a given graph, say $f(x) = x^2$, change when the graph is shifted, but the shape is not changed? For example, in the following multiplot (several graphs plotted on the same set of axes), all of the plots have the shape of the graph of $f(x) = x^2$.

Shifting Graphs

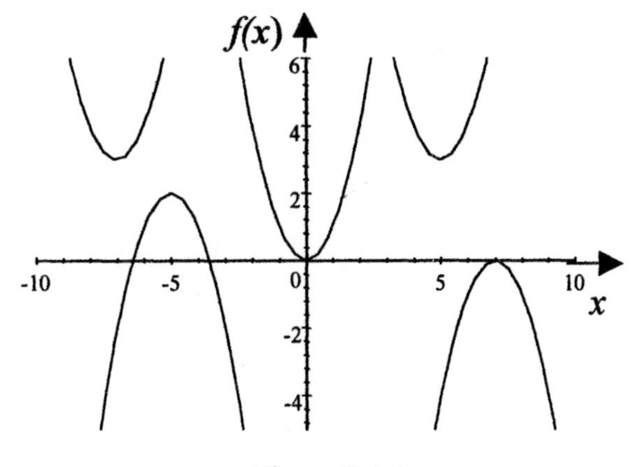

Figure 3.4.1

Small-Group Activity

Each group reproduces Figure 3.4.1 on their calculators and then group representatives explain to the class what they did and how they did it.

Shifting Graphs

Our purpose is to learn how to transform a function expression in order to shift the graph of the function left or right and up or down without changing the shape. There are two rules:

1. Horizontal Shift:

   To shift the graph of a function p units to the right without changing the shape, subtract a positive constant, p, from the independent variable, x (that is, replace x by $x - p$). See plot C in Figure 3.4.2.
   To shift the graph of a function p units to the left without changing the shape, add a positive constant, p, to the independent variable, x (that is, replace x by $x + p$). See plot B in Figure 3.4.2.

   We illustrate this rule with the function $f(x) = x^2$ and $p = 3$.

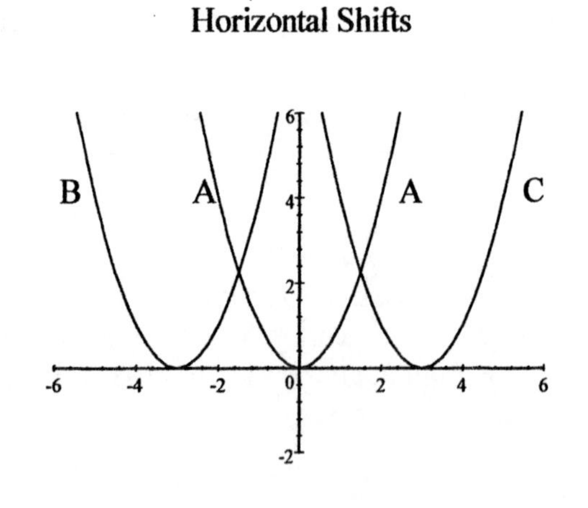

Legend

A: $f(x) = x^2$

B: $g(x) = f(x+3) = (x+3)^2$

C: $h(x) = f(x-3) = (x-3)^2$

Figure 3.4.2

2. **Vertical Shift:**

To shift the graph of a function down c units without changing the shape, subtract a positive constant, c, from the dependent variable, $f(x)$, (that is, replace $f(x)$ by $f(x) - c$). See plot C in Figure 3.4.3.
To shift the graph of a function up c units without changing the shape, add a positive constant, c, to the dependent variable, $f(x)$, (that is, replace $f(x)$ by $f(x) + c$). See plot B in Figure 3.4.3.

We illustrate this rule with the function $f(x) = x^2$ and $c = 2$.

Vertical Shifts

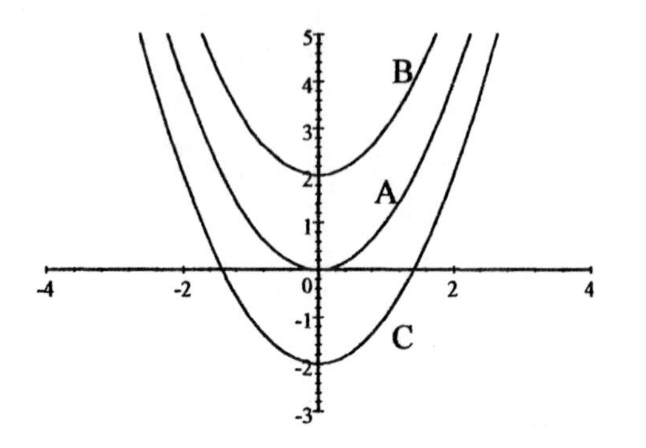

Legend

A: $f(x) = x^2$

B: $g(x) = f(x) + 2 = x^2 + 2$

C: $h(x) = f(x) - 2 = x^2 - 2$

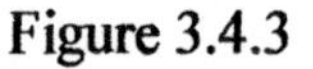
Figure 3.4.3

Query 1.
Transform the function $f(x) = x^3$ so that the graph of the transformed function is the graph of $f(x) = x^3$ shifted two units to the right and three units down.

A second type of transformation involves changing the actual shape of a graph, but not its basic shape. For example, reflecting the graph in the horizontal axis will turn the graph over, but will not change its basic shape (plot D in Figure 3.4.4). Consider the graphs in the following multiplot. Although they are all different, they all have the same basic shape (which is the shape of the graph of $f(x) = x^2$).

Scaled Graphs of $f(x) = x^2$

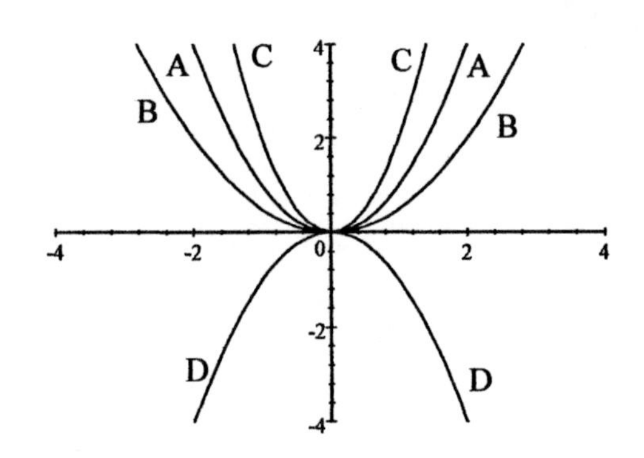

Legend

A: $f(x) = x^2$

B: $g(x) = \frac{1}{2}f(x) = \frac{1}{2}x^2$

C: $h(x) = 2f(x) = 2x^2$

D: $k(x) = -f(x) = -x^2$

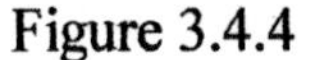

Figure 3.4.4

Scaling Graphs

Scaling the graph of a function means to change the actual shape of the graph, but not its basic shape. There are two possibilities, multiply the independent variable by a constant (plot C in Figure 3.4.5) or multiply the dependent variable by a constant (plot B in Figure 3.4.5). Note that multiplying the dependent variable by a constant is the same as multiplying the function expression by the constant. We give two illustrations of scaling; one with the exponential function $f(x) = 2^x$ and one with the power function $f(x) = x^3$. In both illustrations the constant will be 2.

Scaling $f(x) = 2^x$

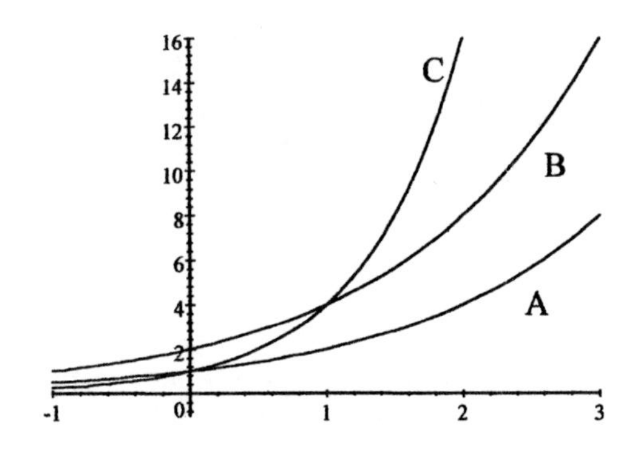

Legend

A: $f(x) = 2^x$

B: $g(x) = 2f(x) = 2 * 2^x$

C: $h(x) = f(2x) = 2^{2x}$

Figure 3.4.5

Scaling $f(x) = x^3$

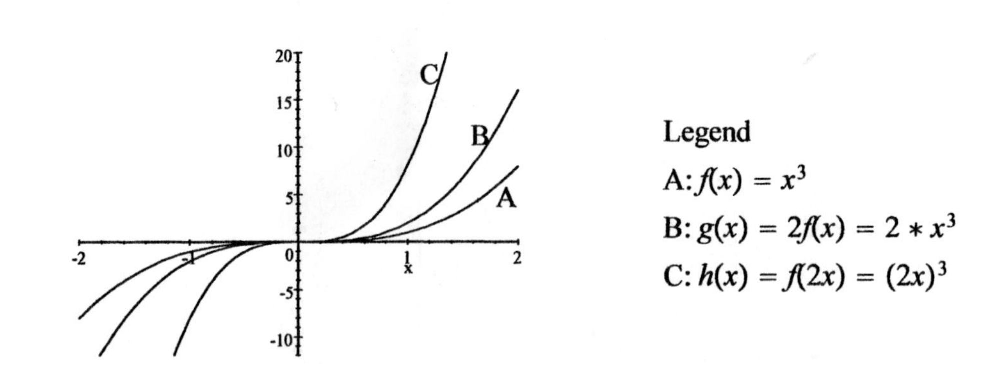

Figure 3.4.6

Query 2.

How would the graph of $f(x) = x^3$ change if the dependent variable was scaled by 0.5? Illustrate your answer with a multiplot.

Example 3.4.1.

Scaling the graph of a function by multiplying the independent variable by -1 reflects the graph in the vertical axis. We illustrate with the exponential function $f(x) = e^x$.

Reflection in the Vertical Axis

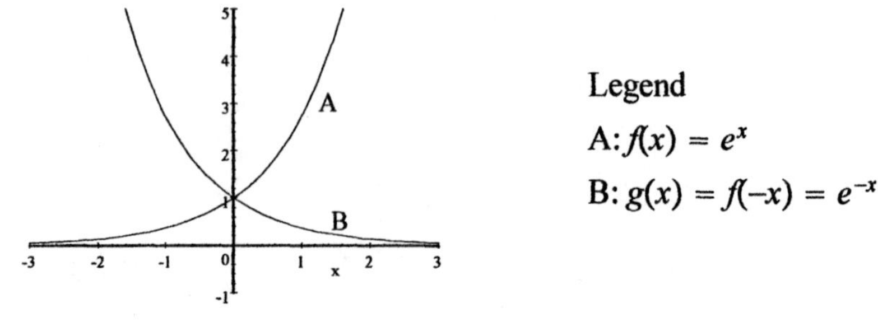

Figure 3.4.7 ▲

An **asymptote** of a graph is a line that the graph approaches and remains arbitrarily close to as either the independent or dependent variable becomes very large in the positive or negative sense. Figure 3.4.7 illustrates that the positive x-axis as an asymptote of the graph of $g(x) = f(-x) = e^{-x}$ and the negative x-axis is an asymptote of the graph of $f(x) = e^x$.

Query 3.

How would the graph of $f(x) = e^x$ change if the dependent variable was scaled by -1? Illustrate your answer with a multiplot.

Query 4.

Can a graph approach a line asymptotically from below the line? What is an asymptote?

The following example demonstrates the art of curve fitting. The reader is encouraged to visualize each of the steps by plotting the functions involved.

Example 3.4.2.

In Example 3.1.2, Margaret recorded temperature data of a can of cold soda as it warmed sitting on a counter in a room with temperature of 75°F. The following is the data and a line plot of the data.

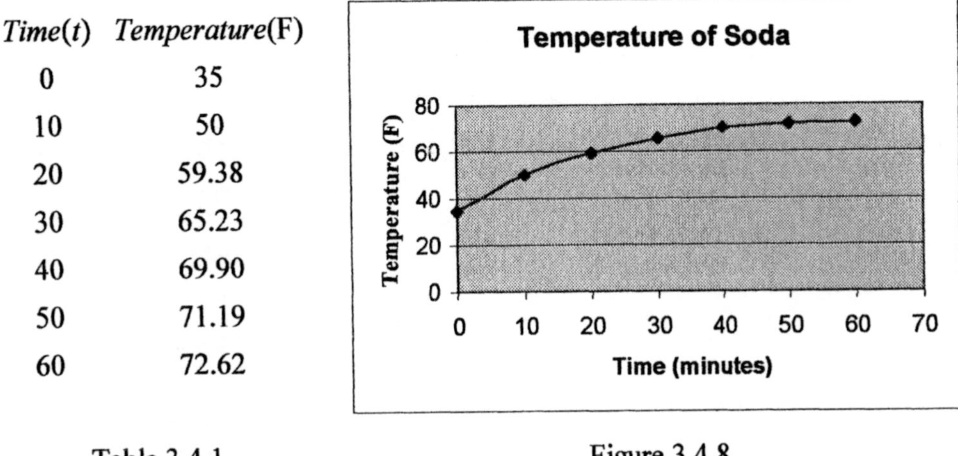

Time(t)	*Temperature*(F)
0	35
10	50
20	59.38
30	65.23
40	69.90
50	71.19
60	72.62

Table 3.4.1

Figure 3.4.8

How can we identify a basic function whose graph can be shifted and/or scaled to fit Margaret's data?

Solution

A first response might be to think about shifting and scaling the graph of $f(t) = \sqrt{t}$. (We use the input variable, t, to denote time.) In order to fit the curve to the first data point $(0, 35)$, we shift the graph vertically by 35 units. That is, consider $g(t) = f(t) + 35 = \sqrt{t} + 35$. As can be observed from the following multiplot, the plot of g does not give a suitable approximation to the data. After a few more scaling attempts, we produce $h(t) = f(23t) + 35 = \sqrt{23t} + 35$.

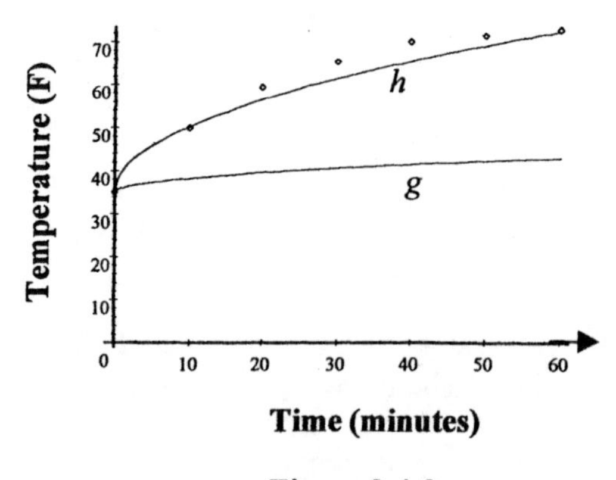

Figure 3.4.9

Although the graph of h approximates the data much better than does the graph of g, the multiplot shows that the shape of h is not the correct basic shape of the data. The data points appear to be leveling off while the graph of h appears to be increasing almost linearly. A little more thought enables us to explain what is happening. The temperature of the soda is bounded above by the room temperature (75°) whereas the function h is unbounded (meaning that the values of h can become arbitrarily large).

We need to consider how to obtain an increasing function whose graph approaches the line $y = 75$ arbitrarily closely (that is, is asymptotic to the line, $y = 75$). From Example 3.4.1, we know that the graph of e^{-t} decreases asymptotically to the positive t-axis. Thus the graph of $-e^{-t}$ must increase asymptotically to the positive t-axis. (See Query 4.) Therefore, to obtain a graph that increases asymptotically to the line $y = 75$, we shift the graph of $-e^{-t}$ vertically 75 units. Hence we consider $f(t) = 75 - e^{-t}$. Figure 3.4.10 indicates that we have the correct asymptotic behavior. However, the initial temperature, $f(0) = 74$, is not correct.

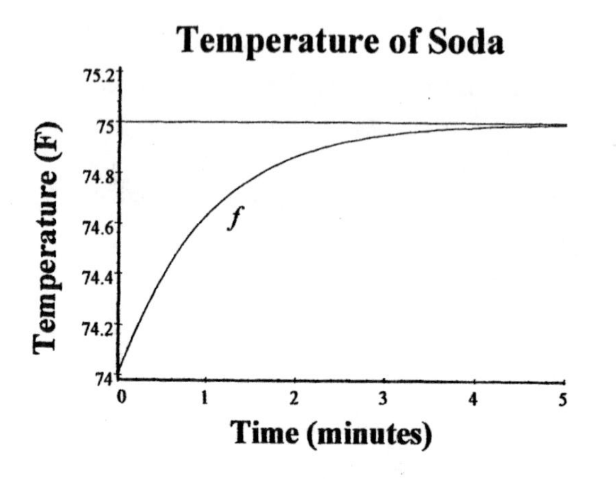

Figure 3.4.10

Because the temperature of the soda in the refrigerator is 35°, we want $f(0) = 35$. In order to obtain this, we scale the graph of f by multiplying the e^{-t} term by 40. (Why? Why not shift downward by 39, that is, write $f(t) = 36 - e^{-t}$?) We now have the function $g(x) = 75 - 40e^{-t}$. We show the multiplot of the data points, the line $y = 75$, and the graph of $g(x) = 75 - 40e^{-t}$.

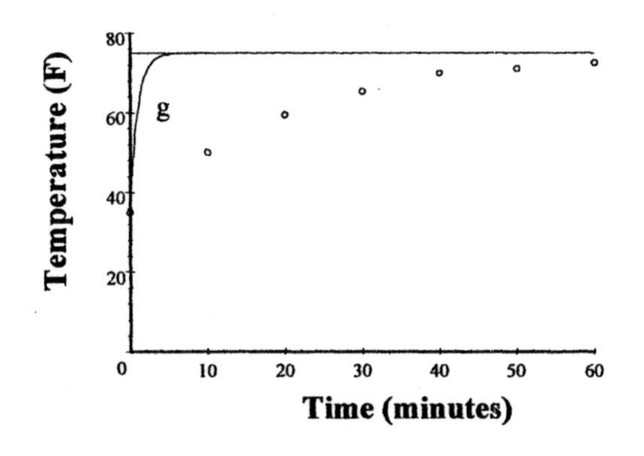

Figure 3.4.11

The graph of g has the correct basic shape, and it crosses the temperature-axis at the correct point (that is, 35), but it needs to be scaled. (Should g be scaled by multiplying t by a number greater than one or less than one?) The following plots show the effects of multiplying t first by a scale factor of 0.1 and then by 0.05.

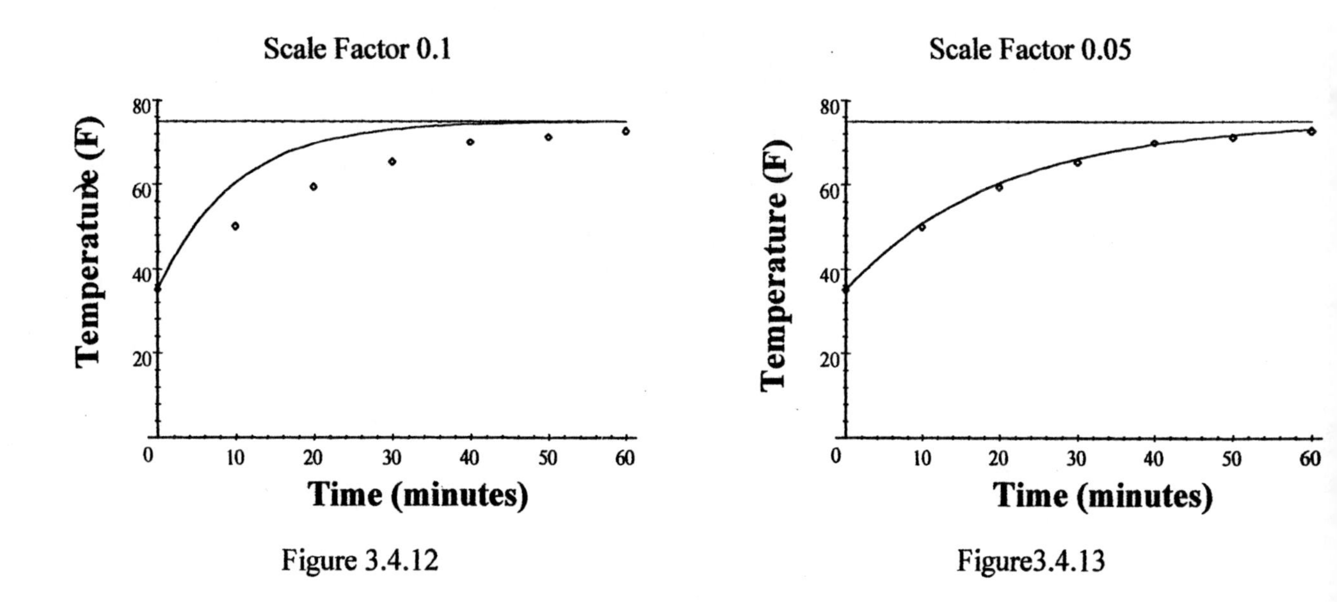

Figure 3.4.12

Figure3.4.13

We accept $h(x) = g(.05x) = 75 - 40e^{-.05x}$ as a good temperature model for this example. ▲

Asymptote

We conclude this section by commenting on the term *asymptote* and defining a second function whose graph has asymptotic properties. ($f(x) = e^{-x}$ was the first function we encountered with asymptotic properties (see Example 3.4.1). As stated earlier, a line is an asymptote of the graph of a function provided the distance between the graph and the line becomes arbitrarily small as either the independent or dependent variable becomes large in the positive or negative sense. For example, the graph of $f(x) = e^{-x}$ comes arbitrarily close to the positive x-axis as x becomes arbitrarily large. In this case, we say that the x-axis is an asymptote of the graph of the function and that the function is asymptotic to the x-axis. Thus shifting the graph two units upward gives a function, $g(x) = e^{-x} + 2$, which is asymptotic to the line $y = 2$.

The two most important asymptotic functions for our work are: $f(x) = e^{-x}$ and $h(x) = \frac{1}{x}$. Figure 3.4.14 shows that both the x and y axes are asymptotes of h.

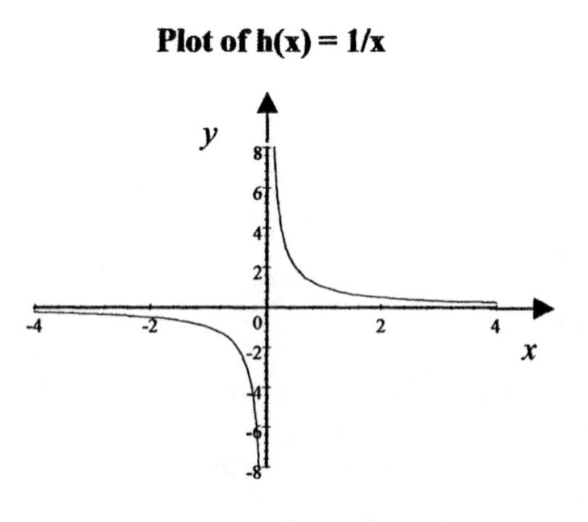

Figure 3.4.14

Figure 3.4.15 shows that the horizontal line $y = -2$ and the vertical line $x = 1$ are asymptotes of h when it is shifted one unit to the right and two units down.

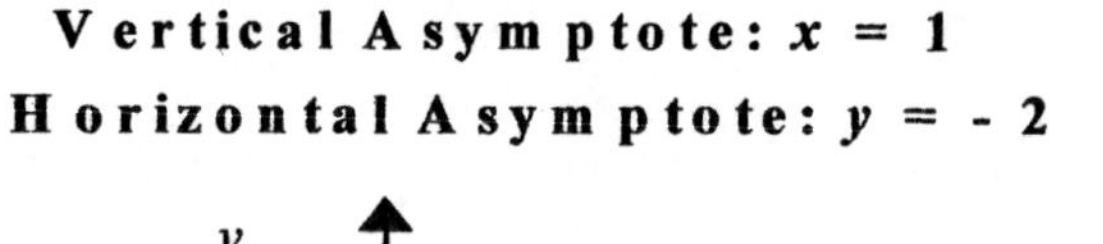

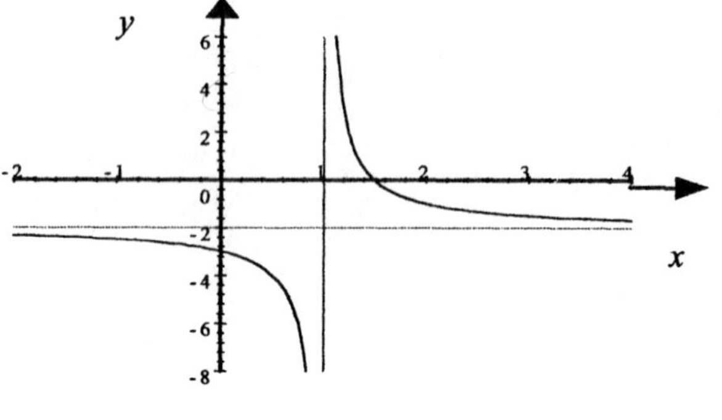

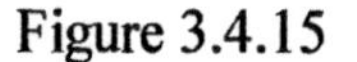
Figure 3.4.15

Note that $f(x) = e^{-x}$ has only a horizontal asymptote whereas $h(x) = \frac{1}{x}$ has both a horizontal and a vertical asymptote.

Query 5.

How should the graph of $h(x) = \frac{1}{x}$ be shifted so that the lines $y = 3$ and $x = -2$ are asymptotes?

Exercises 3.4

1. (Computational Skill) Describe how you would shift the graph of $f(x) = x^2$ so that the resulting graph passes through the point.
   a. $(0, 10)$
   b. $(2, 4)$
   c. $(-1, 5)$

2. (Computational Skill) Describe how you would scale the graph of $g(x) = x^3$ so that the resulting graph passes through the point.
   a. $(1, 0.5)$
   b. $(-1, 1)$
   c. $(1, 8)$

3. (Computational Skill) Make up and solve three exercises similar to Exercises 1a–c, 2a–c.

4. Let f be the absolute value function, $f(x) = |x|$. Form a multiplot of $f(x)$, $f(.5x)$, and $f(-x)$.

5. Shift the absolute value function, $f(x) = |x|$, such that the vertex is at:
   a. $(3, 2)$
   b. $(-1, -2)$
   c. $(-2, 3)$

6. Shift the graph of $f(x) = -x^2$ so that the maximum point is at $(3, 2)$.

7. Shift the graph of $f(x) = x^2$ so that the minimum point is at $(3, 2)$.

8. Shift the graph of $f(x) = e^{-x}$ so that $x = -3$ is an asymptote of the shifted graph.

9. Shift the graph of $h(x) = \frac{1}{x}$ so that $x = -3$ and $y = 4$ are asymptotes of the shifted graph.

In Exercises 10–16, you are asked to plot several conjectures. After each one, analyze the difference between the data points and the curve to determine if scaling or shifting is needed to improve the conjecture.

10. Shift and/or scale the graph of $f(x) = x^2$ such that the resulting graph approximates the following data points. Hint: plot f and the data points.

x	y
–1	1.5
0	2
.5	1.5
2	0

Table 3.4.2

11. Shift and/or scale the graph of $f(x) = x^3$ such that the resulting graph approximates the following data points. Hint: Plot f and the data points.

x	y
–1	2
0	1
1	–2

Table 3.4.3

12. Shift and/or scale the graph of $f(x) = 2^x$ such that the resulting graph approximates the following data points. Hint: Plot f and the data points.

x	y
–1	3
0	2
1	1.5
4	1.03

Table 3.4.5

13. Shift and/or scale the graph of $f(x) = e^x$ such that the resulting graph approximates the following data points. Hint: Plot f and the data points.

x	y
–1	–3.5
0	–3
2.5	0
4	5

Table 3.4.5

14. Shift and/or scale the graph of $f(x) = \log_{10}(x)$ such that the resulting graph approximates the following data points. Hint: Plot f and the data points.

x	y
1	3
2	3.6
3	4
4	4.2

Table 3.4.6

15. Shift and/or scale the graph of $f(x) = \sin(x)$ such that the resulting graph approximates the following data points. Hint: Plot f and the data points.

x	y
$-\pi/2$	3
0	1
2.6	0
$3\pi/2$	3

Table 3.4.7

16. Shift and/or scale the graph of $f(x) = \cos(x)$ such that the resulting graph approximates the following data points. Hint: plot f and the data points.

x	y
–2	–0.65
0	–0.42
2	1
4	–.42

Table 3.4.8

17. Given the function $f(x) = 3\sqrt{x-1}$, match the expressions in the following table. (There may be entries that have no match.)

A.	$f(x)+1$	a.	$6\sqrt{x-1}$
B.	$f(x+1)$	b.	$3\sqrt{-x-1}$
C.	$-f(x)$	c.	$3\sqrt{x}$
D.	$f(-x)$	d.	$3\sqrt{x-1}+1$
E.	$f(2x)$	e.	$6\sqrt{x-1}$

Table 3.4.9

18. Given the functions $f(x) = x^2 - 3e^x$ and $g(x) = \sqrt{x+1}$, match the expressions in the following table. (There may be entries that have no match.)

A.	$f(g(3))$	a.	$\sqrt{6e-3}$
B.	$f(g(5))$	b.	$\sqrt{\frac{1+x}{x}}$
C.	$f(g(\frac{1}{x}))$	c.	$4-3e^2$
D.	$g(f(-2))$	d.	$\frac{\sqrt{5e-3}}{e}$

Table 3.4.10

ᐧtch the functions *a* through *d* with their graphs.

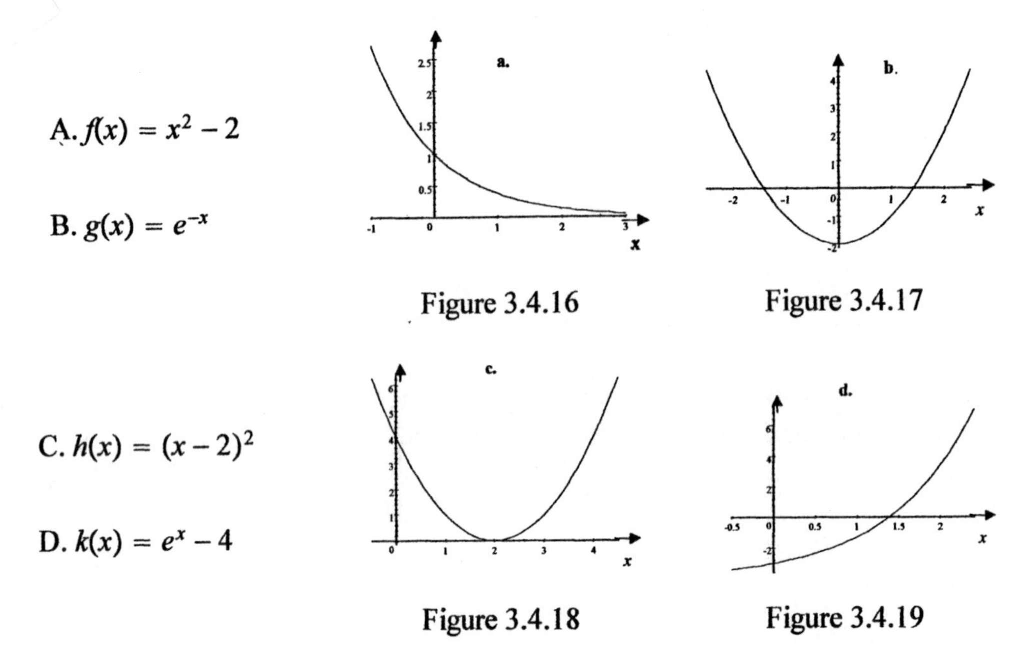

A. $f(x) = x^2 - 2$

B. $g(x) = e^{-x}$

C. $h(x) = (x - 2)^2$

D. $k(x) = e^x - 4$

Figure 3.4.16 Figure 3.4.17

Figure 3.4.18 Figure 3.4.19

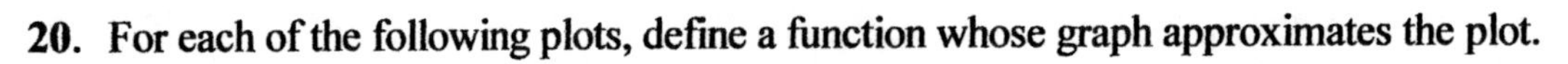

20. For each of the following plots, define a function whose graph approximates the plot.

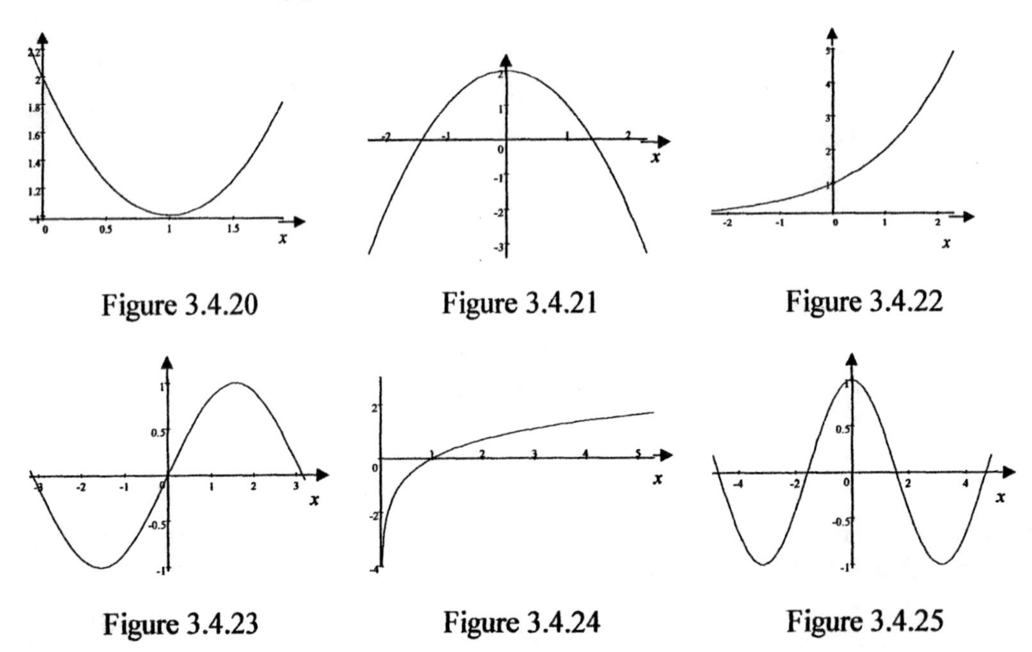

Figure 3.4.20 Figure 3.4.21 Figure 3.4.22

Figure 3.4.23 Figure 3.4.24 Figure 3.4.25

21. Write a paragraph explaining the reasoning involved in a horizontal shift (e.g., Why should replacing x by $x - 2$ shift the graph two units to the right?).

3.5 Algebra of Functions

In Sections 3.2 and 3.3, we defined the function relation and introduced five basic categories of functions. In this section, we will show how functions from these basic categories can be combined to form additional functions. The basic function operations are ***addition, subtraction, multiplication, division, composition,*** and ***taking inverses.*** We shall consider each of these operations from the graphic, numeric, and symbolic points of view.

Addition of Functions

The sum of two functions f, g defined over the same domain is a function h defined over their domain by the equation

$$h(x) = (f+g)(x) = f(x) + g(x).$$

Example 3.5.1 (Graphic Addition of Functions).

Let functions f and g be defined by their graphs.

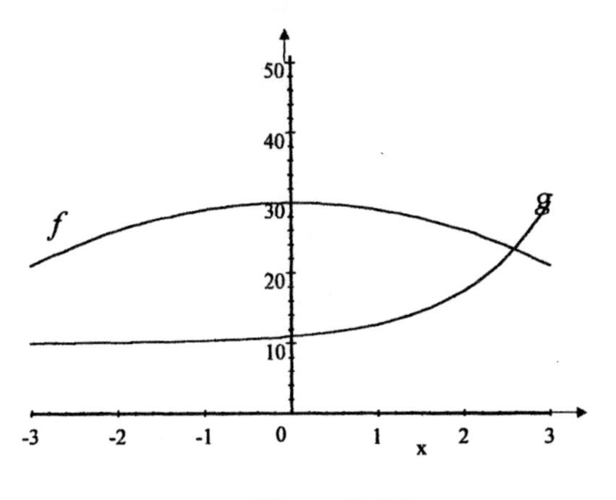

Figure 3.5.1

The process of graphically adding two functions is to select an input value, approximate the corresponding output values for each of the functions, add the two approximate output values, and then plot this new ordered pair. The first coordinate of the new ordered pair is the selected input value and the second coordinate is the sum of the output values of the two functions. We do this for several input values and then draw a smooth curve through the plotted points. We illustrate this process by first determining and then plotting six points that lie on the sum curve, $h(x) = f(x) + g(x)$.

x	$\sim f(x)$	$\sim g(x)$	$(x, \sim f(x) + \sim g(x))$
-3	20	10	$(-3, 30)$
-2	27	10	$(-2, 37)$
-1	30	10	$(-1, 40)$
0	30	12	$(0, 42)$
2	27	18	$(2, 45)$
3	20	30	$(3, 50)$

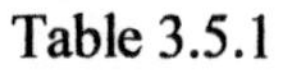
Table 3.5.1

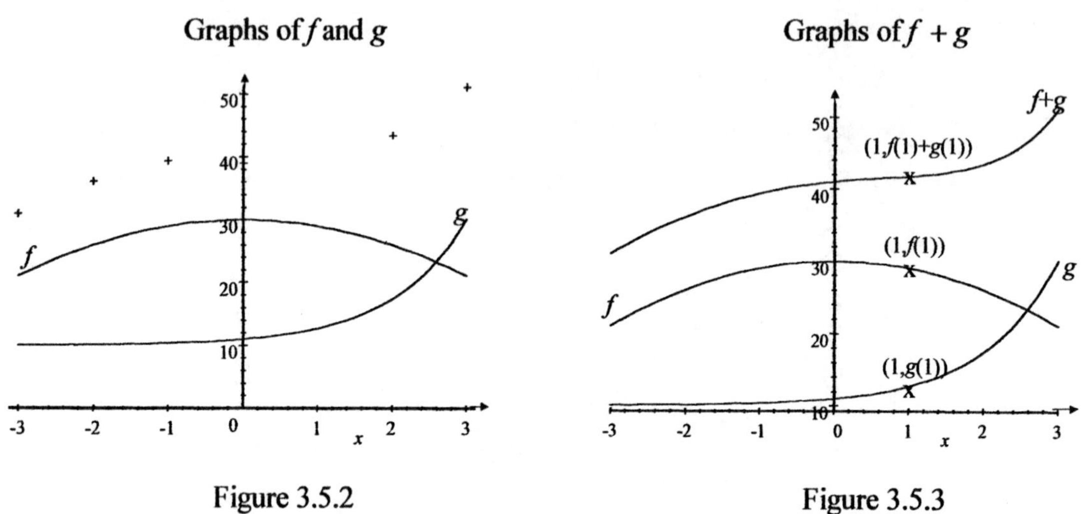

Figure 3.5.2

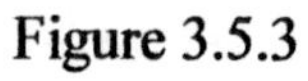
Figure 3.5.3

▲

Example 3.5.2 (Numeric Addition of Functions).

Let functions f and g be defined by the following table.

x	$f(x)$	x	$g(x)$
-2	4	-10	10
0	1	-2	-3
0.5	5	0	6
1	3	1	5
3	1	3	7
4	6	4	2

Table 3.5.2

The functions f and g can only be added over common values in their domains. Thus, although g is defined for $x = -10$, f is not, and therefore the sum of the two functions is not defined when $x = -10$. A similar statement is true when $x = 0.5$. The sum of these two functions is given by the following table.

x	$(f+g)(x) = f(x) + g(x)$
-2	$4 - 3 = 1$
0	$1 + 6 = 7$
1	$3 + 5 = 8$
3	$1 + 7 = 8$
4	$6 + 2 = 8$

Table 3.5.3 ▲

Example 3.5.3 (Symbolic Addition of Functions).

Let functions f and g be defined symbolically as

$$f(x) = x^3 - 4x + 5 \quad \text{and} \quad g(x) = x^2 + 7x - 6.$$

Their sum is defined as

$$h(x) = (f+g)(x) = (x^3 - 4x + 5) + (x^2 + 7x - 6) = x^3 + x^2 + 3x - 1.$$ ▲

Query 1.

How do you define subtraction of functions? Illustrate your definition graphically, numerically, and symbolically.

Example 3.5.4.

Sketch and label the graphs of $h(x) = (f+g)(x) = f(x) + g(x)$ and $k(x) = (f-g)(x) = f(x) - g(x)$, where f and g are defined by the following graphs.

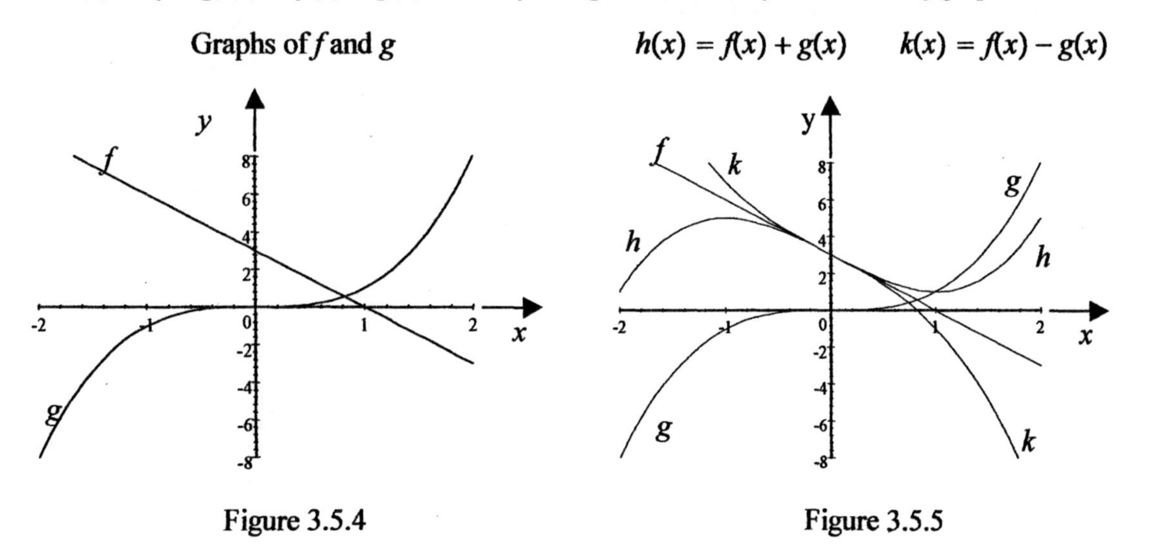

Figure 3.5.4 Figure 3.5.5 ▲

Multiplication of Functions

The product of two functions f, g defined over the same domain is a function h defined over the same domain by the equation $h(x) = (f * g)(x) = f(x) * g(x) = f(x)g(x)$.

(The asterisk indicates the multiplication operation.)

Example 3.5.5 (Graphic Multiplication).

Let functions f and g be defined by the following graphs.

Graphs of f and g

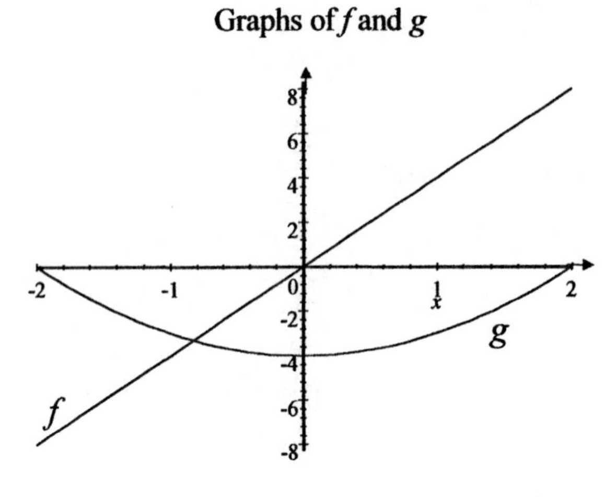

Figure 3.5.6

The process of graphically multiplying two functions is similar to the process of graphically adding two functions except that the output values of the two functions are multiplied rather than added. That is, select an input value and then form and plot the ordered pair (input value, product of the two output values). We do this for several input values and then draw a smooth curve through the plotted points. We illustrate this process by first determining and then plotting five points that lie on the product curve.

x	$\sim f(x)$	$\sim g(x)$	$[\sim f(x)] * [\sim g(x)]$
−2	−7	0	(−2,0)
−1.5	−5	−2	(−1.5,10)
−1	−4	−3	(−1,12)
−.5	−2	−4	(−.5,8)
1	4	−3	(1,−12)
1.5	6	−2	(−1.5,−12)

Table 3.5.4

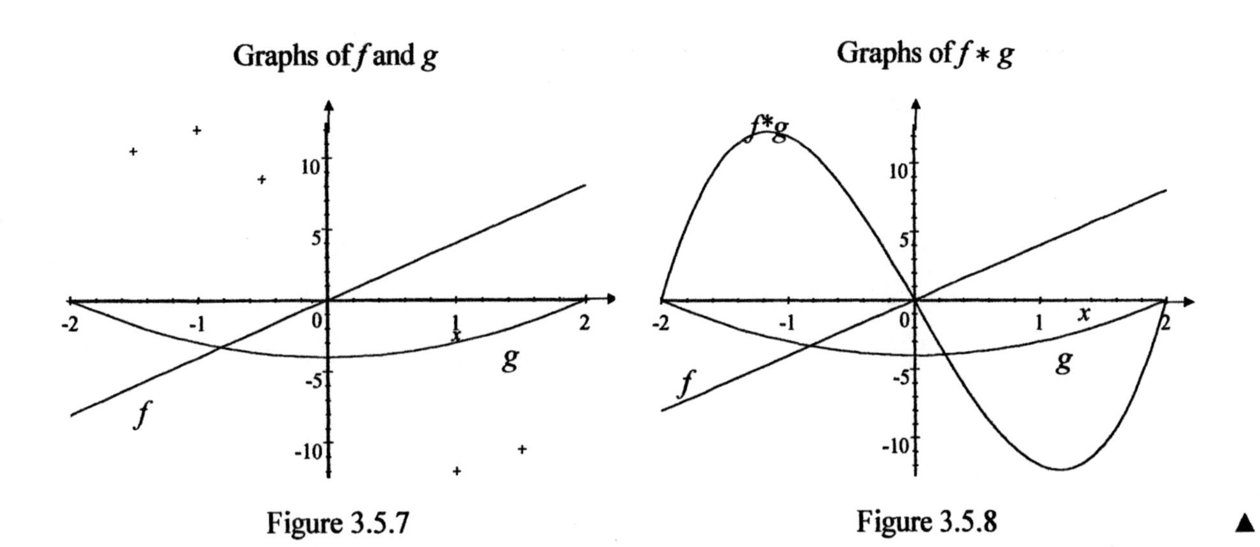

Figure 3.5.7

Figure 3.5.8 ▲

Example 3.5.6 (Numeric Multiplication of Functions).

Let functions f and g be defined by the following table.

x	$f(x)$	x	$g(x)$
−2	4	−10	10
0	1	−2	−3
0.5	5	0	6
1	3	1	5
3	1	3	7
4	6	4	2

Table 3.5.5

The functions f and g can only be multiplied over common values in their domains. Thus although g is defined for $x = -10$, f is not, and therefore the product of the two functions is not defined when $x = -10$. A similar statement is true when $x = 0.5$. The product of these two functions is given by the following table.

x	$(f * g)(x) = f(x)g(x)$
−2	$(4)(-3) = -12$
0	$(1)(6) = 6$
1	$(3)(5) = 15$
3	$(1)(7) = 7$
4	$(6)(2) = 12$

Table 3.5.6 ▲

Example 3.5.7 (Symbolic Multiplication of Functions).

Let functions f and g be defined symbolically as

$$f(x) = 4x \quad \text{and} \quad g(x) = x^2 - 4.$$

Their product is defined as

$$h(x) = (f * g)(x) = (4x)(x^2 - 4) = 4x^3 - 16x. \qquad ▲$$

Query 2.

How do you define division of functions? Illustrate your definition graphically, numerically, and symbolically.

Example 3.5.8.

Sketch and label the graphs of $h(x) = (f * g)(x) = f(x)g(x)$ and $k(x) = (f/g)(x) = f(x)/g(x)$, where the functions f and g are defined by the following graphs.

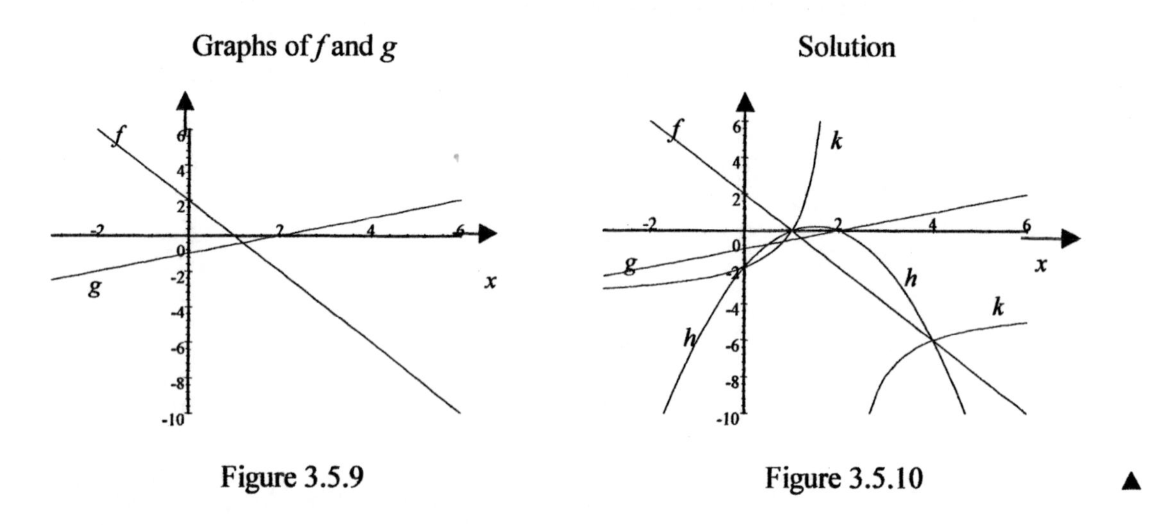

Figure 3.5.9 Figure 3.5.10 ▲

Query 3.

a. Did you expect that the graph of $h(x) = f(x) * g(x)$ would be a parabola? Why?

b. What happened to the graph of $k(x) = \frac{f(x)}{g(x)}$ near $x = 2$?

Note that if the domains of the functions f and g are different, then the operations of addition and multiplication of f and g are only defined over the common portion of the two domains. The same is true of their inverse functions, subtraction and division. Furthermore, division is not defined wherever the denominator function is zero.

Composition of Functions

Example 3.5.9 (Composition of Functions).

A Dairy Queen's profit function on selling ice cream is $p(d) = d^{3/2}$, where d represents the demand for ice cream cones. (The more ice cream cones sold, the greater the profit margin because of volume purchases.) The demand for ice cream cones is dependent on the weather: the warmer the weather, the greater the demand. If the demand function is $d(t) = 2t + 10$, where t represents the temperature (Celsius), how can the profit be expressed as a function of temperature?

Solution

A solution is to compose the profit function with the demand function. That is, for each input value of t, evaluate $d(t)$, and then use that value of $d(t)$ as input to evaluate $p(d(t))$. For example if $t = 20$, then $d(t) = d(20) = 50$ and the profit is

$$p(d(t)) = p(d(20)) = p(50) = 50^{3/2} = \$353.33.$$

Function f is composed with function g by first evaluating function g and then using its output values as input values for evaluating f. That is, $(f \circ g)(x) = f(g(x))$. The notation for composition is the small circle. Be careful to distinguish between the notation for multiplication and the notation for composition of functions. ▲

We will reverse the order of illustrations used in the previous examples and begin with a symbolic illustration.

Example 3.5.10 (Symbolic Composition of Functions).

Let functions f and g be defined as follows

$$f(x) = x^2 - 5 \text{ and } g(x) = 2x.$$

The composition of f with g is

$$(f \circ g)(x) = f(g(x)) = f(2x) = (2x)^2 - 5 = 4x^2 - 1.$$

Considering g to be the inside function and f to be the outside function, the composition $(f \circ g)(x) = f(g(x))$ is evaluated by using the output of the inside function as input for the outside function. ▲

Query 4.

Using the definitions for f and g in the preceding example, is $f(g(x)) = g(f(x))$?

Example 3.5.11 (Symbolic Composition of Functions).

Let functions h and k be defined as follows

$$h(x) = x^2 - 5 \text{ and } k(x) = 3^x + 2.$$

The composition of h with k is

$$(h \circ k)(x) = h(k(x)) = h(3^x + 2) = (3^x + 2)^2 - 5 = 3^{2x} + 4 * 3^x - 1. \quad ▲$$

Example 3.5.12 (Numeric Composition of Functions).

Let functions f and g be defined by the following table.

x	$f(x)$	x	$g(x)$
−2	4	0	10
0	1	2	3
0.5	5	4	0
1	3	1	4
3	1	3	−2
4	6	7	2

Table 3.5.8

The function f can only be composed with the function g at points where $g(x)$ is in the domain of f. Thus although g is defined for $x = 0$, $g(0) = 10$ is not in the domain of f, and so $f(g(0))$ does not exist. A similar statement is true when $x = 7$. The composition of f with g is given by the following table.

x	$g(x)$	$f(g(x))$
2	3	1
4	0	1
1	4	6
3	−2	4

Table 3.5.9 ▲

Example 3.5.13 (Graphic Composition of Functions).

Composing functions graphically is more complicated than adding, subtracting, multiplying, or dividing functions graphically. We illustrate the process by composing the function f with g where the functions are defined by the following graphs.

Plots of f and g

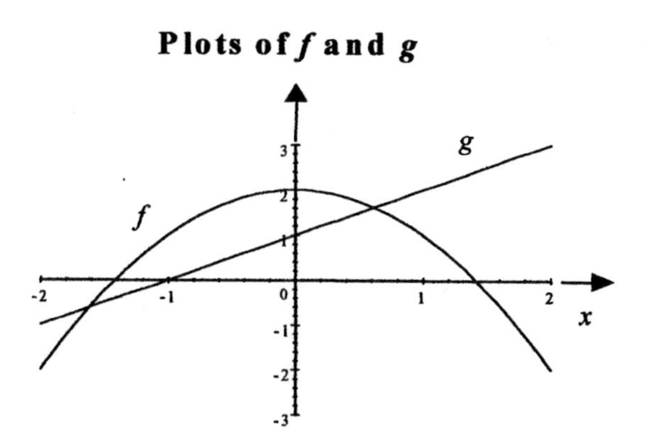

Figure 3.5.11

The procedure, as with addition and multiplication, is to plot several points and then connect the points with a smooth curve. The difficulty is that two functions need to be evaluated in order to determine the output value for a given input value. The points in the right-hand column of the following table were approximated from the plots of f and g as shown in the Figure 3.5.11. The plot in Figure 3.5.12 was obtained by drawing a smooth curve through these points.

x	~g(x)	~f(g(x))	(x, f(g(x))
−2	−1	$f(-1) = 1$	(−2, 1)
−1	0	$f(0) = 2$	(−1, 2)
0	1	$f(1) = 1$	(0, 1)
0.5	1.5	$f(1.5) = -.25$	(.5, −.25)
1	2	$f(2) = -2$	(1, −2)

Table 3.5.9

Composition of f with g

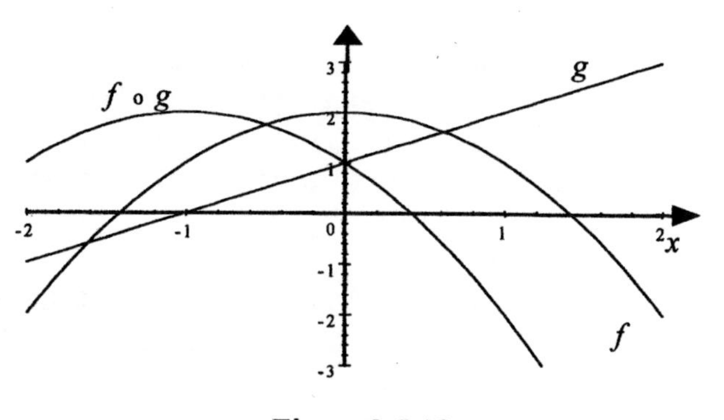

Figure 3.5.12

Query 5.

Complete the following table:

x	$f(x)$	$g(x)$	$f(x)+g(x)$	$f(x)-g(x)$	$f(x)(g(x))$	$f(g(x))$
0	0	3				
1	2	2				
2	4	1				
3	5	0				

Table 3.5.10

Small-Group Activity

Hurrying to catch a flight, Elcie walks on the moving walkway in the airport. If she walks at 3 miles per hour and the walkway is moving at 2 miles per hour, how fast is Elcie moving relative to the shops along the side of the concourse?

Each group should discuss how they would answer the question and then present their reasoning to the class. In particular, each group should decide if Elcie's walking speed and the speed of the moving walkway should be added, multiplied, or composed.

We use the algebra of functions to form complicated functions from simple functions. We also reverse this process to decompose functions into combinations of simpler functions.

Example 3.5.14.

The function $h(x) = 2^{\cos(x)}$ can be expressed as a composition of $f(x) = 2^x$ and $g(x) = \cos(x)$. ▲

Inverse Functions

The mapping interpretation of function is that of an input –> output process. That is, a function maps an input (independent variable) to an output (dependent variable). The inverse of a function undoes the function mapping.

Purchasing an item from a catalog is an example of a mapping function. The inputs (that is, domain) are the items listed in the catalog. (Each item is uniquely identified by catalog number.) The mapping operation is submitting an order (sending money) and the company sending you the purchased item. The inverse of this function is the company returning your money when you return the item.

In a store, think of the checkout counter as a function and the return counter as the inverse function. Note that the domain of the inverse function consists of only the items purchased in that store. That is, the range of a function is the domain of its inverse function.

In Section 3.2, we noted that a two-column table defines a relation (see Query 1, Section 3.2). Thus a data set expressed as a two-column table defines a relation that maps the elements in the first column to the corresponding elements in the second column. The inverse relation reverses this mapping. That is, the inverse relation maps the elements in the second column to the corresponding elements in the first column. Review Section 3.2 to determine what conditions must be satisfied in order for these relations to be functions.

Example 3.5.15.

The average value of homes (1940–1990) function, V, is defined by data given in Section 2.1, Exercise 5. The mapping tables for V and its inverse function, denoted by V^{-1} are shown in the following tables.

V : Year -> Average Value of Homes

Year (input)	– >	Average Value (output)
1940	– >	$27,400
1950	– >	$39,900
1960	– >	$52,500
1970	– >	$57,300
1980	– >	$74,900
1990	– >	$79,100

Table 3.5.11

V^{-1} : Average Value of Homes -> Year

Average Value (input)	– >	Year (output)
$27,000	– >	1940
$39,900	– >	1950
$52,500	– >	1960
$57,300	– >	1970
$74,900	– >	1980
$79,100	– >	1990

Table 3.5.12 ▲

Query 6.

The following home run function is defined from data on the top ten home run hitters as of January 1, 2002. Is the inverse relation a function? Explain.

Player	– >	Home Runs
Hank Aaron	– >	755
Babe Ruth	– >	714
Willie Mays	– >	660
Barry Bonds	– >	658
Frank Robinson	– >	586
Mark McGwire	– >	583
Harmon Killebrew	– >	573
Reggie Jackson	– >	563
Mike Schmidt	– >	548
Mickey Mantle	– >	536

Table 3.5.12

Determining the inverse of a relation that is defined graphically is almost as easy as when the relation is defined numerically (by a two-column table). Because the input values of a relation are plotted on the horizontal axis and the output values on the vertical axis, the graph of the inverse relation can be obtained by interchanging the horizontal and vertical axis. This is done graphically by reflecting the graph of the relation in the line that is the graph of $y = x$.

Example 3.5.16.

The following plots graphically illustrate obtaining the graph of an inverse function (f^{-1}) by reflecting the graph of the function (f) in the line $y = x$.

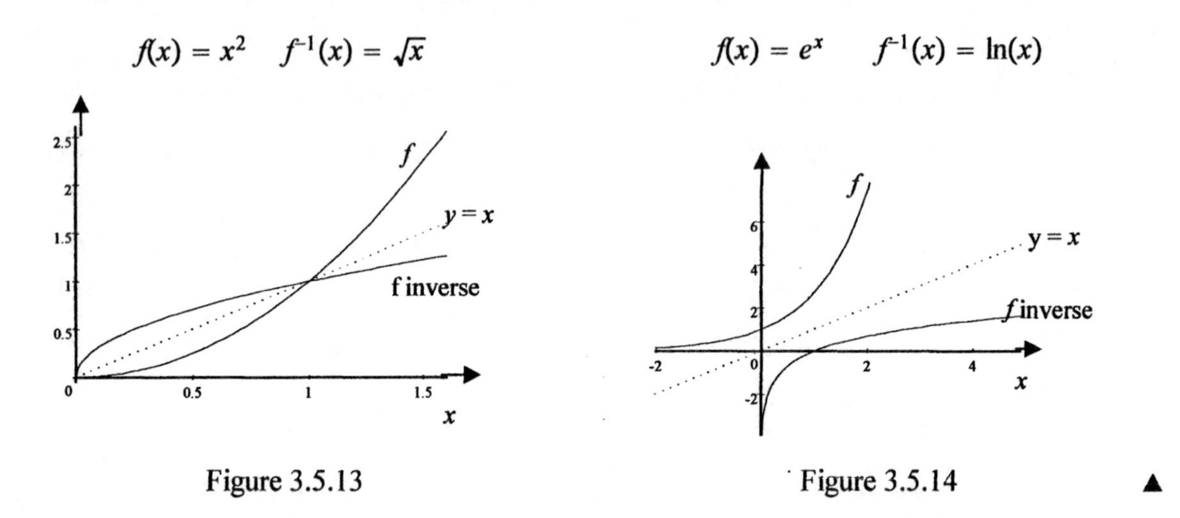

Figure 3.5.13 Figure 3.5.14 ▲

A **horizontal line test** makes it easy to graphically determine if the inverse of a relation is a function. Recall from Section 3.2 that the vertical line test provides a graphical method for determining if a relation is a function. Similarly the horizontal line test says that the inverse of a relation is a function provided any horizontal line intersects the graph of the relation in at most one point. (Consider how the horizontal line test would be applied to the graph of the function f in Figure 3.5.13 or Figure 3.5.14.

Query 7.

Is it true that the inverse relation of any increasing (decreasing) relation is a function? Explain.

Note that the notation: f^{-1} denotes the inverse of f, not the reciprocal of f.

Query 8.

Just as not all relations are functions, not all inverse relations are inverse functions. Recall that a relation must satisfy a uniqueness property for it to be a function. What must be true of a function if its inverse relation is also a function? Give some examples of functions whose inverse relations are not functions and also some examples of functions whose inverse relations are functions.

Query 9.

What is the result of composing a function with its inverse function? What is the result of composing an inverse function with its function? Illustrate.

The algebra of functions allows us to consider individual terms in a function expression as functions. For example, the quadratic function $f(x) = 3x^2 + 6x - 4$ can be considered as the sum of the three functions $g(x) = 3x^2$, $h(x) = 6x$, and $k(x) = -4$. This allows us to apply the shifting and scaling operations (Section 3.4) to individual terms. We will use this flexibility in the next section when we fit curves to plots of data sets.

Exercises 3.5

1. (Computational Skill) Simplify each of the following.
   a. $(x+4)^2 - (2-x)^2$
   b. $\dfrac{(x+2)(x-2)}{x-3}$
   c. $2^3(2^2 - 4^2)$

2. (Computational Skill) Let functions f and g be numerically defined by the following tables of data.

x	$f(x)$
−1	3
0	1
2	2
4	−1
5	0

Table 3.5.13

x	$g(x)$
0	2
4	−1
−2	5
3	3
−4	−2

Table 3.5.14

   a. Compute $f(x)g(x)$
   b. Compute $f(x)/g(x)$
   c. Compute $f(g(x))$
   d. Compute $g(f(x))$

3. (Computational Skill) Make up and solve three exercises similar to Query 4.

4. Sketch and label the graphs of $h(x) = f(x) + g(x)$ and $k(x) = f(x) - g(x)$ where functions f and g are defined by the following graphs.

Graphs of f and g

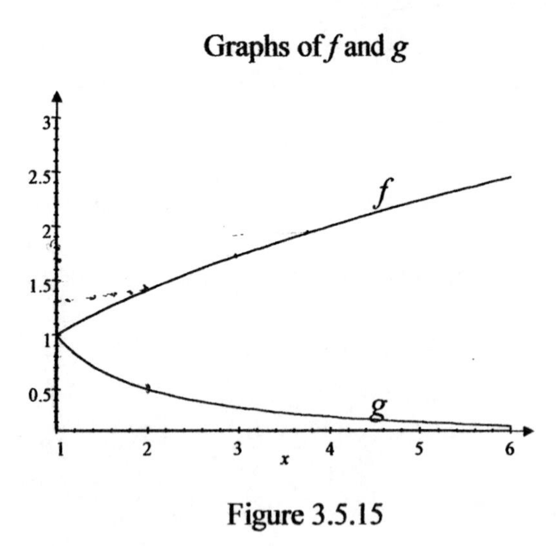

Figure 3.5.15

5. Sketch and label the graphs of $h(x) = f(x) + g(x)$ and $k(x) = f(x) - g(x)$ where functions $f(x) = \sin(x)$ and $g(x) = \cos(x)$ as shown in the following graphs.

Graphs of f and g

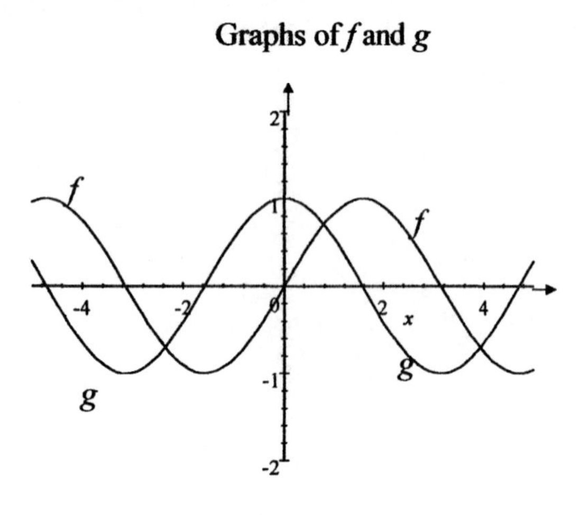

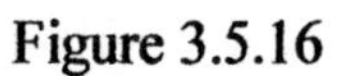

Figure 3.5.16

6. Sketch and label the graphs of $h(x) = f(x) * g(x)$ and $k(x) = g(x)/f(x)$ where functions f and g are defined by the following graphs:

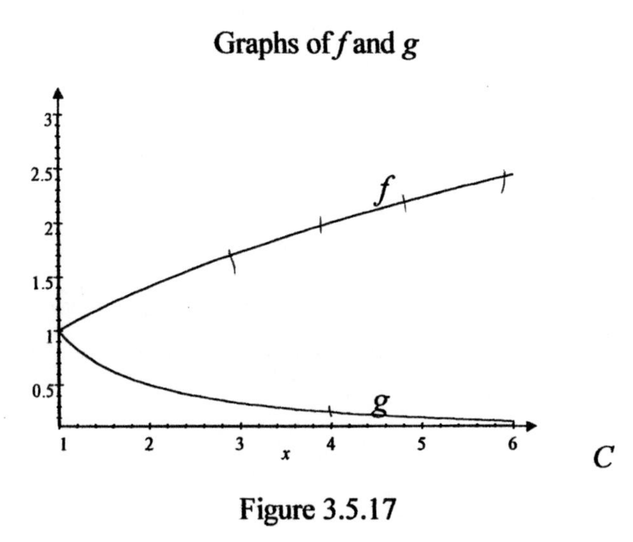

Figure 3.5.17

7. Plot the graphs of $f(x) = \sin(x)$ and $g(x) = \cos(x)$ and then sketch the graph of $h(x) = 2f(x) + g(x)$.

8. Sketch and label the graphs of $h(x) = f(x)g(x)$ and $k(x) = f(x)/g(x)$ where functions $f(x) = \sin(x)$ and $g(x) = \cos(x)$ as shown in the following graphs.

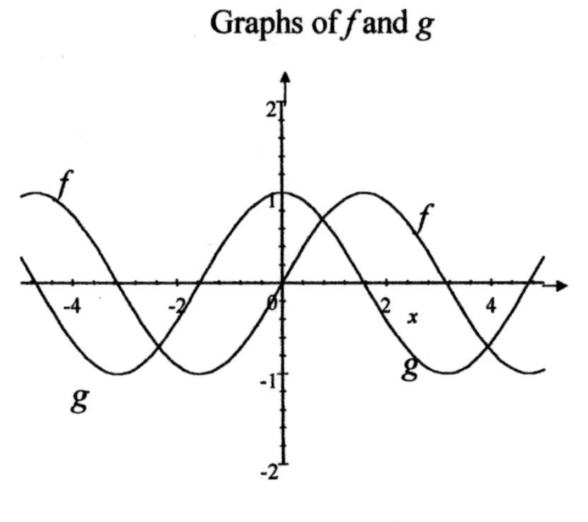

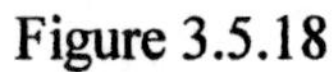
Figure 3.5.18

9. Sketch and label the graphs of $h(x) = f(g(x))$ and $k(x) = g(f(x))$ where functions f and g are defined by the following graphs.

Graphs of f and g

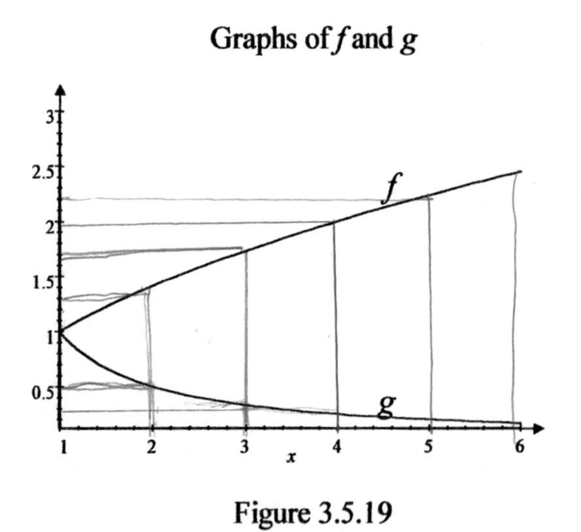

Figure 3.5.19

10. Sketch and label the graphs of $h(x) = f(g(x))$ and $k(x) = g(f(x))$ where functions $f(x) = \sin(x)$ and $g(x) = \cos(x)$ as shown in the following graphs.

Graphs of f and g

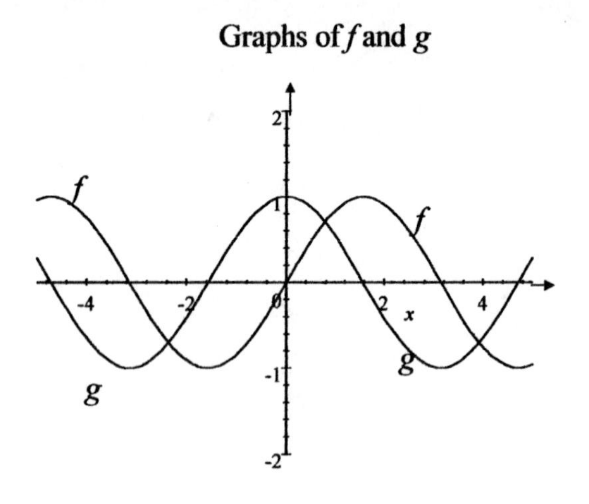

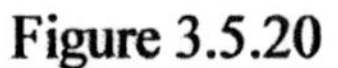

Figure 3.5.20

11. Complete the following table.

x	$f(x)$	$g(x)$	$f(x)+g(x)$	$f(x)-g(x)$	$f(x)(g(x))$	$f(g(x))$
0	4	3				
1	2	0				
2	4	1				
3	1	4				

Table 3.5.14

12. For each of the following, define functions f and g such that.

a. $f(x)g(x) = \sqrt{x(x+1)}$

b. $f(g(x)) = \sqrt{x(x+1)}$

c. $f(g(x)) = (x+2)^3$

d. $f(g(x)) = 2^{4x}$

e. $f(g(x)) = \sin(3x)$

13. Plot the graphs of $f(x) = x^2$ and $y = x$. Then obtain the graph of the inverse of f, that is $f^{-1}(x) = \sqrt{x}$, by reflecting the graph of f in the line $y = x$.

14. Repeat Exercise 13 for the cubic function: $f(x) = x^3$.

15. A Celsius to Fahrenheit function maps temperature in the Celsius system to the Fahrenheit system. This function is defined by $F(C) = \frac{9}{5}C + 32$.

a. Define a Fahrenheit to Celsius temperature function.
b. Compose the Celsius to Fahrenheit function with the Fahrenheit to Celsius function.
c. Compose the Fahrenheit to Celsius function with the Celsius to Fahrenheit function.
d. Do the results of parts b and c prove that the two functions are inverses of each other? Explain.

16. Write a paragraph explaining why reflecting the graph of a function in the line $y = x$ gives the graph of the inverse function.

17. Plot $f(x) = \sin(x)$ and then use the horizontal line test to determine if.
a. The inverse of $f(x) = \sin(x)$ is a function.
b. The inverse $f(x) = \sin(x)$ for $-\frac{\pi}{2} \leq x \leq \frac{\pi}{2}$ is a function.
(This exercise illustrates the common practice of restricting the domain of a function so that its inverse is a function.)

3.6 Graphical Approximations

In this section, we show how to graphically approximate solutions of equations and then we return to the topic of graphically modeling data sets. With our knowledge of the five basic function categories, how to shift and scale graphs, and the algebra of functions, we are prepared to resume the study of graphically fitting a curve to a scatter plot of data that we began in Section 3.3.

Solving Equations

Solving equations is often part of a mathematical solution process. Sometimes this is easy to do (for example, solve for x in the equation $3x = x - 4$), and sometimes it is difficult (for example, solve for x in the equation $x^2 = 2^x$). When solving an equation is easy to do, we will use algebraic methods. When solving is not easy to do, we will approximate the solutions using graphical methods.

Example 3.6.1.

Solve for x in the equation $6x = 3x - 12$.

Solution

$6x = 3x - 12$	Original equation	
$3x = -12$	Subtract $3x$ from both sides of the equation.	
$x = -4$	Divide both sides of the equion by 3.	▲

This was easy to do, and thus we solved it by hand as illustrated. Now we consider an equation that is not easy to solve by hand.

Example 3.6.2.

Solve for x in the equation $x^3 - 6x + 4 = 0$.

Solution:

We plot $y = x^3 - 6x + 4$ and then approximate the values of x where the graph touches the x-axis.

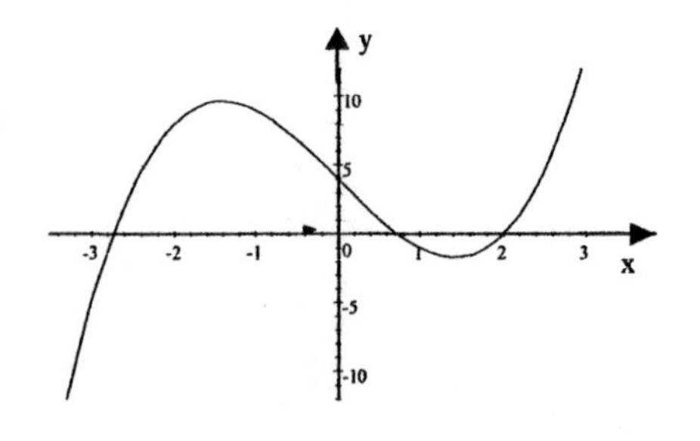

Figure 3.6.1

We approximate the zeros to be -2.8, 0.7, and 2. To get more accurate approximations, we would zoom in on the points where the curve touches the x-axis. ▲

The **solution** of an equation in one variable is a value of the variable that when substituted into the equation reduces the equation to an identity (the same numerical value on both sides of the equal sign). If the equation is linear, there will be at most one solution. If the equation is nonlinear, there may be more than one solution. For example, $x = 1$ and $x = 2$ are solutions of the equation $x^2 = 3x - 2$. The solutions of an equation in one variable are also called **root(s)** of the equation.

The **zero(s) of a function** are the values of the independent variable that the function maps into zero. Graphically, these are the points where the graph of the function touches the horizontal axis. If a function is defined by an equation, the zeros of the function are the roots of the defining equation. That is, the values of the independent variable that make the dependent variable equal to zero. Thus the procedure is to set the dependent variable equal to zero, reducing the problem to the previous case of an equation in one variable. A function may have more than one zero as illustrated as illustrated in Figure 3.6.1.

Example 3.6.3.

A ball is fired straight upward from the ground with initial velocity of 48 feet per second. The height of the ball is modeled by the function $h(t) = -16t^2 + 48t$ where t, measured in seconds, is the time since the ball was fired. The height of the ball after t seconds is $h(t)$, measured in feet. How long will the ball be in the air before it hits the ground?

Solution

The ball is at ground level when $h(t) = 0$. Thus we want to solve the equation

$$-16t^2 + 48t = 0.$$

We may rewrite the equation as $-16t(t-3) = 0$. (This is called factoring out the common factor, $-16t$.) Because $h(t) = 0$ when $t = 0$ and when $t = 3$, 0 and 3 are the zeros of the height function h.

Because $t = 0$ is when the ball is fired, $t = 3$ is the time when the ball returns to earth. That is, the ball lands after 3 seconds. ▲

In the preceding Example, 0 and 3 are called **zeros** of the function h. (They are the input values that yield a zero output value.) They are also referred to as the **roots** or **solutions** of the equation $-16t^2 + 48t = 0$.

Example 3.6.4.

A diver springs off a diving board that is 16 feet above the water with a velocity of 24 feet per second. The height of the diver above the water is modeled by the function $h(t) = -16t^2 + 24t + 16$, $t \geq 0$ where t, measured in seconds, is the time since the diver leaves the board. The height of the diver above the water after t seconds is $h(t)$, measured in feet. How long before she hits the water?

Solution:

The diver hits the water when $h(t) = 0$. Thus we want to compute the zero(s) of h. That is, solve for the values of t that will make

$$-16t^2 + 24t + 16 = 0.$$

We check (by multiplication) that

$$-16t^2 + 24t + 16 = -8(2t^2 - 3t - 2) = -8(2t + 1)(t - 2).$$

This is called factoring. (Note that factoring is the inverse of expanding. For information on factoring, see Appendix, Computational Skills and Basic Functions.) Thus

$$h(t) = -16t^2 + 24t + 16 = -8(2t + 1)(t - 2)$$

and so

$$h(t) = 0 \text{ implies that } -8(2t + 1)(t - 2) = 0 \text{ or that } 2t + 1 = 0 \text{ or } t - 2 = 0.$$

Hence the zeros of h are $t = -\frac{1}{2}$ or $t = 2$.

The value $t = -\frac{1}{2}$ does not make sense for the model even though $t = -\frac{1}{2}$ is a root of $-16t^2 + 24t + 16 = 0$. (Why?) Thus the diver hits the water 2 seconds after she leaves the diving board. ▲

Example 3.6.5.

Suppose the situation in Example 3.6.4 was modified so that the diver leaves the diving board with a velocity of 12 feet per second. The model is then $h(t) = -16t^2 + 12t + 16$, $t \geq 0$. As before, we want to compute the zero(s) of h in order to determine how long before the diver hits the water.

Solution.

The diver hits the water when $h(t) = 0$. Thus we want to compute the zero(s) of h. That is, solve for the values of t that will make

$$-16t^2 + 12t + 16 = 0.$$

Factoring the polynomial $-16t^2 + 12t + 16$ is not easy, and therefore we will graphically approximate the zero(s) of h. Recall the zeros of a function are the points on the horizontal axis where the graph of the function touches the horizontal axis. (Why?) Thus we plot h and zoom in on the point(s) where the graph touches the horizontal axis.

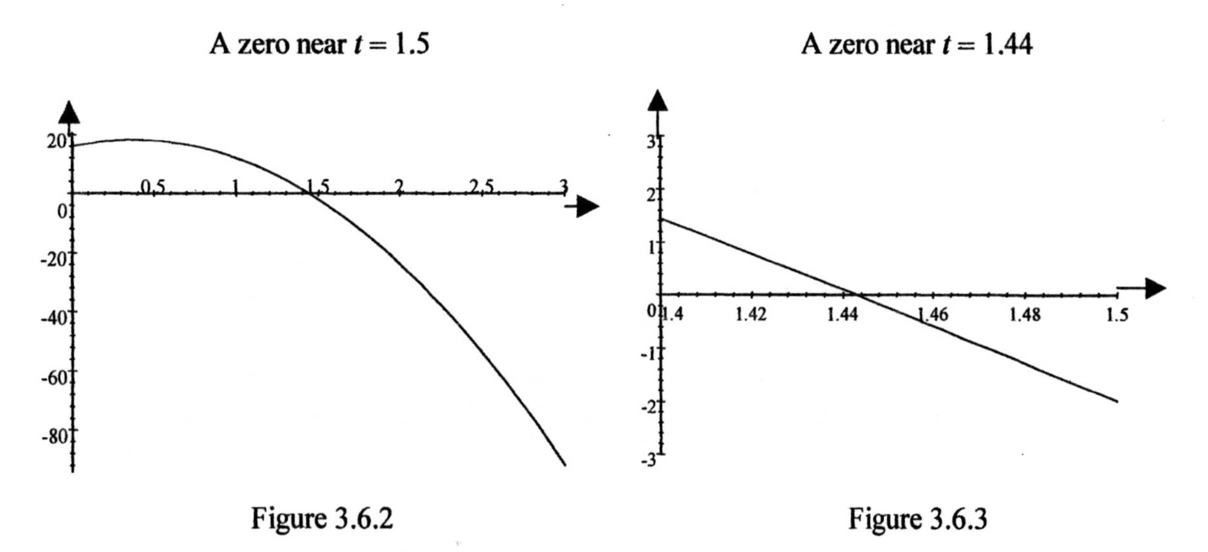

Figure 3.6.2 Figure 3.6.3

Zooming again gives an approximation for the solution of $t = 1.443$, as shown by the following plot.

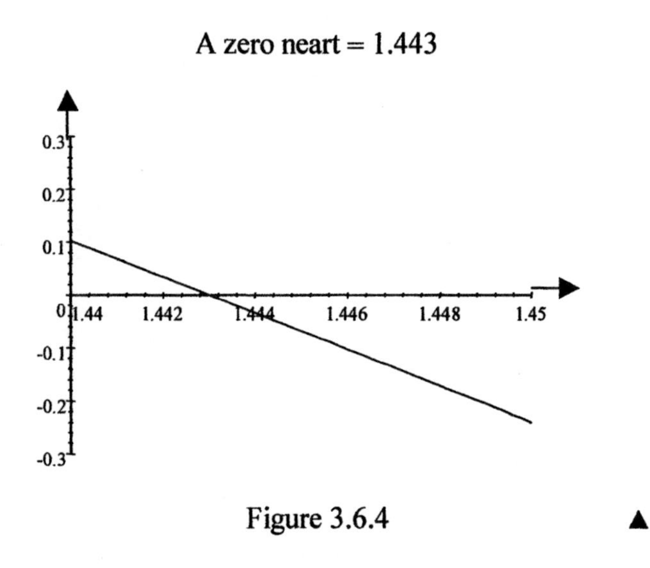

Figure 3.6.4 ▲

Because of the importance in understanding how to solve or at least approximate solution(s) of equations, especially those used to define functions, we restate the graphical process.

Process for Graphically Approximating the Solution of an Equation

Rewrite the equation of one variable, if necessary, so that all of the terms are on the same side of the equal sign, say the left side. The equation is then in the form "expression equals zero." Use the expression to define a function and then plot the function. The points where the graph touches the horizontal axis are the zeros of the function, that is, the solutions of the equation. Zooming in on these points allows for more accurate approximations of the solutions.

Caution: When approximating the zeros of a function or the solutions of an equation, you must always ask the question: Have I found them all? This can be a difficult question to answer as all that you can see is what is in the plotting window. Because the window is always finite, you need to analyze what the graph looks like outside the window.

Example 3.6.6

What are the zeros of the function $f(x) = 2^{-\sin(x)}$?

Solution

Plotting the function over the intervals [-5,5] and [-5,25] shows that the function has no zeros in [-5,25]. Are there any zeros for $x < -5$ or $x > 25$? How can we tell?

Figure 3.6.6 indicates that f is a periodic function. This observation is confirmed by recalling that the sin(x) function is periodic with period 2π, that is sin(x) takes on the same set of values over any interval of length 2π. How does knowing that f is periodic help us answer the question? What is the answer ?

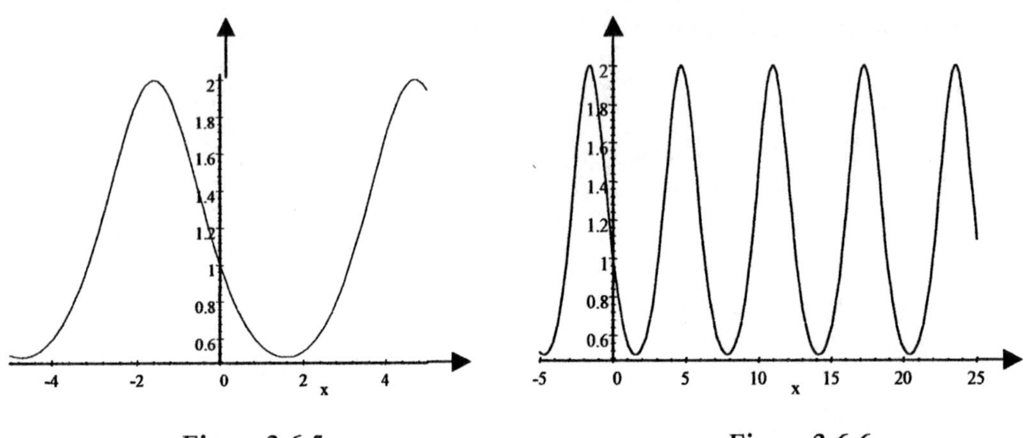

Figure 3.6.5

Figure 3.6.6 ▲

Query 1.

Find all the solutions of $2^x = x^2$ and explain how you know that you have found *all* the solutions.

Graphically Approximating a Scatter Plot of Data

Graphically approximating functions (for example, a data set) is an **exploratory activity**. The objective is to obtain a reasonable approximation, not an exact fit. Quantifying "reasonable" depends on the scenario of the problem and any accuracy bounds stated in the problem. The approach is an iterative one. It consists of making an approximation based on the shape of the data or graph, modifying the approximation to obtain a better fit, and then repeating the modifying process until a reasonable approximation has been found. We illustrate the process with several examples.

Query 2.

In the preceding paragraph, what does the sentence, "The approach is an iterative one." mean?

Example 3.6.7.

Biologists have recorded the following data on the root mass (mg) of pea plants. (Source: Richard Burton, *Biology by the Numbers*, Cambridge University Press, 1998, p. 149.) The numbers represent the averages for 12 plants. Graphically approximate the plot of this data with a curve.

Days	Root Mass (mg)
0	7.2
7	14.2
14	33.4
21	68.7
28	116

Table 3.6.1

Solution:

We begin by plotting the data and recognizing the basic shape of the data. We show six iterations leading to one possible solution.

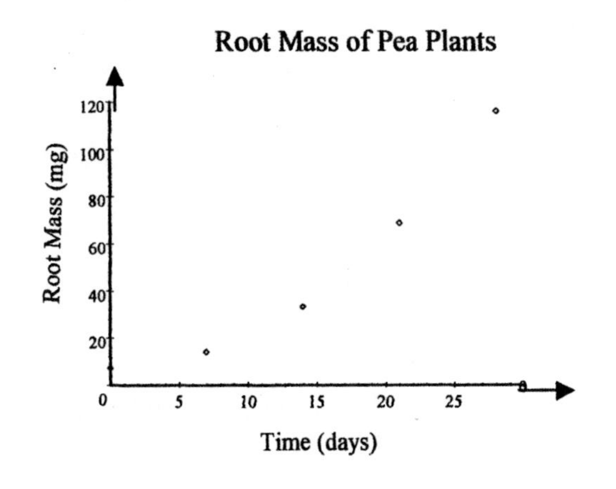

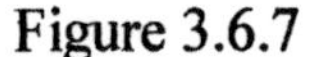

Figure 3.6.7

The basic shape of the set of data points appears to be that of a quadratic (second degree) polynomial or an exponential. We will choose to work with a quadratic polynomial, $ax^2 + bx + c$, because it is easier than working with the general exponential expression, $ae^{bx+c} + d$. (If we are unsuccessful fitting a polynomial to the data, we will then try fitting an exponential expression.)

First Iteration: Because the first data point is (0,7.2), we set $c = 7.2$ (Why?). We will work first with the ax^2 term and then later fine-tune with the bx term. Let *app* denote approximation and define our first approximation to be $app1(x) = x^2 + 7.2$. The plot of *app*1 has the correct basic shape, but needs to be scaled.

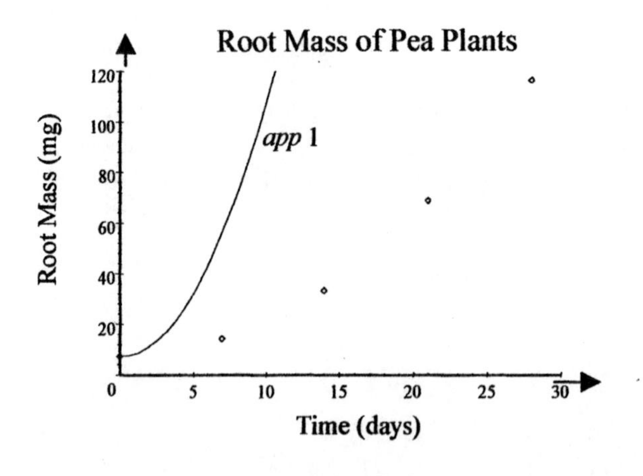

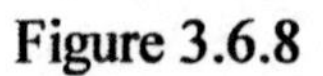
Figure 3.6.8

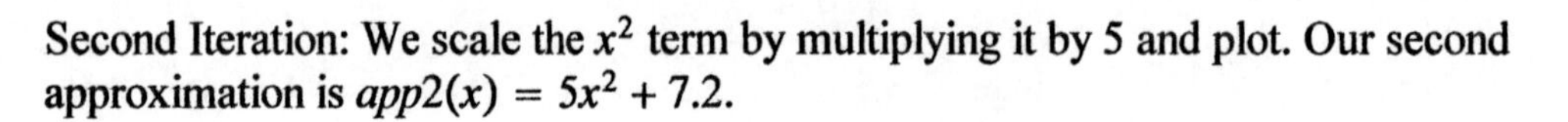
Second Iteration: We scale the x^2 term by multiplying it by 5 and plot. Our second approximation is $app2(x) = 5x^2 + 7.2$.

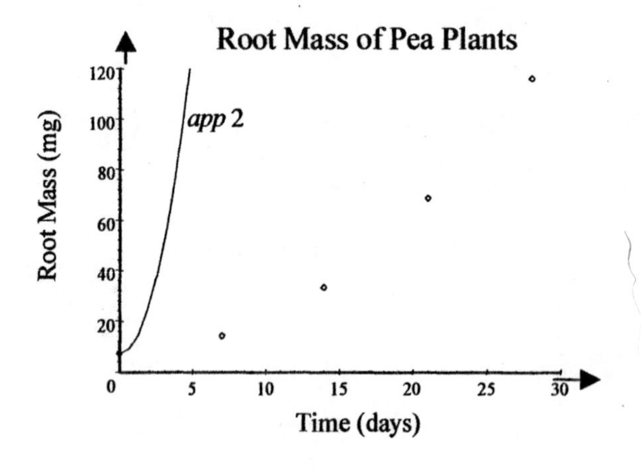

Figure 3.6.9

Oops! We should have divided rather than multiplied. Making that change gives $app2(x) = \frac{x^2}{5} + 7.2$.

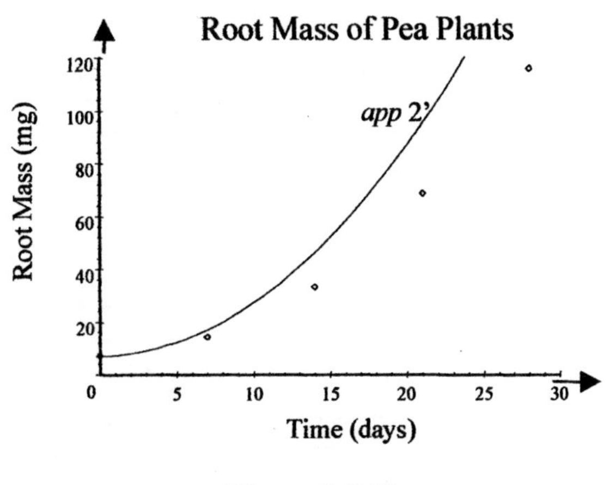

Figure 3.6.10

This looks much better.

Third Iteration: We need to scale the x^2 even more. This time we divide by 6 rather than by 5. So $app3(x) = \frac{x^2}{6} + 7.2$.

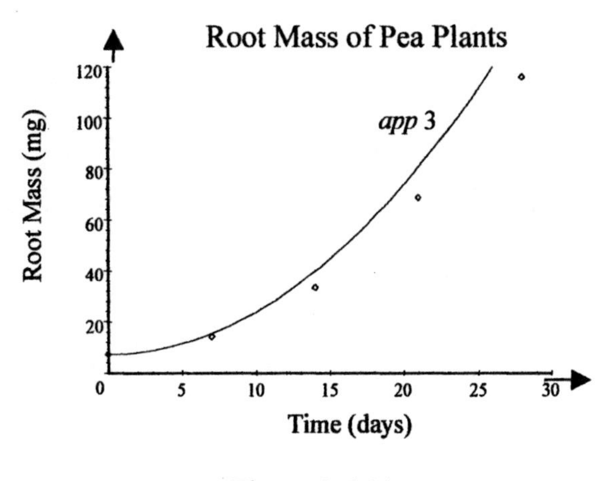

Figure 3.6.11

$app3(x) = \frac{x^2}{6} + 7.2$ looks good for the first two data points, but then the error gets continually worse. Correcting for the later data points by continued scaling of the x^2 term would undo the fit we have on the second data point. Thus we will now begin working with the bx term to fine-tune our approximation.

Fourth Iteration: Because the graph of $app3(x) = \frac{x^2}{6} + 7.2$ lies above the third, fourth, and fifth data points, we want to subtract something from $app3(x)$. We consider $app4(x) = \frac{x^2}{6} - x + 7.2$.

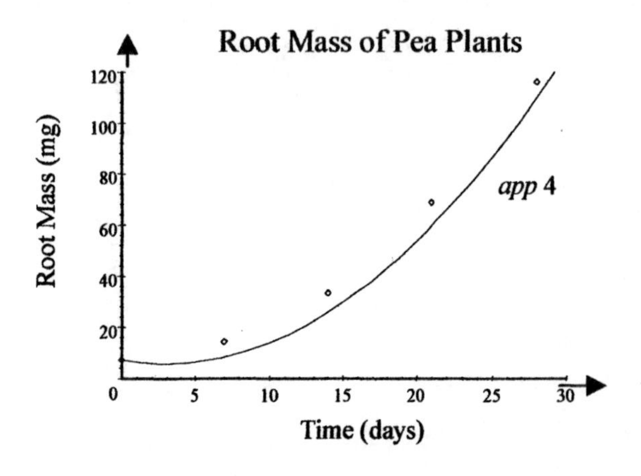

Figure 3.6.12

We subtracted too much in subtracting x. Thus we need to scale the x term.

Fifth Iteration: We scale the x term by dividing it by 2. So $app5(x) = \frac{x^2}{6} - \frac{x}{2} + 7.2$.

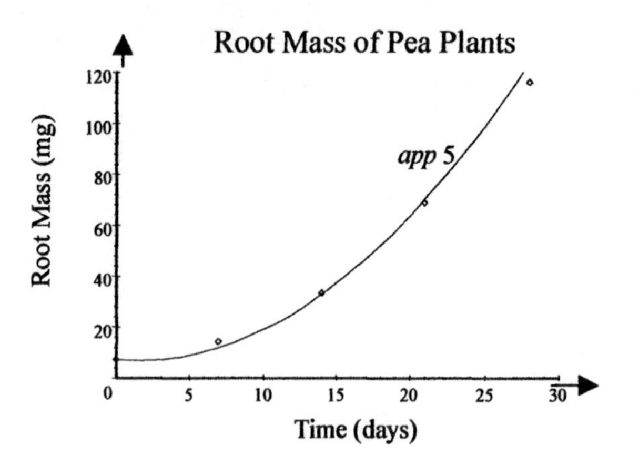

Figure 3.6.13

This looks pretty good, but we can probably do better. Decreasing the factor of $\frac{1}{6}$ in the x^2 term will lower the right portion of the graph, and decreasing the factor of $\frac{1}{2}$ in the x term will raise the left portion of the graph (explain why).

Sixth Iteration: Letting $app6(x) = \frac{x^2}{7} - \frac{x}{7} + 7.2$, gives a reasonable approximation.

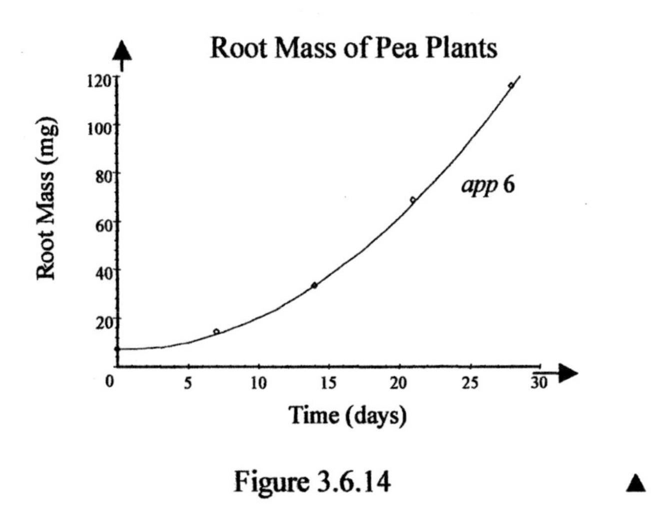

Figure 3.6.14 ▲

Example 3.6.8.

In the final seconds with her team down by two, Paula shoots the basketball from three point range. The bottom of the ball is 7 feet off the floor when it is shot toward the basket that is 30 feet away and is 10 feet off of the floor. Does Paula score?

The following data is given for the location of the ball. The first coordinate is the horizontal distance of the ball from Paula and the second coordinate is the height of the ball.

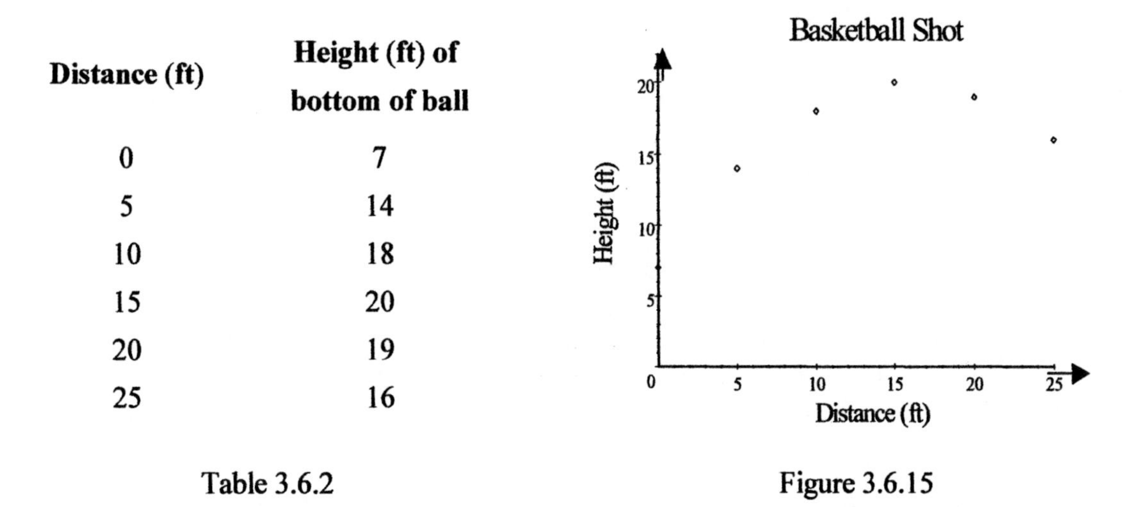

Distance (ft)	Height (ft) of bottom of ball
0	7
5	14
10	18
15	20
20	19
25	16

Table 3.6.2

Figure 3.6.15

Solution:

We will graphically approximate a function fitting the given data and then evaluate the function at $x = 30$.

First Iteration: The basic shape of the set of data points appears to be that of a parabola opening downward that has been shifted upward 20 units and shifted to the right by 15 units. Thus we set our first approximation as $app1(x) = -x^2 + 20$.

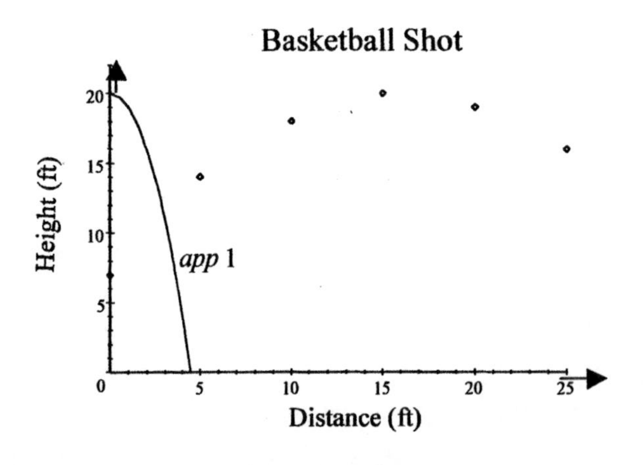

Figure 3.6.16

Oops! We forgot to shift the curve to the right.

Second Iteration: Because the center of the parabola is at approximately (15,20), we shift the curve 15 units to the right. Thus $app2(x) = -(x-15)^2 + 20$.

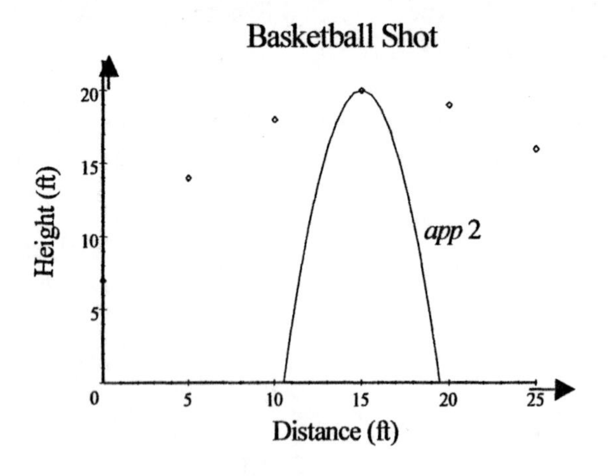

Figure 3.6.17

The basic shape appears to be right, but now we need to scale the $(x-15)^2$ term in order to lift the sides. (Note that multiplying the left-hand side of $app2(x)$ is the same as dividing the right-hand side.) Should we multiply or divide the $(x-15)^2$ term? (Why?)

Third Iteration: Scale the $(x-15)^2$ term by dividing it by 20. This gives $app3(x) = -\frac{1}{20}(x-15)^2 + 20$.

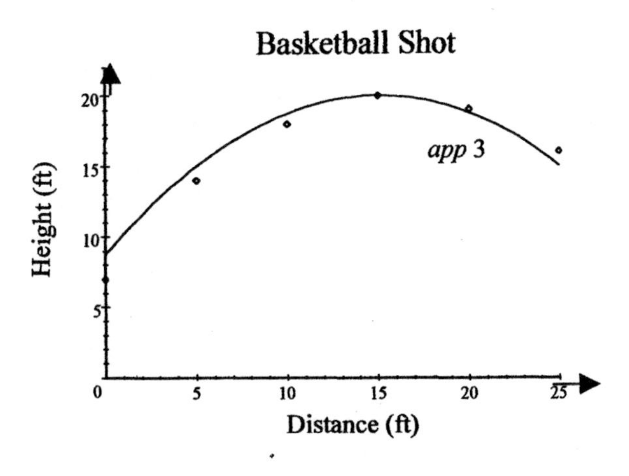

Figure 3.6.18

This is beginning to look good. Now it is time to fine-tune. Adding an $x - 15$ factor will drop the left part of the curve and lift the right part. (Why?) Because we do not want very much change, we scale the $x - 15$ factor, say by $\frac{1}{15}$. (Why does dividing $x - 15$ by 15 reduce the amount of change?)

Fourth Iteration: Consider $app4(x) = -\frac{1}{20}(x - 15)^2 + \frac{1}{15}(x - 15) + 20.$

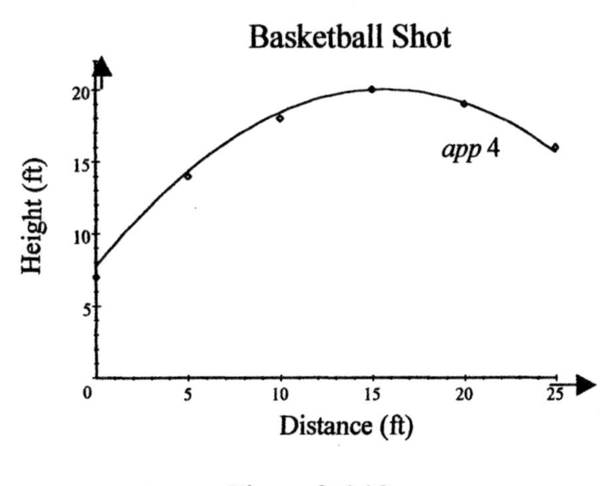

Figure 3.6.19

This looks like a reasonable approximation. Now let's evaluate $app4(x)$ at $x = 30$.

$$app4(30) = -\frac{1}{20}(15)^2 + \frac{1}{15}(15) + 20 = 10.25$$

SCORE !!! The ball just clears the rim, and Paula's team wins by one point. ▲

The results of this last example will be referenced in Section 4.4 in connection with Newton's Second Law of Motion.

Example 3.6.9.

Physicists have recorded the following time-distance data for a ball bearing dropped from the top of a 500-foot-tall building. The zero distance position (that is, the origin) is at the top of the building, and thus the recorded distances are expressed as negative numbers to indicate that the ball bearing is falling toward earth. We plot the data and then determine a model that approximates the data.

Time (sec)	Distance (ft)
.5	−3.96
1	−16.5
1.5	−36.63
2	−64.02
2.5	−100.98
3	−145.53
3.5	−198.99
4	−257.4
4.5	−327.36
5	−405.24

Table 3.6.3

Solution

We begin by plotting the data and recognizing the basic shape of the data

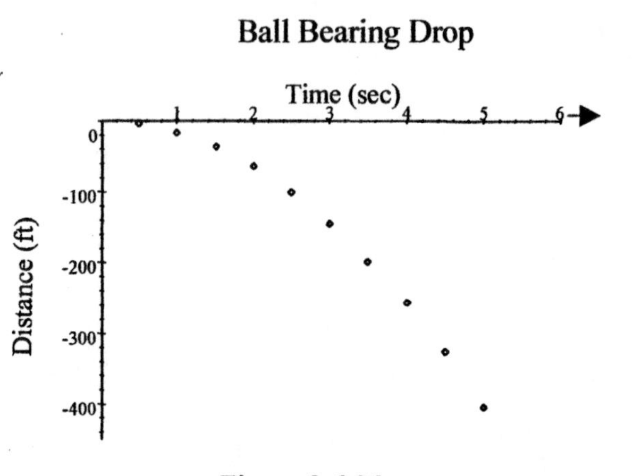

Figure 3.6.20

The basic shape of the data appears to be the right-hand side of a parabola opening downward with the center (that is, vertex) of the parabola at the origin.

First Iteration: We define our first approximation to be $app1(t) = -t^2$. It is not obvious that the plot has the correct shape. What is clear is that the plot needs to be scaled.

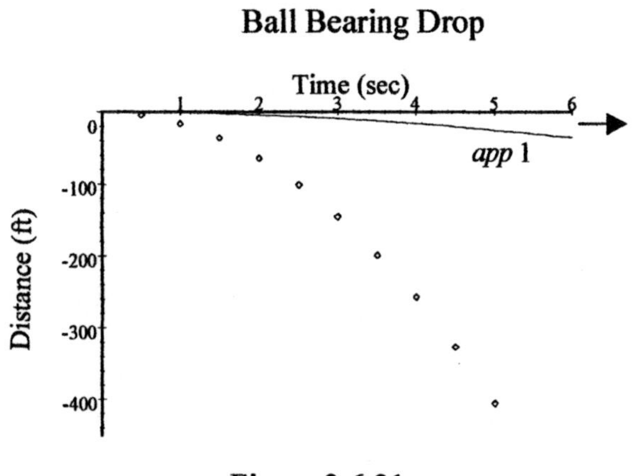

Figure 3.6.21

Second Iteration: Multiplying t^2, by a positive number greater than one will cause the right-hand side of the plot to move downward. Why? Let us try scaling t^2 by multiplying it by 10. Thus $app2(t) = -10t^2$.

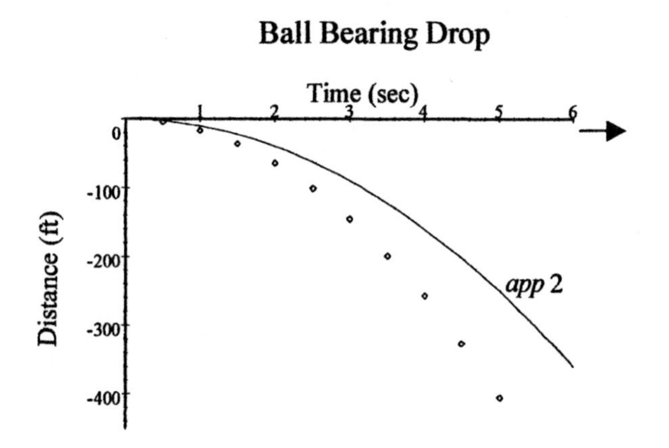

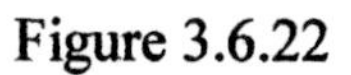
Figure 3.6.22

We are on the right track and only need to increase the scaling factor. Let's try scaling by 20.

Third Iteration: $app3(t) = -20t^2$.

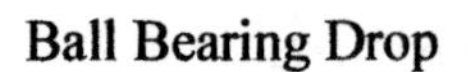

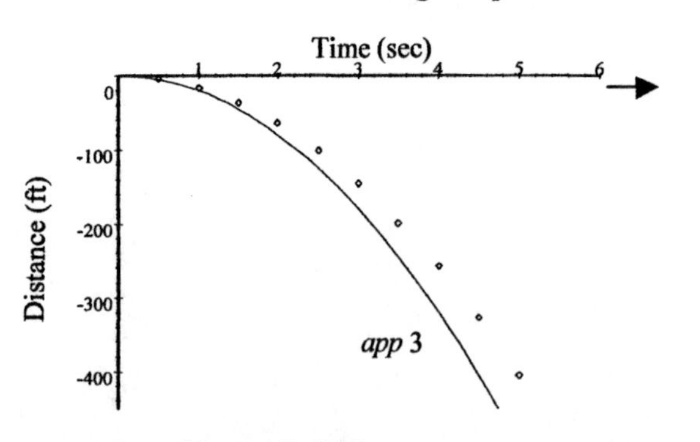

Figure 3.6.23

Too much. We should reduce our scaling factor, say to 15.

Fourth Iteration: $app4(t) = -15t^2$

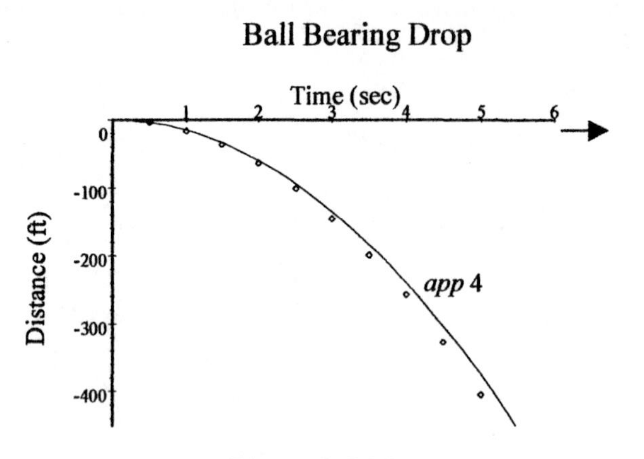

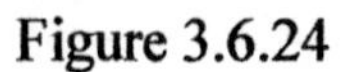

We can probably do a bit better. Let us increase the scaling factor to 16.

Fifth Iteration: $app5(t) = -16t^2$.

Ball Bearing Drop

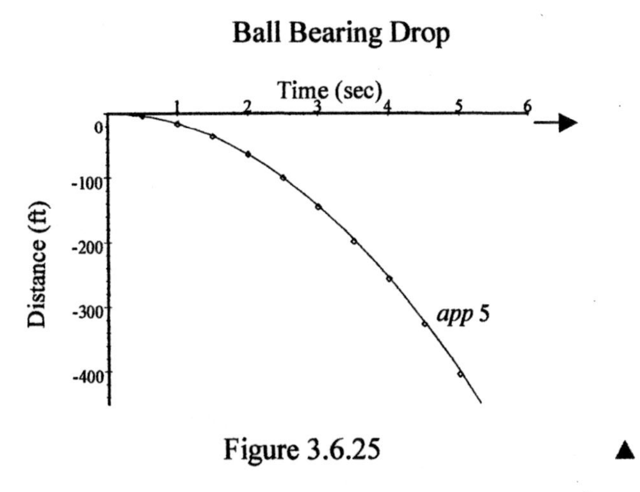

Figure 3.6.25 ▲

This looks like a very good fit. Our desired model is $f(t) = -16t^2$.

Query 3.

What would the model have been in the preceding example if the origin was placed at the foot of the building rather than at its top?

Small-Group Activity

(Students should work in pairs on this activity.) Begin the activity with pairs of students swapping calculators. Each student enters a second-degree polynomial equation in the last equation entry position of the calculator, plots the polynomial equation, and then returns the calculator to his/her partner. (Entering the equation in the last entry position effectively hides the equation.)
The task of each student is to experimentally determine a second-degree polynomial function whose graph is a good approximation to the graph that his/her partner has entered and to do this without looking at the statement of the function that was entered. Each pair must decide on what is meant by a "good approximation."

Hint: An effective procedure is to make up and then plot a second-degree polynomial function whose graph has the same basic shape of the given graph. That is, one that opens in the same direction as the given plot. Consider this to be the first approximation (it will probably not be a very good approximation). Now experiment by changing one of the coefficients (and replotting) to get a better approximation (several attempts will probably be necessary). Repeat the experimentation with each of the other two coefficients.

Small-Group Activity

Develop a model for the sum of the first n positive integers by generating data using a recursive sequence, plotting the data, and then fitting a curve to the scatter plot. (See Section 2.4 for the definition of a recursive sequence.)
Define $s(n) = 1 + 2 + 3 + ... + n$ where n is a positive integer. (Thus $s(3) = 6, s(4) = 10$.)
Complete the following.

1. **a.** Show that $s(n)$ can be expressed as the recursive sequence $s(n+1) = s(n) + (n+1)$.

b. Create a two-column table with column headings n and $s(n)$ for $n = 1,2,3,...,20$.

c. Plot the data from the table in part b.

d. Fit a curve to the data. Hint: Begin by recognizing the basic shape of the scatter plot and scaling the corresponding basic function. Then *fine-tune* (see Example 3.6.8).

We emphasize that different people using this process of developing a sequence of improved approximations will probably obtain a different sequence of approximations. This is typical of modeling real-world problems.

Exercises 3.6

1. (Computation Skills) Expand (multiply) the following factored expressions. Observe aspects of the expanded expression that could be determined by inspection of the factored expression (for example, degree, coefficients).

    a. $(x-2)(x+3)$

    b. $x(x-2)(x-2)$

    c. $x(x-3)(x+3)$

2. (Computational Skill) Factor the following polynomial expressions.

    a. $x^4 + 7x^3$

    b. $x^2 - 4$

    c. $x^3 - 5x^2 + 6x$

3. (Computation Skill) Make up three additional expansion exercises similar to Exercise 1a–c and three additional factoring exercises similar to Exercise 2a–c.

4. Graphically approximate the zeros of $f(x) = x^3 - 7x + 5$. How many zeros are there? Explain.

5. Graphically approximate the zeros of $f(x) = x^4 - 7x^3 + 5$. How many zeros are there? Explain.

6. Make up two functions, each of which has thee zeros.

7. Make up two functions, neither of which has a zero.

8. Plot the following data on stopping distances as given in the Texas Drivers Handbook (1995), and then graphically determine a model that approximates the data.

Speed (mph)	Stopping Distance (ft)
20	45
30	78
40	125
50	188
60	272
70	381

Table 3.6.4

9. In Section 2.1, Exercise 6, we were given the following data on average freeway speeds in Harris Country. Plot the data and then graphically determine a linear model that approximates the data. Using your model, approximate the average freeway speed in 1997.

Year	Miles per Hour
1982	38.3
1985	41.1
1988	45.6
1991	47.6
1994	49.0

Table 3.6.5

10. Plot the following data on senior populations and then graphically determine a model that approximates the data.

Year	Population (millions)
1900	3.1
1992	32.3
2000	35.3
2010	40.1
2020	53.3
2030	70.2

Table 3.6.6

11. The following estimates are given of the growth in the number of pagers from 1990 to 1997 (*New York Times*, May 21, 1998). Plot this data, graphically determine a quadratic model that approximates the data, and then approximate the number of pagers in 1999.

Year	# Pagers (million)
1990	10
1991	12
1992	15
1993	20
1994	26
1995	33
1996	41
1997	49

Table 3.6.7

12. In searching through some past records, Scott found an old savings bank book showing how a \$100 deposit made by his parents when he was born grew until he cashed it in on his tenth birthday. Scott's bank book gave the following data (balance amounts rounded up to full dollars).

No. of Years	Balance (\$)
0	100
2	123
3	137
4	152
5	169
6	187
8	230
9	256
10	284

Table 3.6.8

Plot this data, graphically determine an exponential model that approximates the data, and then approximate what the balance would have been if Scott had left the account alone until his twenty-first birthday. Hint: Consider an exponential model in the form $f(x) = a * b^t$ where t represents years and a and b are parameters to be determined.

13. The Austin Community Profile published by the Greater Austin Chamber of Commerce (1996) gave the following population data for the City of Austin.

Year	Population (thousands)
1910	29
1930	53
1950	132
1970	252
1995	512

Table 3.6.9

Plot the population data, graphically determine a model that approximates the data, and then, using your model, approximate the population for the year 2000. (Suggestion: Plot the data point (1910,29) as (10,29) and likewise for the other four points.)

Hint: The data suggests that the population is doubling about every 20 years. Thus a reasonable model to consider is an exponential model of the form $p(t) = a * 2^{bt}$ where t represents years. Let 1900 correspond to $t = 0$ and thus $p(0) = a$. Estimate what the population was in 1900, set that value equal to a, and then experimentally determine a suitable value for b.

14. The World Health Organization published the following data on the 1995 Ebola outbreak in Africa.

Date	Deaths
5/11/95	27
5/13/95	57
5/14/95	59
5/17/95	77
5/19/95	89
5/20/95	97
5/24/95	108
5/26/95	121
5/30/95	153
6/1/95	164

Table 3.6.10

Plot this data, graphically determine a linear model that approximates the data, and then, using your model, approximate the number of deaths on 6/5/95 if the outbreak had continued. Fortunately the outbreak was contained in early June 1995. (Suggestion: Plot the data point (5/11/95, 27) as (11, 27) and similarly for the other points.)

15. Plot $f(x) = \cos(x)$ over the interval $[-1.4, 1.4]$. Graphically determine a polynomial function whose graph is a good approximation of the graph of f.

16. Develop a model for summing the squares of the first n positive integers. That is, $s(n) = 1^2 + 2^2 + 3^2 + \ldots + n^2$. Hint: See the Small-Group Activity in this section on summing the first n positive integers.

3.7 Symbolic Approximation of Data

In the last section, we developed a process for graphically determining a model that approximates the scatter plot of a data set. In this section, we look at an analytical method, the least squares method, for determining a model to approximate a given data set. This method produces the best model in the sense of minimizing the sum of the squares of the point errors.

Example 3.7.1.

Determine the best linear approximation of the following data set such that the resulting line passes through the origin.

x	y
1	2
3	2
4	5
5	4

Table 3.7.1

The problem is to determine the equation of a line that best fits the scatter plot of the data, that is, an approximation that has the least error.

Solution

We start by plotting the data and drawing a line that looks reasonable, call it $y = mx$. We then draw a vertical line segment from each data point to the line. The lengths of these line segments represent the point errors. We have four point errors, one for each of the four data points. Our objective is to determine an m, the slope of the line, that will minimize the sum of the squares of these four point errors.

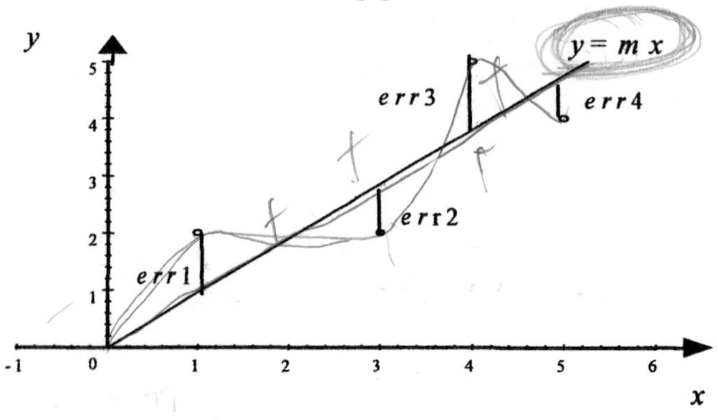

Figure 3.7.1

Because the error segments are vertical line segments, the endpoints of the error segments have the same first coordinates. Thus the point error for a data point is the y coordinate of that data point minus the y coordinate of the corresponding point on the line $y = mx$.

(Explain how the corresponding point on the line $y = mx$ is determined.) We use the square of a point error rather than the point error itself in order to avoid a negative error value canceling a positive error value. In Figure 3.7.1, point *err*1 and point *err*3 are positive whereas point *err*2 and point *err*4 are negative.

For each data point, we now compute the corresponding point on the line $y = mx$ and the square of the point error.

Data Point	Line Point	Square of the Point Error
(1,2)	(1,m)	$err1^2 = (2 - 1m)^2$
(3,2)	(3,3m)	$err2^2 = (2 - 3m)^2$
(4,5)	(4,4m)	$err3^2 = (5 - 4m)^2$
(5,4)	(5,5m)	$err4^2 = (4 - 5m)^2$

Table 3.7.2

Note that each of the point errors is a function of m, the slope of the line. Thus the sum of the squares of the point errors will also be a function of m. We define the error function, $error(m)$, as the sum of the squares of the 4 point errors.

$$error(m) = (2 - 1m)^2 + (2 - 3m)^2 + (5 - 4m)^2 + (4 - 5m)^2$$

In order to find the value of m that minimizes $error(m)$, we plot $error(m)$ and note the value of m that gives the lowest point on the plot. We then zoom in on that point.

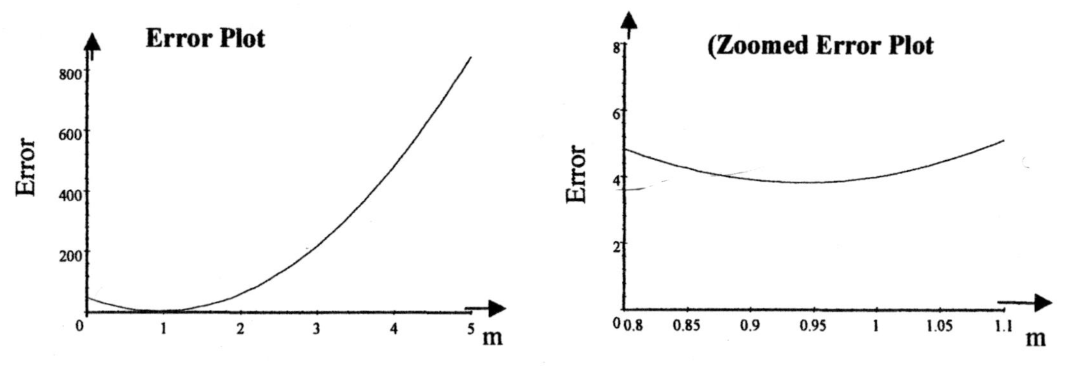

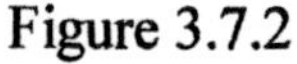
Figure 3.7.2

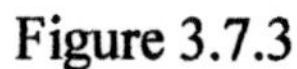
Figure 3.7.3

It appears that $error(m)$ has a minimum value at $m = 0.95$. Thus the line of best fit, called the regression line, to the data points is approximately $y = 0.95x$. (This answer is approximate because we only obtain an approximate value in reading the coordinates of the minimum point on a graph.)

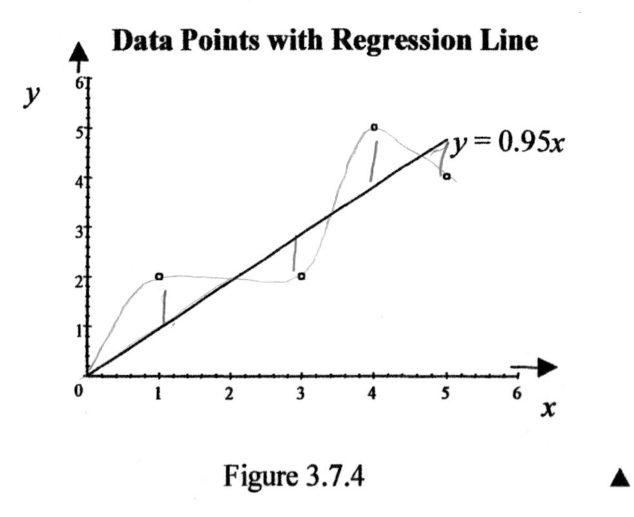

Figure 3.7.4 ▲

Example 3.7.2.

Exercise 14 in Section 2.6 gives data on family income eligibility for free school lunch programs from July 1, 1996, through June 30, 1997. A portion of that data is reproduced here.

Number in Household	Income per Week ($)
0	0
2	259
4	325
6	456
8	587

Table 3.7.3

Using the method illustrated in Example 3.7.1, determine the best linear approximation of this data and then approximate the weekly income figure for a household of nine.

Solution

We begin by plotting the data and visually estimating a line $y = mx$, where x represents the number of persons in the household and y represents the income per week. The task is then to determine a value for m that will minimize the sum of the squares of the point errors.

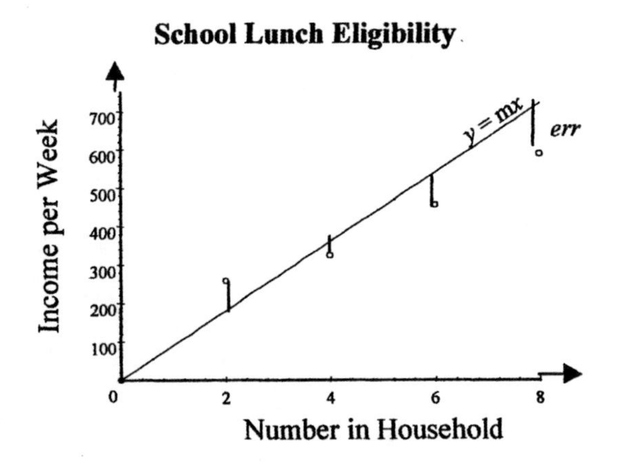

Figure 3.7.5

The error function is

$$error(m) = (259 - 2m)^2 + (325 - 4m)^2 + (456 - 6m)^2 + (587 - 8m)^2.$$

Because the slope of the estimated line appears to be about 80, we would expect the slope of the desired line to be in the range 65 to 85. Thus we plot the error function over that range.

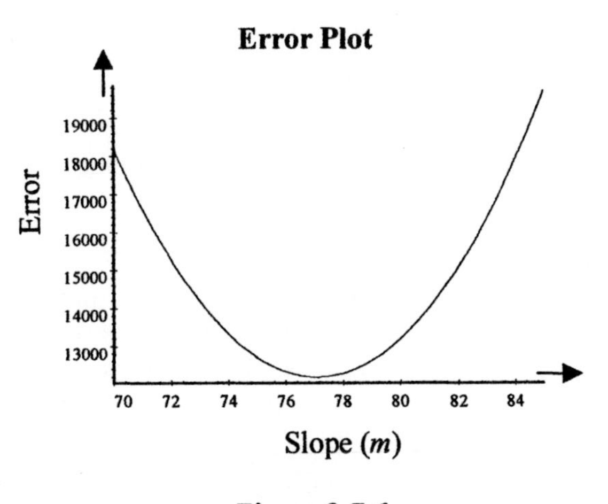

Figure 3.7.6

Zooming in near the low point, it appears that $error(m)$ has a minimum at $m = 77.1$.

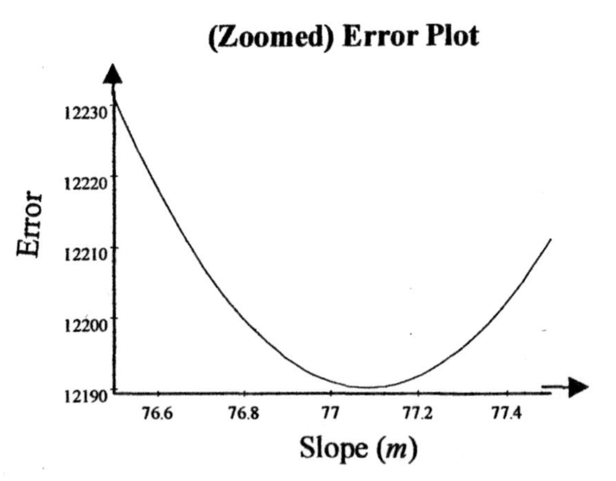

Figure 3.7.7

Thus the best linear approximation of this data is $y = 77.1x$, and so the eligible weekly income for a household of nine is approximately $694. ▲

Regression Analysis

The previous two examples were special cases of **regression analysis**, which is the name for determining a function whose graph best fits the scatter plot of a given data set. The most frequently used regression method is that of minimizing the sum of the squares of the errors.

Examples 3.7.1 and 3.7.2 illustrate the process of minimizing the sum of the squares of the errors to determine the equation of a line passing through the origin that best approximates a scatter plot. Because the equation of a line through the origin, $y = mx$, has only one independent variable, m, the error function is just a function of m. We were thus able to determine the appropriate value for m by plotting the error function and determining the value of m that gives the lowest point on the error curve.

If we drop the restriction that the desired line pass through the origin, then the equation of the line is $y = mx + b$. There are now two independent variables, m and b. Thus the error function will be a function of the two variables, m and b. Minimizing a function of two or more variables is beyond the content of this course (calculus is required), and thus we will turn to technology (graphing calculators, computers). In addition to linear regression, most graphing calculators and computer algebra systems also contain built-in programs for computing exponential, logarithmic, quadratic, and cubic regression. The regression process in each case is the same as was illustrated in Examples 3.71 and 3.72 where the regression line passed through the origin. The difference in the computation is in the number of independent variables in the error function. As illustrated, we can graphically approximate where a function of one variable assumes its minimum. For functions of more than one variable, we need to lean on our friends in calculus or use technology.

Exact Polynomial Fit

A general polynomial of degree n has exactly $n + 1$ parameters or coefficients. For example, a second-degree polynomial has the form $p(x) = ax^2 + bx + c$ and a third-degree polynomial has the form $q(x) = ax^3 + bx^2 + cx + d$. To determine a polynomial whose graph passes through each point of a data set means to determine appropriate values for the parameters or coefficients. In the exact polynomial fit method, a unique polynomial of degree one less than the number of data points is determined whose graph passes through the data points. We illustrate the method with the following

9 < n

example.

Example 3.7.3 (Exact Polynomial Fit).

Determine a second-degree polynomial function whose graph passes through the points $(0,1),(3,3),(6,0)$. That is, determine the parameters (coefficients) of a general second-degree polynomial function, $p(x) = ax^2 + bx + c$, that satisfy the conditions $p(0) = 1$, $p(3) = 3$, and $p(6) = 0$.

Solution

We start with a general second degree polynomial function

$$p(x) = ax^2 + bx + c$$

and then determine the parameter values. The condition that the graph pass through a given point yields a linear equation in the three unknown coefficients: a, b, c. For example, requiring the graph to pass through the point $(3,3)$ means that $p(3) = 3$ and thus $9a + 3b + c = 3$. Thus the three conditions yield a system of three linear equations in three unknowns (the coefficients a, b, c).

Condition		Equation
$p(0) = 1$	=>	$c = 1$
$p(3) = 3$	=>	$9a + 3b + c = 3$
$p(6) = 0$	=>	$36a + 6bc = 0$

To solve this system, we reduce it to a system of two equations in two unknowns, a and b, by substituting $c = 1$ in the second and third equations.

$$9a + 3b + 1 = 3 \quad \Rightarrow \quad 9a + 3b = 2$$
$$36a + 6b + 1 = 0 \quad \Rightarrow \quad 36a + 6b = -1$$

To solve the reduced system, we multiply the first equation by –4 and then add the two equations.

$(-4)(9a + 3b = 2)$	=>	$-36a - 12b = -8$	
second equation	=>	$36a + 6b = -1$	
	=>	$-6b = -9$	Add the two equations.
	=>	$b = \frac{3}{2}$	Divide by -6.

To solve for a we substitute $c = 1$ and $b = \frac{3}{2}$ into either the second or third original equation, say the third.

$$36a + 6b + c = 0 \qquad \text{Third equation}$$

$$36a + 6(\tfrac{3}{2}) + 1 = 0 \qquad \text{Substitute for } b \text{ and } c.$$

$$36a = -10$$

$$a = \frac{-10}{36} = -\frac{5}{18}$$

The desired second degree polynomial function is

$$p(x) = -\tfrac{5}{18}x^2 + \tfrac{3}{2}x + 1.$$

We check our computations symbolically by evaluating the solution function at $x = 0, 3, 6$ and graphically by plotting the given points and the graph of our solution function.

$$p(0) = 1, \text{OK}$$

$$p(3) = (-\tfrac{5}{18})(3^2) + (\tfrac{3}{2})(3) + 1 = 3, \text{ OK}$$

$$p(6) = (-\tfrac{5}{18})(6^2) + (\tfrac{3}{2})(6) + 1 = 0, \text{ OK}$$

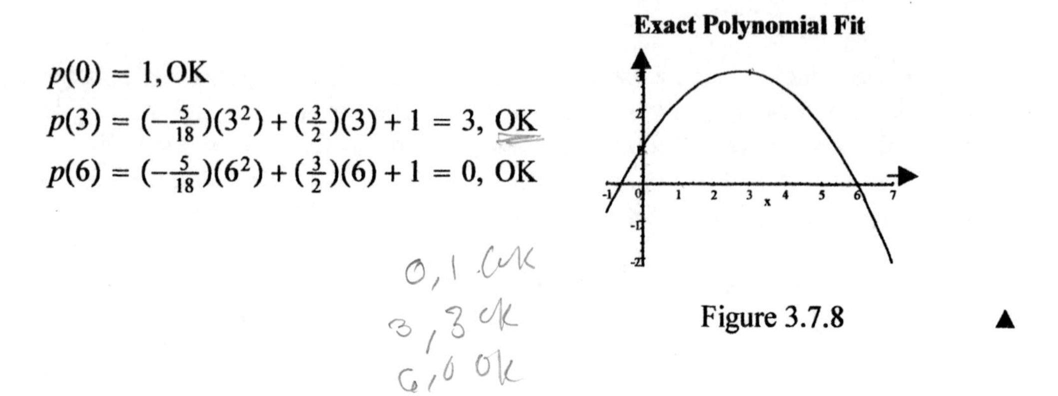

Figure 3.7.8 ▲

An alternative method for solving the system of equations

$$\begin{cases} c = 1 \\ 9a + 3b + c = 3 \\ 36a + 6b + c = 0 \end{cases}$$

is to express the information in the system as an augmented matrix (see Section 2.8)

$$\begin{bmatrix} 0 & 0 & 1 & 1 \\ 9 & 3 & 1 & 3 \\ 36 & 6 & 1 & 0 \end{bmatrix}$$

and then use a calculator to reduce the augmented matrix to its row reduced echelon-form (rref).

$$\begin{bmatrix} 1 & 0 & 0 & -.2778 \\ 0 & 1 & 0 & 1.5 \\ 0 & 0 & 1 & 1 \end{bmatrix}$$

Thus the coefficients are $a = -0.2778, b = 1.5$, and $c = 1$. This matrix approach is the preferred approach for most systems of three or more equations.

Although the exact polynomial fit method theoretically works for any number of data points, it is seldom used in cases of more than five data points for two reasons.

1. The oscillations in the graphs of high-degree polynomials can be very large and thus do not serve as good models of reality.
2. To solve for the parameters or coefficients of a polynomial of degree n involves solving a system of $n + 1$ linear equations. Solving large systems of linear equations involves much greater amounts of computing than a corresponding regression analysis and thus is subject to much greater round-off error than for regression analysis.

Exercises 3.7

1. **(Computational Skill)** Solve the following systems for the variables a and b. (Refer to Section 2.8.)

   a. $\begin{cases} a + 3b = 2 \\ 2a - b = -1 \end{cases}$

   b. $\begin{cases} 3a + 2b = 4 \\ 2a - b = 2 \end{cases}$

   c. $\begin{cases} 2a + 4b = 2 \\ a + 2b = 1 \end{cases}$ What is wrong? Explain.

2. **(Computational Skill)** Determine the distance between each of the following pairs of points.
   - **a.** $(1,2)$ and $(3,4)$
   - **b.** $(3,2)$ and $(-2,1)$
   - **c.** $(-2,5)$ and $(-1,-2)$

3. **(Computational Skill)** Make up and solve three exercises similar to Exercises 1a–c, 2a–c.

4. Use the method of Example 3.7.1 to determine the linear function of the form $f(x) = mx$ that best fits the following data.

x	y
2	1
3	5
4	5

Table 3.7.4

5. Salt is used in old-fashioned ice cream churns and on icy highways to lower the freezing point. The following table gives data on the effect of salinity (the weight in grams of dissolved salts in one kilogram of water) on the freezing point of water.

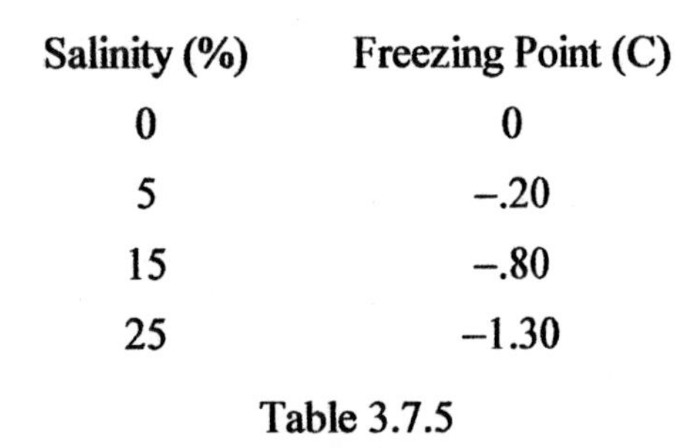

Salinity (%)	Freezing Point (C)
0	0
5	–.20
15	–.80
25	–1.30

Table 3.7.5

Plot this data and then, using the method of Example 3.7.1, determine the linear function that best fits the scatter plot.

6. Determine a second-degree polynomial function (i.e., $p(x) = ax^2 + bx + c$) whose graph passes through the points $(-\frac{\pi}{2}, 0), (0, 1), (\frac{\pi}{2}, 0)$ and then form a multiplot over the interval $[-\frac{\pi}{2}, \frac{\pi}{2}]$ of your function and $f(x) = \cos(x)$. Comment on the result.

7. Determine a second-degree polynomial function (i.e., $p(x) = ax^2 + bx + c$) whose graph passes through the points $(0, 0), (\frac{\pi}{2}, 1), (\pi, 0)$ and then form a multiplot over the interval $[0, \pi]$ of your function and $f(x) = \sin(x)$. Comment on the result.

8. Use the built-in linear regression program in your calculator to determine the linear function that best fits the 1995 Ebola outbreak data given in Exercise 14 in Section 3.6.

9. Use the built-in quadratic regression program in your calculator to determine the quadratic function that best fits the data given in Example 3.6.8 in Section 3.6.

10. Use the built-in quadratic regression program in your calculator to determine the quadratic function that best fits the data given in Exercise 10 in Section 3.6.

11. Use the built-in exponential regression program in your calculator to determine the exponential function that best fits the data on the population of Austin, Texas, given in Exercise 13 in Section 3.6.

12. Apply the what-if technique to Example 3.7.3 to determine a third-degree (cubic) polynomial, $p(x) = ax^3 + bx^2 + cx + d$, that passes through the points $(-2, -2), (0, 3), (3, 2), (6, 5)$. Hint: Form a system of four equations in the four unknowns a, b, c, d and then use the matrix method to solve the system (see Section 2.8). Compare your answer with the cubic equation found using the built-in cubic regression program in your calculator applied to the scatter plot of the four data points.

13. Apply the what-if technique to both Example 3.7.3 and Exercise 12 by determining a fourth-degree (quartic) polynomial, $p(x) = ax^4 + bx^3 + cx^2 + dx + e$, that passes through the points $(-2, -2), (0, 3), (3, 2), (6, 4), (9, -2)$ by forming and then solving (matrix method) a system of five equations in five unknowns. Compare your answer with the quartic equation found using the built-in quartic regression program in your calculator applied to the scatter plot of the five data points.

3.8 Optimization

Optimization problems involve finding maximum and minimum values. In the past, optimization problems required calculus for solutions. However, today's technology, such as a graphing calculator, allows us to find maximum and minimum values graphically and numerically without relying on calculus. In an optimization problem, the function to be maximized or minimized is called the **objective** function. Frequently the objective function will be a function of two or more variables. In these instances, it is necessary to find additional relations between the variables that can be used to transform the objective function into a function of a single variable. These additional relations are called **constraint** equations.

Example 3.8.1

Mary Lou wants to build a rectangular-shaped dog pen. She has 300 feet of fencing, and she wants the pen to be as large as possible. What should be the dimensions of Mary Lou's dog pen?

We first note that Mary Lou can make several different size, rectangular dog pens with her 300 feet of fencing. Here are three possibilities.

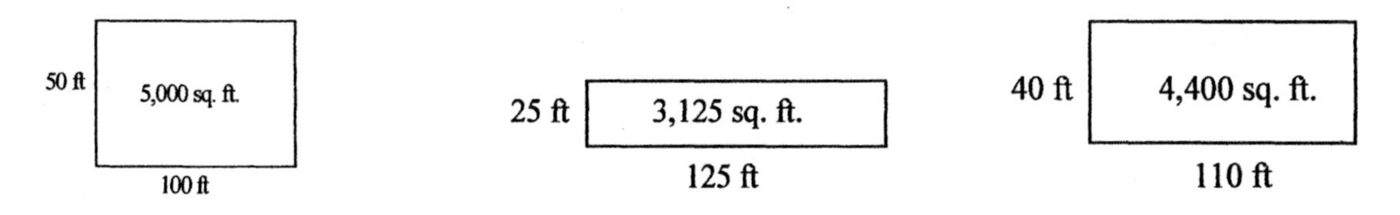

Figure 3.8.1

Solution.

We begin by sketching and labeling a picture.

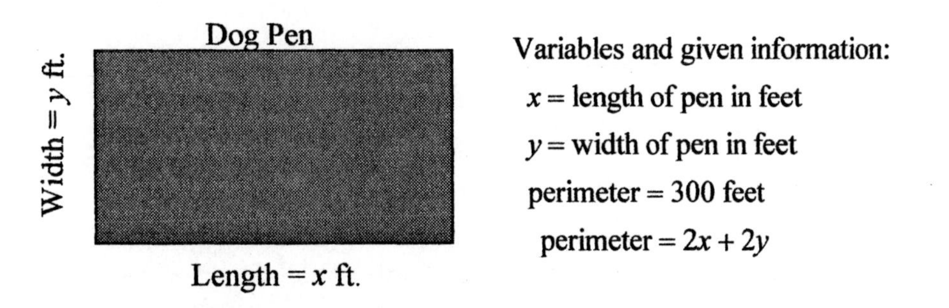

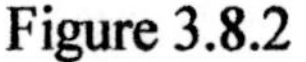
Figure 3.8.2

Because Mary Low wants to maximize the area of the dog pen, we define the objective function as

$$A(x,y) = xy$$

and the constraint equation as

$$2x + 2y = 300.$$

We solve the constraint equation for y and then substitute for y in the objective function.

$$y = \frac{300-2x}{2} = 150 - x.$$

Thus $A(x) = (150 - x)x$.

Now we can approximate the value of x that will maximize the objective function by looking at the graph of the objective function.

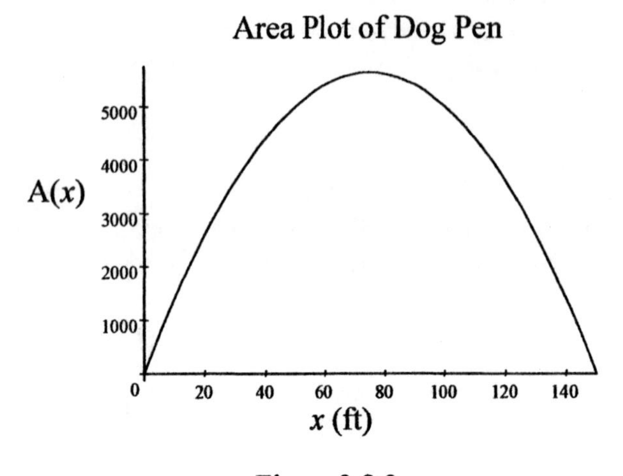

Figure 3.8.3

The maximum value for the area appears to occur for x between 70 and 80. Zooming in on this interval would show that the maximum occurs at $x = 75$ ft. Thus $y = 150 - 75 = 75$ ft also. Hence, the largest (in area) dog pen that Mary Lou can build is a square 75 feet on a side. ▲

Query 1.

Example 3.8.1 showed that the largest rectangle having a fixed perimeter is a square. Is it also true that the longest perimeter of a rectangle with fixed area is the perimeter of a square having that area?

Query 2.

Could Mary Lou have enclosed more area if she had built a circular dog pen?

Example 3.8.2.

Jed is planning the most efficient route to a bus stop. He lives diagonally across the street from a large vacant and partially overgrown city block that is 2400 feet long and 1200 feet wide. The bus stop is at the furthest corner of the block from Jed's house. He estimates that he can walk at 4 mi/hr along the sidewalks surrounding the block, but that his rate cutting through the block would be only 3 mi/hr. What is Jed's most efficient route from his home to the bus stop, and how many minutes will it take him to cover the route?

Solution

We assume that efficiency refers to time, and thus we want to find the route that would minimize time. The choice of routes include walking along the sides of the block on the sidewalk, cutting diagonally through the block, or a combination of these two as shown in the following sketch.

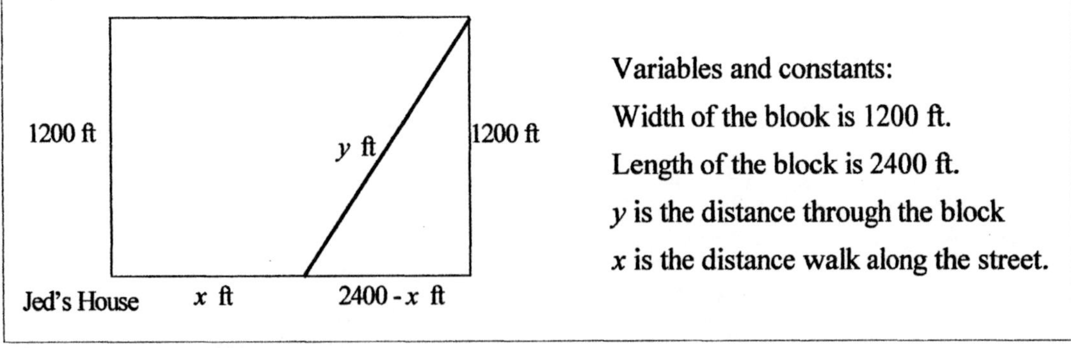

1200 ft
y ft
1200 ft
Jed's House
x ft
2400 - x ft
Variables and constants:
Width of the block is 1200 ft.
Length of the block is 2400 ft.
y is the distance through the block
x is the distance walk along the street.

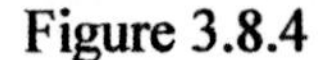
Figure 3.8.4

The given rate information is 4 mi/hr walking on the sidewalks and 3 mi/hr cutting through the block. Because the distances are given in feet and one answer is to be in minutes, we convert the rates to ft/min.
4 mi/hr* 5280 ft/mi * 1/60 hr/min = 352 ft/min
3 mi/hr *5280 ft/mi * 1/60 hr/min = 264 ft/min

Because the combination route includes the other two choices as special cases, we form an objective function to represent the time for the combined route. The objective function will be the time to walk the diagonal (y ft.) crossing the block plus the time to walk the x ft. along the street. From the formula: distance equals rate times time, we have time is equal to distance over rate. Thus the objective function is

$$f(x,y) = \frac{y}{264} + \frac{x}{352} \text{ min.}$$

Because the objective function involves two variables, we look to our sketch to recognize a constraining relationship between the variables. The right triangle in the sketch suggests the constraint equation $y^2 = (2400 - x)^2 + 1200^2$ or $y = \sqrt{(2400 - x)^2 + 1200^2}$. Substituting for y in the objective function reduces the objective function to a single variable function, x.

$$f(x) = \frac{\sqrt{(2400-x)^2+1200^2}}{264} + \frac{x}{352} \text{ min.}$$

We now plot the objective function and identify where the lowest point on the curve occurs.

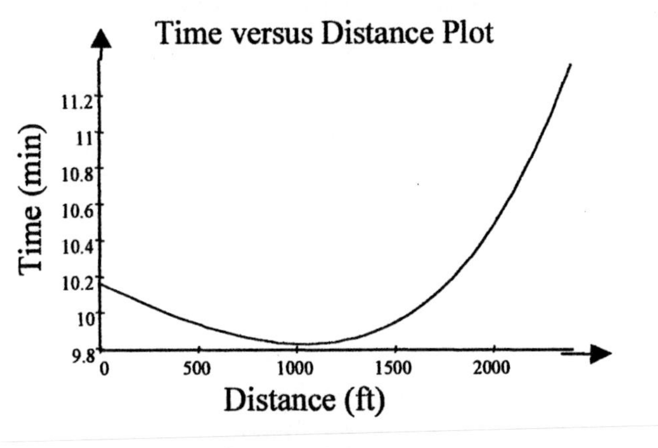

Figure 3.8.5

The minimum point appears to be a little more than 1000 feet. We replot over the interval [1030,1060] to get a clearer picture.

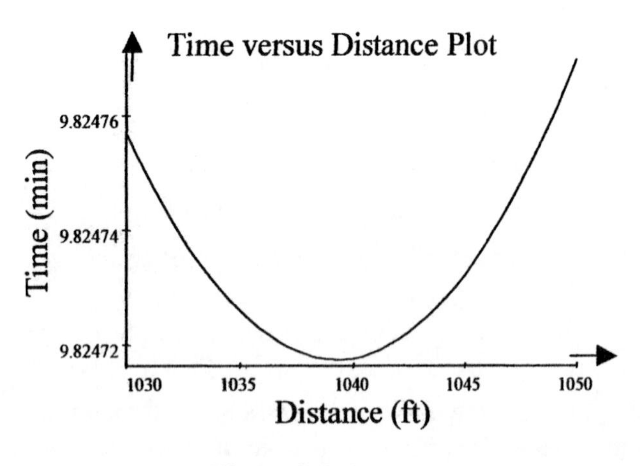

Figure 3.8.6

We estimate the minimum occurs for $x = 1039$. (We could continuing zooming to obtain more accuracy.) Thus Jed should walk 1039 feet along the sidewalk and then cut diagonally to the bus stop. The length of the diagonal cut is $\sqrt{1200^2 + (2400 - 1039)^2} = 1814.5$ ft, and his minimum time is $f(693) = \frac{\sqrt{1200^2+(2400-1039)^2}}{264} + \frac{1039}{352} = 9.8247$ minutes.

To check if this is a reasonable result, we compute the time to walk diagonally from Jed's house to the bus stop and the time for Jed to walk along the sidewalk to the bus stop.
Time for the diagonal route is $\frac{\sqrt{2400^2+1200^2}}{264} = 10.164$ minutes.
Time to walk along the sidewalk is $\frac{2400+1200}{352} = 10.227$ minutes.
The time to travel the sidewalk/diagonal path is less than the other two options, and thus we accept it as being reasonable. ▲

Query 3.
What would be the result for Example 3.8.2 if Jed could only walk 2 mi/hr through the vacant lot?

The solutions to Examples 3.8.1nd 3.8.2 illustrate a problem-solving process that is effective for solving a wide range of problems, including optimization problems. The process, presented in outline form, is

1. Create a model
   a. Sketch and label a picture illustrating the problem. (Label the parts of the picture that are fixed with constants and the parts that can change with variables.)
   b. Define the variables and list assumptions.
   c. List the given information.
   d. Develop an objective function (the function to be maximized or minimized).
   e. List any constraint relations.
2. Solve the model
   a. Use the constraint relations to express the objective function as a function of a single variable.

   b. Check to see that proper units are used.
   c. Plot or enumerate the objective function (the highest value or the largest number is the maximum value, the lowest value or the smallest number is the minimum value).
3. Interpret the results in terms of the original problem setting.
   a. Do the results make sense?
   b. Do the results provide a satisfactory solution?

Exercises 3.8

1. Determine the maximum value of the objective function $f(x,y) = x(y^2 - x^2)$ over the domain $-2 \leq x \leq 2$, $-2 \leq y \leq 2$, given the constraint equation $x^2 + y^2 = 4$.
2. Determine the minimum value of the objective function $g(x,y) = xe^{\sin(y)}$ over the domain $-2 \leq x \leq 2$, $-2 \leq y \leq 2$, given the constraint equation $y = x$.
3. Make up an exercise to determine the maximum and the minimum values of an objective function of two variables given a constraint equation.
4. A rectangle is formed in the first quadrant with two sides lying on the x and y-axes and one corner on the graph of $y = \frac{1}{x}$. Determine the coordinates of the corner point on the curve that yields the maximum area of the rectangle.
5. Dr. Lei Yu (Texas Southern University) conducted a study of vehicle emissions on a highway in Houston, Texas. He collected data on carbon monoxide (CO), a dangerous tasteless, orderless, and colorless gas that reduces the delivery of oxygen to body organs. Plotting the data and fitting a curve to the resulting scatter plot yielded the function $CO_s(x) = e^{-2.465+0.028x}$ where x is the speed in miles per hour. The units for $CO_s(x)$ are grams per second (g/sec). Convert the CO emissions to grams/mile and then determine the speed that minimizes the CO emissions in terms of grams/mile. Hint. Use dimensional analysis as a guide to determine the conversion factors. That is, $(\frac{\text{grams}}{\text{second}})$ (conversion factors) $= (\frac{\text{grams}}{\text{mile}})$.
6. The school's Sports Club is organizing a bus trip to Albany to see the Crows and Groundhogs baseball teams compete for the championship. The trip is both a social event and a fund-raiser for the club. For a 49-passenger bus, the school charges a fixed charge of $400 plus $5 per person. The school requires a minimum of 30 people. To encourage people to participate, the club decides to reduce the $45 per person fee by $1 for each person over the required minimum number. Determine the maximum amount the club can realize and the number of passengers required to obtain this amount.

In Exercises 7–16, explicitly follow the problem-solving process outlined in this section.

7. Determine the dimensions of a rectangle with perimeter 100 m whose area is as large as possible. Does the result imply that when a given length is divided into two lengths such that the product of the two lengths is a maximum, the two lengths must be equal? (Think of the two lengths as representing the sides of a rectangle.) If the problem were modified so that the area was fixed (rather than the perimeter), would the resulting rectangle still be a square?
8. Consider a cardboard box with a square base and an open top that has a volume of 3,375 cm^3. Determine the dimensions of the box that minimize the amount of cardboard used.
9. Determine the dimensions of an open (no top) rectangular box of maximum volume that can be formed by cutting out the corners of a 16-in by 20-in sheet of cardboard.

10. A pizza box is formed from a rectangular sheet of cardboard by "folding in the corners." Determine the dimensions of the minimum-size sheet of cardboard needed to form a box for a 16-in pizza if there is to be at least one-half inch between the pizza and the edge of the box and the inside height of the box is to be 2 inches.
11. Consider a circle of radius 1 centered at the origin and a line segment in the first quadrant that is tangent to the circle and has slope = −1. Determine the dimensions of the rectangle of maximum area that lies in the first quadrant with one corner at the origin and the diagonally opposite corner on the line segment. Hint: The slope of a line tangent to a circle is perpendicular to the radius drawn to the point of tangency. Two lines are perpendicular provided their slopes are negative reciprocals of each other.
12. Venus and Oscar live in houses that are 200 feet apart when the distance is measured along the straight road in front of the houses. However Venus's house sits 100 feet from the road and Oscar's sits 50 feet from the road. The school bus driver only wants to make one stop to pick up Venus and Oscar. Assuming that Venus and Oscar will make direct paths from their houses to the bus stop, where should the bus stop to minimize the total distance that Venus and Oscar walk.
13. An equipment rental business purchased a Quonset-type building for its storehouse. The front of the building is parabolic in shape with the roof line defined by the equation $y = 16 - x^2/20$ with x and y measured in feet. The owners plan to install the largest possible set of rectangular doors on the front. Determine the dimension of the opening for the doors.
14. A 20-foot rain gutter is formed from a 10-inch-wide strip of aluminum 20 feet long by bending up the sides of the strip to form a *U* shape 20 feet long.(The sides are bent perpendicular to the bottom.) Determine the width of the sides and bottom that will maximize the volume of the rain gutter.
15. A 20-foot rain gutter is formed from a 10-inch-wide strip of aluminum 20-feet-long. The strip is marked linearly into three 20 foot long segments with the two outside segments having the same width. One outside segment is bent perpendicular to the middle segment and the other outside segment is bent to form an angle of 120 degrees with the middle segment. If the width of the middle segment is twice as wide as either outside segment, what is the width of the middle segment that will maximize the volume of the gutter?
16. A pup tent is made from three pieces of canvas—front, back, and top. The front and back pieces hang vertically to close the ends of the tent. If the top piece measures 8.5 ft by 11 ft, determine a suitable height for the tent poles. Clearly state the assumptions you make in modeling and then solving this problem. (This problem is ill posed. Students must define the problem, e.g. determine an objective function, and establish assumptions allowing for a model to be created.)

3.9 Summary

The function concept is **one of the most important concepts in mathematics** and thus should be part of every college mathematics course. In *Contemporary College Algebra*, the fundamental reason for studying functions is to transform data into information and information into action. Section 3.1 (Displaying Functions) emphasizes the importance of being able to recognize and display functions in graphic, symbolic, numeric, and written form. Advantages and disadvantages of each form (graphic, symbolic, numeric) are discussed in a manner similar to the topic of displaying data (Section 2.1). Section 3.2 (Definitions) consolidates the informal treatment of functions, begun in Section 2.4 and continued in Section 3.1, into rigorous definitions of terms used in discussing functions. Definitions are stated and illustrated for relations, functions, domain, range, independent variable, and dependent variable. Functions are defined as **sets of ordered pairs** and as **input-output** relations satisfying a uniqueness relation that requires exactly one output value (dependent variable) for each input value (independent variable).

A person studying data is usually interested in questions such as: What is the trend? How can I make predictions based on the data (that is, interpolate, extrapolate)? What are the long-term results? How can errors in the data be determined? The answers to these types of questions depend on extracting a function relation from the data. The development of such functions often depend on recognizing the basic shape of the data, as this suggests the type of function to be developed. Basic shapes include the shapes of the graphs of power functions (degrees 0, 1, 2, and 3), radicals (square and cube roots), exponentials, logarithms, periodic functions (sine, cosine), and asymptotic behavior. Section 3.3 provides background for this type of data analysis.

Sections 3.4 and 3.5 address the theme of doing more with less. Shifting and scaling are the two basic graph transformations for changing the actual shape and location of a graph without changing the basic shape. A person who understands these two transformations and is able to recognize the basic shapes, is able to analyze a broad spectrum of graphs. Being able to mix and match basic shapes of graphs is as important in analyzing graphs as it is in planning a person's wardrobe. The process of expanding a basic set into a broad spectrum is carried into functions in Section 3.5 (Algebra of Functions). In this section, the fundamental operations of addition (subtraction), multiplication (division), and composition are applied to functions allowing for new functions to be formed from combinations of the basic functions. These operations are presented graphically, symbolically, and numerically underscoring the major ways in which data is presented. The graphical and numerical presentations of functions facilitate finding inverse relations. The reflection of a graph in the line $y = x$ yields the graph of the inverse relation. For a function defined numerically by a table, reversing the columns in the table yields the inverse relation. We recall that not all inverse relations are functions. (The study of the inverse operation is always included with the study of the operation.) The goal of Sections 3.4 and 3.5 is to develop the ability to combine functions using the fundamental algebraic operations and to decompose functions into basic parts along with the ability to recognize basic shapes and the inverses of basic shapes.

The modeling theme, initiated in Chapter 2 with linear models, is expanded in Chapter 3 to include curve fitting. A primary objective of this course is to provide students with a realization and experience in modeling real-world situations by collecting and plotting data and then fitting a curve to the data. The resulting function (whose graph fits the data) is the model and can be used for predictive purposes. Sections 3.6 and 3.7 focus on developing a function whose graph approximates or fits a given data set. This is done graphically by first recognizing the basic shape of the data and then shifting and scaling the corresponding basic function until a suitable fit is obtained. The shifting and scaling is an iterative process that usually requires several iterations until a suitable fit is found. The symbolic **regression analysis** method of least squares is illustrated in Examples 3.7.1 and 3.7.2. In this approach, the first step is again to recognize the basic shape of the data and determine the corresponding function. The sum of the squares of the errors between the data points and the points on the graph of the corresponding function constitute the error function. The value of the variables that minimize the error function substituted into the corresponding function yield the best fit function. This process is illustrated with the one variable case in which the graph of the corresponding function is a straight line passing through the origin. Thus slope is the only variable involved, and its value can be obtained by graphing the error function. When two or more variables are involved, calculus is required to carry out the minimization process. For these situations, we rely on the built-in regression programs in our calculators and computers.

Determining the basic shape of data (the first step in the procedure described in the previous paragraph) is not always easy. For example, the shape of the data on masses of roots of pea plants (Example 3.6.7) could suggest either a quadratic polynomial or an exponential shape. When there is a question among two or more possible basic shapes, one should consider the physical aspects of the situation generating the data. If this does not help, then choose the simplest option. If this option does not work out well, then choose the next simplest option, and so forth. In general, polynomials are simpler options than are exponential or trigonometric functions.

An additional method of fitting data (Section 3.7) is to determine a polynomial whose graph contains each of the data points. This method requires solving a system of linear equations. Because the number and size of fluctuations in the graphs of polynomials increase with the degree of the polynomial, this method is usually used only with polynomials of degree four or less.

Optimization, finding a maximum or minimum value, is a central feature of many real-life applications of mathematics. Once the purview of calculus courses, today's technology (graphing calculators) have made optimization a college algebra topic. A key aspect of solving optimization problems is to understand the roles of the objective function and the constraint equations. Section 3.8 concludes with an outline of a process that summarizes the aspects of problem-solving developed in Chapters 2 and 3.

Fun Projects

(See Fun Projects in Chapter 2, for suggestions on assigning Fun Projects, formatting of the project reports, and assigning student responsibilities related to their groups.)

1. Comfort Function for Stairways

(Purposes: Model a real-life situation; collect data; create a function of more than one variable; provide a writing exercise; provide a small-group work experience.)

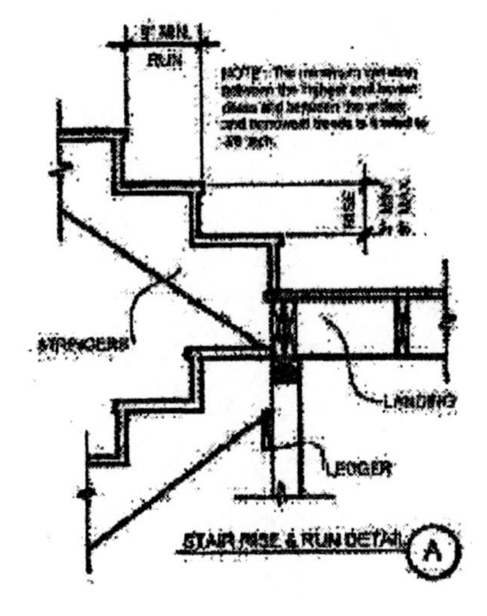

Introduction: Some stairways feel more comfortable to walk up and down than others. What factors help make a stairway comfortable? For example, are circular stairways more comfortable than straight stairways? Does the stairway material (wood, iron, concrete, etc.) affect the comfort? Are railings and lighting comfort factors in a stairway? Furthermore some stairways are enclosed and others are open, some stairways are steeper than others, and some stairways are longer than others.

Your task is to identify several factors related to stairways and then develop a comfort function for stairways. Your function needs to have at least three different types of inputs. (That is, your function must be a function of three or more variables.) The output of your function must be an element in the set {dangerous, comfortable, uncomfortable, convert to a ramp}. (That is, the range of your function consists of this output set of four items.) You should provide a written explanation of your reasoning for each of the following.

1. Assumptions and variables
   a. List your assumptions (for instance, average leg length).
   b. Define your variables (for instance, the pitch or slope of the stairway).
   c. Assign a range of weights to each variable that will allow you to quantify each variable for a given stairway.
2. Analyze at least ten different stairways, collecting data on each of your variables. Display your data in a table.

3. Define your stairway comfort function.
4. Apply your stairway comfort function to each of the stairways you analyzed. For each stairway, explain whether or not your function output coincides with your intuitive feeling about the stairway.

2. Determining the Dimensions of a Soda Can

(Purposes: Provide opportunity for interdisciplinary cooperation with the English department; model a real-world situation; graphically analyze a function; determine function optimization graphically; provide a writing exercise; provide a small-group work experience.)

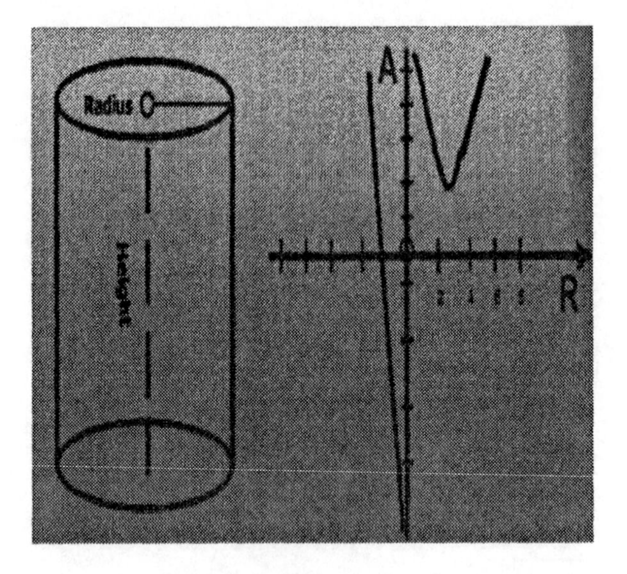

Introduction: The article "The Aluminum Beverage Can" in the September 1994 issue of *Scientific American* stated that approximately 100 billion drink cans are produced every year. With this number of cans, a small reduction in the cost of producing a drink can would result in a sizeable increase in profits. Today's 12-ounce beverage can weighs approximately 0.48 ounces compared to 0.66 ounces when the cans were first introduced in the 1960s. The savings from a further reduction in weight of 1% is approximately $20 million dollars per year.

In this Fun Project, you are asked to optimize the dimensions of a 12-ounce aluminum beverage can. For this project, assume that a soda can is a closed cylinder. (That is, ignore the neck, domed base, and pull tab of an actual soda can.) Also assume that the idealized soda can holds 12 ounces. (1 ounce = 1.8047 cubic inches.)

Complete the following tasks:.

1. Write a letter to a soft-drink company, say *Coke-Cola* or *Pepsi-Cola*, asking for the approximate number of 12-ounce aluminum beverage cans the industry produced during the previous year. Include a copy of your letter and the response received in your project report.
2. Carefully measure the diameter and height of a 12-ounce soda can.

3. Assume that the material of the can has a uniform thickness (bottom, sides, top). Let the thickness be one unit in order to treat the volume of the material as the outside surface area of the can. Express the surface area of a 12-ounce soda can as a function of the radius of the can and then plot the function.

4. Using the plot from Task 3, graphically determine the radius and from that the height of a 12-ounce soda can that has the minimal surface area.

5. Compare your results against the dimensions of an actual soda can. If the results are considerably different, explain why the soft-drink company would not use your size can in order to save money.

6. Change the assumption in Task 3 to be that the thickness of the top is 3 times the thickness of the sides and bottom (to allow for the pull tab). Now repeat Task 3. Compare your results against the dimensions of an actual soda can. Comment on the comparison.

Instructor Notes.

(1) This project offers an opportunity for interdisciplinary cooperation with the English department. In preparation for Task 1, an English professor could be invited to come to the mathematics class to discuss how to write a business letter. Or, each group could be instructed to submit their letter to an English instructor before mailing it. Sufficient lead time needs to be provided in assigning this project to allow time for students to receive replies from their letters.

(2) Tasks 5 and 6 illustrate the modeling process of modifying the original model when the interpretation of the results are not in accord with the real situation. (In this case, the dimensions of the soda can with minimum surface area were very different from the actual dimensions of a soda can.) The modifications usually involve making changes in the assumptions, as was done in this problem.

3. Postage Stamps

(Purposes: Model a real-world situation; analyze data; graphically fit a curve to a scatter plot; research on the Internet; apply regression analysis; provide a writing exercise; provide a small-group work experience.)

Introduction: The United States Post Office was established in 1775 when the Continental Congress appointed Benjamin Franklin as Postmaster General. He had previously served, since 1753, as the Crown's Postmaster General. In 1753, he established the first weekly mail wagon. It operated between Philadelphia and Boston. In 1765 Franklin established the first postal rate chart, although the first postal stamp was not issued for many years later. The following table shows that it took almost 80 years for the United States Post Service to issue its first 600 different stamps, approximately 30 years for its next 600 stamps, a little more than 20 years for its third 600 stamps, and then less than 10 years for its fourth 600 stamps.

This project focuses on developing a function that models the issuance of new stamps. We call it the stamp issue function and denote it by *SI*. The *SI* function is then compared to stamp issue functions for other countries. The following table shows the cumulative number of United States postage stamps issued by ten-year intervals. (The source is the *Scott Standard Postage Stamp Catalogue* (1988).) Only regular and commemorative issued stamps are included. Excluded are air-mail, special delivery stamps, and postcard stamps.

Year	Number of U.S. Stamps Issued	Year	Number of U.S. Stamps Issued
1848	2	1928	647
1858	30	1938	838
1868	88	1948	980
1878	181	1958	1123
1888	218	1968	1364
1898	293	1978	1769
1908	341	1988	2400
1918	529		

Table 3.FunProject 3.1

Part 1. U.S. postage stamps.

1. Plot the data in the preceding table.
2. Graphically develop an *SI* function by
   a. Recognizing the basic shape of the data
   b. Graphically fitting a curve to the scatter plot (Include in your project report the sequence of equations of the curves you considered in graphically fitting a curve to the scatter plot.)
3. Analytically develop an *SI* function by using a suitable regression program in your graphing calculator.
4. Compare the results from parts 2 and 3 and explain the difference, if there is one.
5. Use your *SI* function to approximate how many years it took to issue the fifth 600 stamps. That is, determine $n - 1988$ where $SI(n) = 3,000$.
6. Research to find the actual year in which the cumulative number of U.S. postage stamps was $\geq 3,000$. Using this result, determine how many years it took to issue the fifth 600 stamps. How does this result compare to your result in part 5?
7. Research to find the year and the denomination of the first stamp issued by the U.S. Postal Service.

Part 2. Foreign Postage Stamps.

The following table shows, by ten-year intervals, the cumulative number of postage stamps issued by Great Britain, France, Norway, and the Soviet Union. (Source is the same as for the preceding table.)

Year	**Great Britain**	**France**	**Norway**	**Soviet Union**
1840	2	–	–	–
1850	6	8	–	–
1860	29	21	5	8
1870	58	47	15	25
1880	81	99	34	30
1890	122	102	43	52
1900	126	132	53	54
1910	145	160	80	86
1920	181	160	91	150
1930	209	261	153	440
1940	255	400	186	816
1950	280	642	309	1536
1960	369	983	387	2417
1970	634	1291	567	3811
1980	926	1718	771	4896

Table 3.FunProject 3.2

1. Compute an *SI* function for each of the four countries whose postage stamp data is given in Table 3.Fun Project 3.2 by:
    a. Plotting the data
    b. Recognizing the basic shape of the data
    c. Fitting a curve to the data (either graphically or using a regression program in the calculator)
2. Do all of the plots, including the plot for the United States, have the same basic shape? If so, explain why it is reasonable to expect this basic shape. If not, discuss reasons why the shapes are different.

Part 3.

Write a one or two-page essay discussing the following points.

a. Is an *SI* function always an increasing function. (Why?)

b. What factors (political, social, etc.) would cause the rate of issuance of new stamps to increase?

4. The Bones Know

(Purposes: Model a real-life situation; collect and analyze data; apply regression analysis; research on the Internet; provide a writing assignment; provide a small-group work experience.)

Introduction: Dr. Mildred Trotter (1899–1991), a distinguished physical anthropologist, worked for the U.S. government to identify skeletal remains of soldiers killed in World War II. She developed mathematical models to predict a person's height based on the lengths of certain bones. Her models are still in use today by archaeologists, police departments, and forensic scientists.

In this project, you will rediscover some of Dr. Trotter's results by analyzing data that you collect. In particular, you will analyze height versus skull length, height versus forearm length, and height versus stride length. You will do this for both a male and a female population. Using linear regression to analyze your data, you will develop six mathematical models for predicting a person's height: two based on skull length, two based on forearm length, and two based on stride length.

The validity of the models produced is dependent on the accuracy of the measurements taken and the variability of the data. (The larger the difference between the smallest and largest item, the less representative is the average or mean. This is why outliers can distort a statistical analysis.) Each group should devise its own procedures for obtaining accurate measurements. (For example, use averages to minimize distortion due to a faulty measurement.) In order to minimize the variability issue, the regression analysis for each category of data will be done twice. The first time, all the data will be used. For the second analysis, the largest two readings and smallest two readings will be removed from the data set. The group will then determine which of the two models is best.

Complete the following tasks.

1. Develop procedures for measuring head length, forearm length, stride length, and height. Your procedures should include a self-correcting provision in order to minimize the distortion of the data from a faulty measurement. All measurements are to be made in centimeters. Include a brief written description of your methods. (Your methods may be different for the different measurements.)
   a. The head length is the vertical distance from the bottom of the chin to the top of the head.
   b. The forearm length is the length from the elbow to the knobby bone at the wrist when the person's arm lies flat on a table with palm placed down.

c. The stride length is the distance from toe to toe for one step. Because a person's step lengths may vary, consider having the person take several steps and average the results. How many steps should the person take?

d. Remove shoes before measuring heights.

2. Measure head length, forearm length, stride length, and height for ten males and ten females. Record your measurements in two tables (one for males and one for females) with the column headings:

Name	Head Length (cm)	Forearm Length (cm)	Stride Length (cm)	Height (cm)

3. Use the linear regression program in your calculator to determine the line of best fit for each of the three categories: height versus head length, height versus forearm length, and height versus stride length. The equations of these lines are graphical models for predicting a person's height.

4. Repeat part 3 after first removing the two largest and two smallest measurement readings for head length, forearm length, stride length, and the corresponding height readings.

5. Check your models by measuring at least three people not in your population samples.

6. Decide which of your six models is the best predictor. Explain your reasoning.

7. Write a one-page essay on some aspect of Dr. Mildred Trotter's life and work.

5. Homecoming Parade

(Purposes: Experience designing and pricing materials for a homecoming float; application of college algebra to a real-life situation; provide a writing assignment; provide a small-group work experience.)

Introduction: This year's homecoming queen is a math major! To help her celebrate, the Mathematics Club has decided to construct a float for their Queen and her court of eight young ladies to ride on in the homecoming parade. The students have decided that to follow the theme of the homecoming they will construct a gazebo on top of a flatbed truck. The base of the gazebo will be in the shape of a regular hexagon. The ladies of the court will be sitting on chairs on the truck bed. In order to give maximum visibility to the queen, the floor of the gazebo should be level with the top of the head of the tallest lady when she is sitting in the queen's court.

There need to be steps built from the truck bed to the floor of the gazebo in order for the queen to make a stately entrance and departure. For decorative purposes, there is to be lattice work around the bottom of the gazebo except for the stair side, and for safety purposes, there is to be a railing around the outside of the gazebo except for the stair side.

The We-Haul trucking company has agreed to donate a flatbed truck for the parade. The Cover-Up carpet company has volunteered to lend a green carpet to cover the flatbed, the Bud Florist has offered to loan eight large pots of flowers, and Shade, the local umbrella company, has donated a large beach umbrella for the roof of the gazebo. The queen will be able to hold onto the umbrella pole to steady herself. The local lumber yard has agreed to provide the structural base for the gazebo provided the club designs the gazebo and buys the lumber needed for the floor, railing, lattice, and steps from them.

The lumber for the floor is to be 6-inch boards that are placed 1/2 inch apart. (A 6-inch board is actually 5.5 inches wide.) Boards come in lengths that are multiples of 2 feet. The railing should be made with 2-by-4s that are cut lengthwise. (A 2-by-4 is actually 1.5 inches thick and 3.5 inches wide.) The lumber for the latticework should be 1.8 inches wide (A 6-inch board cut lengthwise into 3 equal width strips would give strips that are 1.8 inches wide.)

Two coats of paint should be applied to the gazebo floor, railing, and steps.

Your task is to do the following.

1. Design. The flatbed of the truck measures 8 feet wide and 40 feet long. For safety purposes, the gazebo should be designed so that there is exactly one foot between the edge of the gazebo and the edge of the truck bed.
   a. Determine the dimensions of the gazebo. Draw a picture with the dimensions noted.
   b. Determine the total floor area of the gazebo.
   c. Determine the height of the gazebo floor from the truck bed.
   d. Draw a diagram to scale showing the location of the gazebo (with the stairs), the chairs for the ladies of the queen's court, and the flower pots. Indicate the dimensions on your diagram.
2. Materials.* Determine the lumber that you need in order to complete the gazebo with the minimum amount of lumber left over.
   a. The number and lengths of 6-inch boards (The boards for the floor must be at least two feet long.)
   b. The number and lengths of 2-by-4s
   c. The number and lengths of 6-inch boards for the steps
3. Diagrams. Draw diagrams for the following.
   a. The arrangement of the boards on the floor of the gazebo
   b. The latticework around the base of the gazebo
   c. The steps for the gazebo
4. Waste. Determine the amount and cost of the lumber that is left over.

5. Paint. Determine the amount of paint needed.

6. Cost
   a. Obtain prices for the lumber and paint. Determine the lumber price for two different types of lumber. Indicate the name of the lumber yard(s) or store(s) from whom you obtained your prices.
   b. Draw up a sales slip, including tax, for the materials which, includes the following:
      i. For each length of 6-inch board ordered: The number of boards, the type of lumber (pine, hemlock, spruce, etc.), price per board foot, number of board feet, and the cost.
      ii. For each length of 2-by-4 ordered: The number of 2-by-4s, the type of lumber (pine, hemlock, spruce, etc.), price per board foot, number of board feet, and the cost.
      iii. The amount of paint (quarts, gallons), color, and price.

*Lumber is measured (for sale) in board feet. The number of board feet in a piece of lumber is the surface area measured in square feet times the thickness measured in inches. That is thickness(inches) x width(feet) x length(feet).
For example,

A 6-inch board, 10-feet long measures 5 board feet (1-in x 0.5-ft x 10-ft).
An 8-foot 2-by-4 measures 16/3 board feet (2-in x 1/3-ft x 8-ft).

(Instructor, you may individualize this project to your class situation by omitting some of the requirements in order to adjust the time required.)

6. Income Tax

(Purposes: Analyze a contested political issue—taxes; develop and compare tax functions; research the present state tax schedule, conduct a numerical and graphical analysis; provide a writing assignment; provide a small-group work experience.)

Introduction: The politicians of your state are vigorously debating replacing the present state income tax with either a flat tax or a graduated tax. The adoption of a change in the income tax plan would require approval by a vote of the state's residents. The three options under consideration are

a. Present state income tax
b. Flat tax in which people pay 8% on all of their income (no deductions allowed)
c. Graduated tax in which persons pay no tax on their first $15,000, 6% on their income over $15,000 to $50,000, and 10% on their income over $50,000.

Your task is to compare the three options for people with incomes of $20,000, $40,000, $50,000, $75,000, $100,000, and $150,000 by doing the following.

1. Construct a table of values for the tax for each of the six levels of income under each of the options. For the state income tax option, deduct the standard depreciation from the six income levels to obtain the taxable income.
2. In order to have a visual comparison of the three tax options, draw (by hand) the graph of the tax dollars versus income for each option. Draw all three graphs on the same pair of axes. (Let income be measured on the horizontal axis and tax dollars be measured on the vertical axis.)
3. Use your multiplot from part 2 to estimate the income level for which the taxes are approximately the same for each of the tax options.
4. For each of the tax options, construct a tax function in terms of income. (You may need to use a piecewise defined function for some of the tax options.) Let the independent variable represent income and the dependent variable represent the corresponding tax. Check your result from part 3 with your functions.
5. If the median income in your state is $42,000 and the mean (average) income is $46,000, determine which tax option would most likely be voted in by the people. Explain your reasoning.

7. Measuring Earthquakes

(Purposes: Provide a modeling opportunity for a real-life situation (earthquakes); provide a real-life application of logarithms; provide an inquiry and writing exercise; provide for a small-group work experience.)

Introduction: A huge earthquake measuring 8.4 on the Richter scale occurred off the coast of Peru on July 23, 2001. On September 21, 1999, Taiwan was struck with a devastating earthquake that measured 7.7. Approximately 2,300 lives were lost, 8,700 were injured, and damage estimates exceeded $10 billion in U.S. dollars. Turkey experienced a 7.4 earthquake the preceding year. (In 1939, Turkey lost over 30,000 people in a giant quake measured at 8.0 on the Richter scale.) The largest quake in the United States measured 9.2 and occurred in Alaska on March 28, 1964. This was the second-largest quake ever recorded. The largest one, which measured 9.5, occurred in Chile on May 22, 1960. What causes earthquakes? What do the earthquake numbers mean? How are earthquakes compared?

Earthquakes are caused by parts of the earth's crust slipping by each other. Geophysicists believe the slippage is the result of stress buildup through the movement of the earth's crust. The outer forty or so kilometers of the earth is called the crust. It is made up of twelve or more tectonic plates that float on the semisolid rocks of the upper mantle of the earth. The tremendous heat that is built up in the earth's core and mantle causes rocks to soften and to lose density. These less-dense rocks attempt to rise to the surface and in doing so create convection currents. These, in turn, cause the plates to move independently, to occasionally collide, and sometimes to override each other at the edges. The colder, firmer rocks making up the crust do not deform like the hot rocks of the core and mantle. Thus when stresses caused by plates colliding or overriding builds to the breaking strength of rocks, fractures, plate slippage, and violent shock waves result. Approximately 6,000 earthquakes occur every year, although about 90% of them are too small to be noticed without special instruments, and only about fifteen cause major damage.

Earthquakes are measured by the length of the rupturing fracture, called the fault, in the earth's surface and the amount of slippage. Small earthquakes have faults that are tens of hundreds of meters long and a few centimeters of slippage, whereas faults for large earthquakes are measured in hundreds of kilometers and slippage in meters. Seismographs record a zigzag pattern showing the amplitude of ground oscillations beneath the instruments. Very sensitive seismographs can detect and measure earthquake motion in any part of the world. In 1935, Charles Richter at the California Institute of Technology developed a magnitude scale for earthquakes. His scale, known today as the Richter Scale, is a function that maps magnitude of an earthquake to the common logarithm of the size,

$$magnitude = \log(size).$$

A better sense of the magnitude of an earthquake is obtained by comparing magnitudes of earthquakes with similar rating. For instance to determine how much larger a 7.9 earthquake is than a 6.6 earthquake, we compare their sizes

$$\frac{10^{7.9}}{10^{6.6}} = 10^{1.3} = 19.95.$$

Thus a 7.9 earthquake is almost 20 times as large as a 6.6 earthquake.

The difference between a 6.6 earthquake and a 7.9 earthquake is even more striking when their strengths (energies) are compared. The conversion function from magnitude to energy for earthquakes is $\log(E) = 1.5M$ or $E = 10^{1.5M}$ where E denotes energy and M magnitude. Thus one unit of magnitude represents $10^{1.5}$ units of energy. Using this conversion, we see that a 7.9 earthquake is 89.13 times as strong as a 6.6 earthquake. The strength or energy rating is more important than the magnitude rating, as it is the energy that causes the damage in an earthquake.

Frequency of Earthquakes (Worldwide)

Based on Observations since 1900

Descriptor	**Magnitude**	**Strength**	**Average Annually**
Great	8 or higher		1
Major	7–7.9		18
Strong	6–6.9		120
Moderate	5–5.9		800
Light	4–4.9		6,2000 (estimated)
Minor	3–3.9		49,000 (estimated)
Very Minor	< 3.0		Magnitude 2–3: about 1,000/day Magnitude 1–2: about 8,000/day

Table 3.FunProject 7.1

Number of Earthquakes in the United States, 1990–1999:

Magnitude	**1990**	**1991**	**1992**	**1993**	**1994**	**1995**	**1996**	**1997**	**1998**	**1999**
8.0–9.9	0	0	0	0	1	0	0	0	0	0
7.0–7.9	0	1	2	0	1	0	2	0	0	3
6.0–6.9	3	6	9	9	5	7	6	6	3	5
5.0–5.9	72	50	84	69	67	49	109	63	62	52
4.0–4.9	283	255	404	269	331	355	621	362	411	360
3.0–3.9	621	701	1713	1115	1543	1050	1042	1072	1053	1388
2.0–2.9	411	555	996	1007	1194	820	652	759	742	814
1.0–1.9	1	3	5	7	2	0	0	2	0	0
0.1–0.9	0	0	0	0	0	0	0	0	0	0

Table 3.FunProject 7.2

Source: U.S. Geological Survey, National Earthquake Information Center.
wwwneic.cr.usgs.gov/neis/eqlists/eqstats

Your task is to do the following.

1. Fill in the strength column in Table 3.FunProject 7.1.
2. Compare the magnitudes and strengths of a 8.5 earthquake to a 5.4 earthquake.
3. How much stronger was the 1960 Chilean earthquake than the 1999 Taiwan earthquake?
4. Determine the seismograph's amplitude reading as a function of an earthquake's magnitude

rating.

5. Determine the seismograph's amplitude reading as a function of an earthquake's energy rating.

6. Complete the following table: Magnitude versus Ground Motion and Energy

Magnitude Change	Ground Motion Change (Displacement)	Energy Change
	10.0 times	
	3.2 times	
	2.0 times	
	1.3 times	

Table 3.FunProject 7.3

Answer the following based on your table.

   a. The ground change for a 7.2 magnitude earthquake is how many times the ground change for a 6.2 magnitude earthquake?
   b. The energy release for a 7.2 magnitude earthquake is how many times the energy release for a 6.2 earthquake?

7. From Table 3.FunProject 7.2, determine the average number of earthquakes from 1990 to 1999 by magnitude. Plot the resulting averages versus magnitude and then fit a curve to your scatter plot.

8. Using the results from part 7, develop a scenario for the number of earthquakes, by size, for the United States in the year 2000.

9. In defining his magnitude scale, Charles Richter wanted to express large numbers in ways that people could readily understand. Write a short essay explaining why you think Charles Richter chose to use the logarithm function in defining his scale rather than some other function, such as a polynomial or exponential.

8. How Many Times Must I Pump?

(Purposes: Apply college algebra to a real-life situation that is familiar to students; work with ratios; convert from one measurement system to another; provide a writing assignment; provide a small-group problem solving experience.

Introduction: Bicycles have provided a means of transportation and sport for hundreds of years. From bike-riding couriers in New York City to bike traffic in Bombay, India, to kids "riding their bikes" to Lance Armstrong winning his fifth Tour De France, bikes play a meaningful role in the majority of the world's cultures. Bike shops as well as discount chains offer a wide variety of bikes—mountain, racing, n -speed, etc. In this Fun Project, you will explore gear ratios of multispeed bikes and use these ratios along with wheel size to model distance traveled in terms of the number of pumps of the pedals. A pump is one revolution of the pedal sprocket.

The group should select a multispeed bike to analyze, preferably one with two pedal sprockets and several rear-wheel sprockets, and then complete the following tasks.

1. Count the number of teeth in each pedal sprocket and each rear-wheel sprocket.
2. Develop a table to display the gear ratios of the number of teeth in a pedal sprocket divided by the number of teeth in a rear-wheel sprocket. Let the row headings be the pedal sprockets and the column headings be the rear-wheel sprockets.
3. What ratio gives the lowest speed? What ratio gives the highest speed? Explain your answers.
4. Measure the circumference of the rear tire.
5. Theoretically determine how far the bike travels with one pump when the gears are aligned for the lowest speed.
   Theoretically determine how far the bike travels with one pump when the gears are aligned for the highest speed.
6. Measure the distance the bike travels with one pump when the gears are aligned for the lowest speed.
   Measure the distance the bike travels with one pump when the gears are aligned for the highest speed.
7. Compare the results for parts 5 and 6. Explain any unexpected differences.
8. Using the third lowest gear, count the number of pumps required to travel 50 feet. Compute the theoretical number of pumps required.

9. Using the third lowest gear, determine the theoretical number of pumps required to travel 50 feet if the rear wheel is 24 inches in diameter. (If the bike you are using has 24-inch-diameter wheels, then change the diameter to 26 inches.)
10. Develop a function to model the distance traveled in terms of gear ratio, rear-wheel diameter, and number of pumps. That is, gear ratio, rear-wheel diameter, and number of pumps are the inputs, and distance traveled is the output.
11. Write an explanation of why there are fewer pedal sprockets than rear-wheel sprockets.
12. Write an explanation of why the pedal sprockets are larger than the rear-wheel sprockets.

9. Water, Water Everywhere, But Will there be Enough to Drink?

(Purposes: develop an awareness of a growing national problem and an appreciation for the importance and difficulty of establishing a water resource policy; provide an inquiry and writing exercise; provide small-group work experience.)

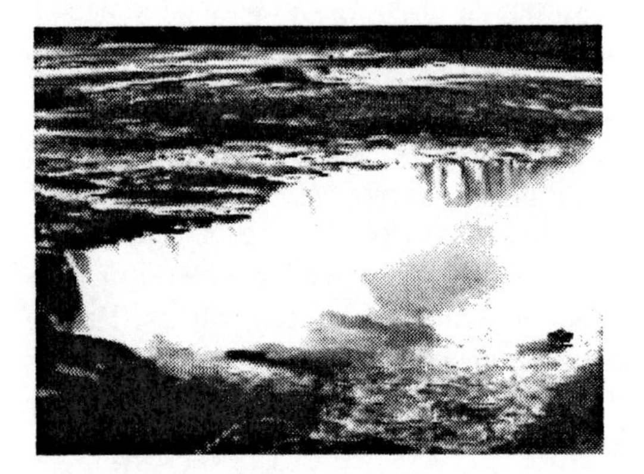

(The information in this article was obtained from the National Geographic, November 1993, Special Edition "Water, The Power, Promise, and Turmoil of North America's Fresh Water" and the August 26, 2002 New York Times article "Saving Water, U.S. Farmers are Worried They'll Parch.")

Water is a renewable, but limited resource. The cycle of evaporation, precipitation, evaporation, precipitation, etc. has existed for the life of our planet. In the United States, precipitation amounts to approximately 4 trillion gallons a day most of which is consumed by evaporation or run-off into reservoirs, rivers, and absorption into vegetation and underground aquifers. On some global scale, the quantity of water contained in vegetation, aquifers, natural and man-made reservoirs, rivers, icebergs, oceans, and in the atmosphere is constant. Of the total quantity, 97% is salt water and two-thirds of the remaining is held in icebergs. Thus only 1% of the world's water is available as fresh water.

Agriculture is the largest user of fresh water. It accounts for 78% of the United State's average usage of 1,300 gallons per person per day. The remaining 22% is consumed by urban needs. Irrigation, the largest component of agriculture's water usage, amounts to approximately 137 billion gallons per day. A frequently used measure in connection with agriculture's use of water is the acre-foot. One acre-foot is the quantity of water needed to cover one acre, one foot deep in water.

California receives, on average, 24 inches of rain per year. This amounts to approximately 193 million acre-feet of water. Of this amount, run-off accounts for approximately 70 million acre-feet, 43 million acre-feet are captured in 1,300 reservoirs, and the remainder evaporates.

Aquifers act like huge underground sponges for storing water. In Florida, 4.5 billion gallons of water are pumped out of aquifers on a daily basis. Nine out of ten Floridians take their drinking water from aquifers, which is twice the national average. Compactification of an aquifer occurs when more water is drawn out than is absorbed, just as a sponge contracts in size when it dries out. When this happens, some of the aquifer's structure breaks down preventing the aquifer from regaining its original size even when sufficient water is available. For example, the Central Valley aquifer in California has lost about 20 million acre-feet of water storage through compactification. The situation in Mexico City provides a more extreme example. The continuing compactification of the aquifer under Mexico City is causing the level of the city to drop between one and two inches per year. A cost of holding water in reservoirs, cementing bottoms and sides of flood drainage systems, and roads, parking lots, etc. of urbanization is to reduce the amount of run-off available for absorption into aquifers.

The increasing demands for water for agriculture, urbanization, and environmental uses that accompany increases in population threaten the aquifer system. For example, in some places in northwest Texas the depth of the Ogallala aquifer has decreased over seventy-five percent. At present the 640,000 acre-feet per year of water drawn from the Edwards aquifer by the city of San Antonio and surrounding agricultural lands is approximately balanced by its present rate of absorption. However, San Antonio is growing and with it the need for more water for both urbanization and agriculture. Thus there is growing pressure to increase the output of the Edwards aquifer while the input is fixed.

Limited quantities of water and increasing population emphasize the importance of regulations, storage, and distribution of fresh water. For example the Texas Water Board estimates the population of Texas will approximately double by 2050, but the amount of water from present sources will decrease by nineteen percent. Resolving water issues and implementing conservation measures is critical for individuals, cities, states, and nations. The difficulty of doing this is compounded by the fact that some aquifers lie under more than one state, such as the Ogallala aquifer that lies under parts of eight states. Other aquifers lie under more than one country. The problem is that an aquifer is like a large can of soda. If everyone who has a straw feels they have right to stick in their straw and start drinking, we may run out of soda.

"When the well's dry, we will know the worth of water." Benjamin Franklin

Questions:

1. What is the average number of gallons of fresh water used in agriculture to support one individual for one day? (.78 x 13000)
2. How many gallons of water are in one acre-foot of water? (~326,000)
3. If the population (1993) of California is 31 million people and 80% of the captured water is used for agriculture, how many gallons of water is used for agriculture per person per day in California? (~1,700 gal.s)
4. Compile a table showing how water is used in your home (or dorm) providing description and amount for each type of usage. (For example, how much water does it take to flush a toilet? How many times is a toilet in your home flushed in a day?)
5. How much water is lost in one year through a faucet that drips every 30 seconds? (1 drop = 1/480 fluid ounces, 16 fluid ounces = 1 pint, 8
6. Write an essay on the problems of restricting the amount of water that individuals, industry, farmers, etc. can withdraw from an aquifer.

David Blackwell

David Blackwell was born in Centralia, Illinois, in 1919. At age sixteen, he entered the University of Illinois, Champaign-Urbana where he received his bachelor's degree in 1938, his master's degree in 1939, and his Ph.D. in 1941, all in mathematics. At the age of 22, he had earned a Ph.D. degree in mathematics and had been awarded a Rosenwald Fellowship to attend the Institute of Advanced Study, Princeton, New Jersey. Thus was launched a career of fifty-plus years as a world-class mathematician.

Dr. Blackwell taught at Howard University where he chaired the Mathematics Department from 1947 to 1954. He also taught at Stanford University, Clark College (now Clark Atlanta University), and Southern University before moving to the University of California to chair the Statistics Department from 1957 to 1961. He has been recognized and acclaimed as a gifted teacher and productive scholar (80 publications prior to retirement in 1954). Dr. Blackwell has received Honorary Doctorates of Science from Harvard, Yale, Howard, Carnegie-Mellon, Michigan State, Syracuse, and Southern Illinois Universities; University of Southern California, University of Warwick, National University of Losotho, and Amherst College. He has served as president of the Institute of Mathematical Statistics, International Association for Statistics in the Physical Sciences, Bernoulli Society, and vice-president of the International Statistical Institute, American Statistical Association, and the American Mathematical Society.

Two of the highest honors bestowed on him have been his election to the National Academy of Science (first African-American mathematician elected) and his election to the American Academy of Arts and Sciences. In 1994 the National Association of Mathematicians (NAM) established the David Blackwell Lectureship program to honor him for his many contributions to education.

Chapter 4 Modeling

Modeling is a means of transforming information into knowledge. A mathematical model of a situation is a mathematical description of the situation. For example, the equation $y = 2x,\ x \geq 0$ is an analytical model of the set of points in the Cartesian plane whose distance from the positive x-axis is twice the distance from the y-axis. The graph of this equation is a graphical model of the specified set. Recall that in Chapter 3 we graphically modeled data sets by fitting a curve to the corresponding scatter plots. We then used the resulting function to interpolate between the given data points as well as to make predictions. For instance, in Example 3.3.2, we extracted a stopping distance function by fitting a curve to the scatter plot of the data taken from a driver's manual. We then used the function to determine a stopping distance for a speed that was not listed in the driver's manual. This was an example of how a model is used to provide insight and understanding beyond the original statement of facts. In this chapter, we focus on analytical rather than on graphical models. Furthermore we intentionally select situations from a wide variety of disciplines to show the wide applicability of mathematics to everyday life. A paradigm for many of our models is

(New Situation) = (Old Situation) + (Change).

4.1 Mathematical Modeling

Example 4.1.1.

Chuck's parents offer him the option of two financial graduation presents.

Option 1: Receive \$1 at graduation, \$2 a month later, \$4 two months later, \$8 three months later, and so on, the amount doubling each month for a total of twelve months.

Option 2: Receive a one-year savings account with an initial investment of \$1,000 that earns 1% interest per month (12% yearly interest).

Which option generates the largest amount of money at the end of one year?

Solution:

Chuck forms the following table to compare the two options.

Month	Monthly Income Under Option 1	Accumulated Value of Option 1	Accumulated Value of Option 2
1	$1	$1	$1,010.00
2	$2	$3	$1,020.10
3	$4	$7	$1,030.30
4	$8	$15	$1,040.60
5	$16	$31	$1,051.01
6	$32	$63	$1,061.52
7	$64	$127	$1,072.14
8	$128	$255	$1,082.86
9	$256	$511	$1,093.69
10	$512	$1,023	$1,104.62
11	$1,024	$2,047	$1,115.66
12	$2,048	$4,095	$1,126.83
Total Amount		**$4,095**	**$1,126.83**

Table 4.1.1

The table clearly indicates that Chuck would be much better off choosing Option 1. ▲

Although the preceding table provides an answer to the stated question, Chuck might like more information than is given by the table. For example, he might like to know the comparisons of the two options for different numbers of months, say twenty-five months; or he might want to know what would happen if the initial investment in Option 2 was different than $1,000; or if Option 1 started with $2 instead of $1; or if the interest rate in Option 2 was different. What Chuck would really like is a mathematical model of each option that would enable him to analyze the situation under different scenarios rather than just answering the specific question.

A mathematical model of a real-world situation is a mathematical description of the situation that enables one to analyze the situation in greater depth than just answering a specific question. For example, when we developed a function from a data set, we formed a mathematical model that would allow for interpolations and predictions. The modeling process consists of three stages as illustrated by the following diagram.

Modeling Process

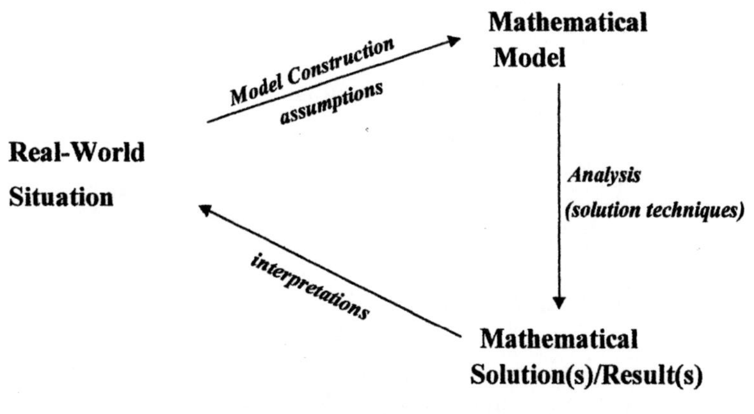

Mathematical Model

Model Construction
assumptions

Real-World Situation

Analysis
(solution techniques)

interpretations

Mathematical Solution(s)/Result(s)

Figure 4.1.1

Because real-world situations are usually far too complicated to be modeled exactly, simplifying assumptions are often made in constructing a model. For example in Chuck's problem, for Option 2 we assumed that each month had the same number of days and we also rounded results to full cents. The analysis stage of the modeling process involves computations similar to those encountered in exercises in mathematics texts. Today, most of these computations are done with a calculator or computer. The results of these computations, although correct mathematically, need to be interpreted in light of the real-world situation. Recall there were two mathematical solutions, $t = -\frac{1}{2}$ and $t = 2$ seconds, to the question in Example 3.6.4 as to how long it would take a diver to hit the water after springing off a diving board. Although both solutions were correct mathematically, the negative solution did not make sense in the given situation and was discarded. Thus the third stage of the modeling process is necessary to interpret the mathematical solutions in the environment of the given situation.

We now return to Example 4.1.1 and use Chuck's option 1 to motivate the development of a geometric series model and his option 2 to motivate a recursive sequence model.

Geometric Series

Chuck's return on option 1 is the sum

$$1 + 2 + 2^2 + 2^3 + 2^4 + \ldots + 2^{11}.$$

This is a special type of summation, called a **geometric series**. A geometric series is a sum of terms in which the first term is one and each term after the first one is a fixed multiple of the preceding term. For example, the third term is two times the second term and the fifth term is two times the fourth term. Thus, denoting the common multiplier by r, a geometric series has the form

$$1 + r + r^2 + r^3 + r^4 + \ \ldots\ + r^{n-1}.$$

We now derive a formula for the sum of a geometric series. Let S denote the desired sum. Thus

$$S = 1 + r + r^2 + r^3 + r^4 + \ \ldots\ + r^{n-1}.$$

(Note that if $r = 1$, then $S = n$. Thus we restrict r to be different from one.)

Multiply both sides of this equation by r obtaining

$$rS = r + r^2 + r^3 + r^4 + \ ... \ + r^{n-1} + r^n.$$

Now subtract the second equation from the first equation:

$$\begin{array}{rl} S = & 1 + r + r^2 + r^3 + r^4 + \ ... \ + r^{n-1} \\ - \ rS = & \quad\ \ r + r^2 + r^3 + r^4 + \ ... \ + r^{n-1} + r^n \\ \hline S - rS = & 1 - r^n \end{array}$$

Factor the S out of the left-hand side to obtain

$$(1 - r)S = 1 - r^n.$$

Divide both sides by $(1 - r)$ to obtain the expression for the sum.

$$S = \frac{1 - r^n}{1 - r} \quad \text{for } r \text{ not equal to one.}$$

We repeat this result for emphasis. The sum of a geometric series with common multiplier r is

$$S = 1 + r + r^2 + r^3 + r^4 + \ ... \ + r^{n-1} = \frac{1 - r^n}{1 - r} \quad \text{for } r \text{ not equal to one}$$

Note that the first term of the series is one and the exponent in the summation formula is one more than the largest exponent in the geometric series. Also the exponent in the summation formula is the number of terms in the geometric series.

Query 1.

What happens to the summation formula for a geometric series when $r = 1$? What is the value of a geometric series when $r = 1$?

We generalize this result by considering series in which each term after the first one is a fixed multiple of the preceding term, but the first term is not equal to one. The procedure is to factor the first term out of each term. This gives a product of two factors: the first factor is the first term of the series, and the second factor is a geometric series. We illustrate with two examples.

Example 4.1.2.

$$\begin{aligned} c + cr + cr^2 + cr^3 + cr^4 + ... + cr^{n-1} &= c(1 + r + r^2 + r^3 + r^4 + \ ... \ + r^{n-1}) \\ &= c\left(\frac{1 - r^n}{1 - r}\right) \end{aligned}$$

▲

Example 4.1.3.

$$\begin{aligned} cr^2 + cr^3 + cr^4 + cr^5 + ... + cr^{n-1} &= cr^2(1 + r + r^2 + r^3 + r^4 + \ ... \ + r^{n-3}) \\ &= cr^2\left(\frac{1 - r^{n-2}}{1 - r}\right) \end{aligned}$$

▲

To develop a geometric series model for Chuck's option 1, we begin by defining variables and stating conditions

n = number of months
$v(n)$ = value after n months
$v(0) = 1$ (initial investment)
$r = 2$ (geometric ratio)

The model is

$$v(n) = c\left(\frac{1-r^n}{1-r}\right) \text{ and } v(0) = 1, \text{ where } c = 1,\ r = 2, \text{ and } n = \text{number of months.}$$

Thus the value of Chuck's option 1 is: $v(12) = \frac{1-2^{12}}{1-2} = 2^{12} - 1 = 4{,}095.$

Query 2.

What is the value of option 1 if the doubling process begins with $0.50 the first month?

Recursive Sequence

Chuck's return on option 2 is the balance in the savings account after 12 months when the initial investment is $1,000 and the monthly interest rate is 1%.

To develop a recursive sequence model, we begin by defining variables and stating the conditions. Let

n = number of months
$a(n)$ = balance in the account after n months.
$a(0) = 1{,}000$ (initial investment)
$r = .01$ (monthly interest rate)

The next part is dependent on the fact that the balance in the account only changes at the end of each month. (Interest is posted to the account after the close of business on the last day of the month.) This means that we can model the situation by expressing next month's beginning balance in terms of this month's beginning balance plus the change that takes place during this month (that is, interest received). Thus our model is

(Next month's balance) = (This month's beginning balance) + (Interest received)

or symbolically

$$a(n) = a(n-1) + ra(n-1) \quad \text{with} \quad a(0) = 1{,}000.$$

This model is a recursive sequence, meaning that the value of each term after the first is dependent on a previous term (see Section 2.4).

The next step in the modeling process is to solve the model. We begin by computing $a(n)$ for the first few values of n. This is called **iterating**. Our purpose is to recognize a pattern and gain a feeling for how the solution should develop. We express each iterate in terms of the initial amount, the only amount that we know.

$a(0) = a(0)$

$a(1) = a(0) + ra(0) = (1+r)a(0)$ Original amount plus interest

$a(2) = a(1) + ra(1) = (1+r)a(1)$ Amount after 1st month plus interest
$= (1+r)[(1+r)a(0)]$ Substitute for $a(1)$
$= (1+r)^2 a(0)$

$a(3) = a(2) + ra(2) = (1+r)a(2)$ Amount after 2nd month plus interest
$= (1+r)[(1+r)^2 a(0)]$ Substitute for $a(2)$
$= (1+r)^3 a(0)$

$a(4) = a(3) + ra(3) = (1+r)a(3)$ Amount after 3*rd* month plus interest
$= (1+r)[(1+r)^3 a(0)]$ Substitute for $a(3)$
$= (1+r)^4 a(0)$

- - - - - - - - - -

- - - - - - - - - -

If the pattern is not clear, the reader should compute $a(5)$, $a(6)$, and so on until the pattern is clear. Assuming that the pattern is clear, we note that the exponent on $(1+r)$ is the same as the input variable (the independent variable). Thus, letting n, represent an arbitrary month, we write

$$a(n) = (1+r)^n a(0).$$

A recursive sequence for Chuck's option 2 where $r = 0.01$ is the monthly interest rate, $a(0) = 1{,}000$ is the initial value, n is the number of the month, and $a(n)$ is the balance in the account after n months is

$$a(n) = (1+0.01)^n 1{,}000 \text{ or } a(n) = 1{,}000(1.01^n).$$

Therefore the value of Chuck's option 2 is $a(12) = 1{,}000(1+.01)^{12} = 1{,}126.83$.

The solution form, $a(n) = 1{,}000(1.01^n)$, is called a **closed form** solution. Note that it allows us to compute $a(n)$ directly without iterating n times. The models for option 1 and option 2 enable us to answer the particular question posed in Chuck's problem and, in addition, allow us to consider the question under alternative conditions (for example, different number of months, different interest rate, different initial investment, different doubling rates).

Query 3.
What is the value of Chuck's option 2, if the initial investment is \$2,000 and the monthly interest rate is 0.75%?

The following pharmacology example involves the local anesthetic Lidocaine. This anesthetic has largely replaced Novocaine in the field of dentistry. Lidocaine is less likely to cause allergic reactions than novocaine and can be dispensed in several formulations (cream, ointment, jelly, solution, aerosol, as well as injection). In contrast, Novocaine is only dispensed by injection.

Example 4.1.4.

Wanda's dentist injects a 500-mg dose of Lidocaine into her gum before removing one of her wisdom teeth. Assume that each hour following the injection, 20% of the Lidocaine present at the beginning of that hour has been eliminated from her body. Develop a recursive sequence model for the amount of Lidocaine remaining in Wanda's body after n hours. How much Lidocaine is in Wanda's system after 5 hours?

Solution:

Although we know the elimination of Lidocaine is continuous, we shall make the simplifying assumption that an hour's elimination takes place at the end of the hour. This allows us to formulate a discrete model using our paradigm

(New Situation) = (Old Situation) + (Change).

We now define our variables and state the initial condition.

n = number of hours
$L(n)$ = amount (mg) of Lidocaine in Wanda's body after n hours
$L(0) = 500$ (initial dose)
$r = -0.2$ (elimination rate)

Using our model paradigm, we have the recursive sequence

$$L(n) = L(n-1) - .2\,L(n-1).$$

This equation can be simplified to $L(n) = (1 - .2)\,L(n-1) = .8\,L(n-1)$. Our model is now

$$L(n) = .8\,L(n-1) \text{ with } L(0) = 500.$$

The next step of the modeling process is to solve the model. We begin by computing $L(n)$ for the first few values of n.

$L(0) =$ initial dose

$L(1) = .8\,L(0)$ — Initial amount less first hour's elimination

$L(2) = .8\,L(1) = .8[.8\,L(0)] = .8^2L(0)$ — First hour's amount less second hour's elimination

$L(3) = .8\,L(2) = .8[.8^2L(0)] = .8^3L(0)$ — Second hour's amount less third hour's elimination

$L(4) = .8\,L(3) = .8[.8^3L(0)] = .8^4L(0)$ — Third hour's amount less fourth hour's elimination

– – – – – – – –

– – – – – – – –

The reader should continue this iteration process until the pattern of $L(n) = .8^nL(0)$ is clear.

Because $L(0) = 500$, a closed form solution to our problem is

$$L(n) = 500\,(.8^n).$$

After 5 hours, Wanda has $L(5) = 500\,(.8^5) = 163.84$ mg of Lidocaine in her system. ▲

The plot of the points $(n, L(n))$ for $n = 1,2,3,\ldots 10$ suggests that $L(n)$ is a decreasing function that approaches zero asymptotically.

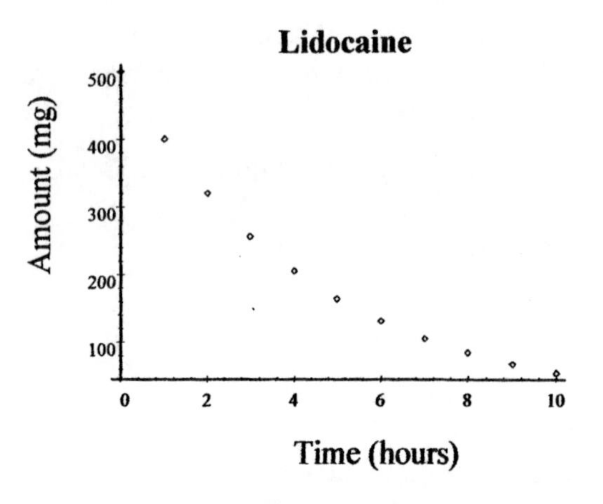

Figure 4.1.2

Let us generalize the discrete closed form solution to a continuous solution by extracting a function from this scatter plot. That is, by fitting a curve to the points as we did in Section 3.6. Explain why choosing $f(t) = 500\,(.8^t)$ is a reasonable conjecture for the function.

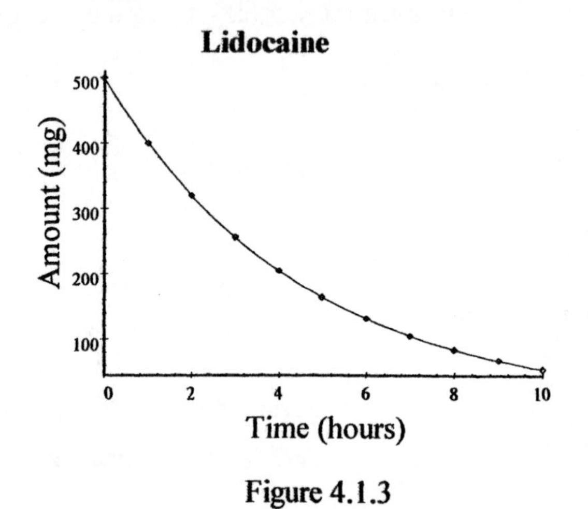

Figure 4.1.3

Suppose the dentist told Wanda that she would not feel any numbness after the lidocaine in her body fell below 150 mg. To determine how long this would be, we superimpose the line $y = 150$ on the lidocaine plot and note where the curves intersect.

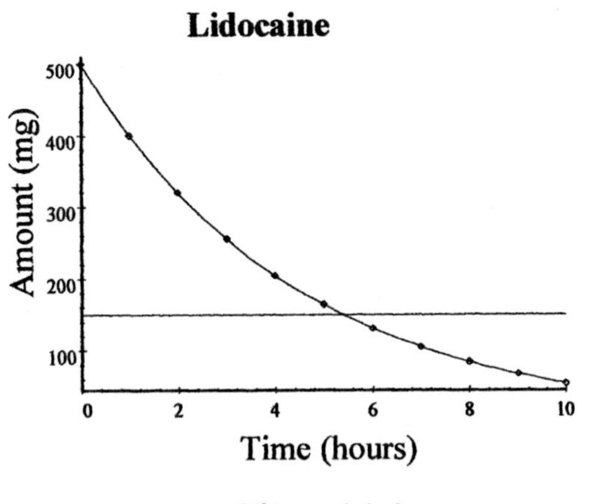

Figure 4.1.4

Zooming in on the intersection gives an approximate answer of 5 hours and 40 minutes. The value can be obtained analytically using logarithms as we show in the next section.

Developing a suitable model is a very creative process. One approach that is particularly effective is to consider the situation to be modeled in terms of a sequence of discrete periods in which change occurring during a period is recorded at the end of the period. The result is a recursive sequence. This approach was illustrated in modeling option 2 of Chuck's problem. An advantage to a recursive sequence model is that a numerical solution can always be found by iterating the recursive sequence, even when a symbolic solution does not exist.

Another useful approach to modeling is to create a physical model and actually take physical measurements. (This would be a reasonable approach in Exercises 12 and 14 of this section.) As noted previously, simplifying assumptions (for example, all months have exactly 30 days) are often made in order to formulate a model. Thus a model is only an approximation of the true situation and therefore it is important to check that results are reasonable in terms of the actual situation.

We now return to geometric series and look at three examples that illustrate summing a generalized geometric series.

Example 4.1.5 (Geometric Series).

Sum the series: $3 + 3 * 2 + 3 * 2^2 + 3 * 2^3 + 3 * 2^4 + \ldots + 3 * 2^{10}$

Solution:

The procedure is to rewrite the given expression in a way that will allow us to use the formula for the sum of a geometric series. The first step in doing this is to factor a 3 out of each term and obtain the expression

$$3(1 + 2 + 2^2 + 2^3 + 2^4 + \ldots + 2^{10}).$$

The expression inside the parenthesis is a geometric series with common multiplier $r = 2$ and largest exponent 10. Thus the sum is

$$3\,\frac{1-2^{11}}{1-2} = 3(2^{11} - 1) = 3 * 2{,}047 = 6{,}141. \qquad \blacktriangle$$

Example 4.1.6 (Geometric Series).

Sum the series $(\frac{1}{2})^3 + (\frac{1}{2})^4 + (\frac{1}{2})^5 + (\frac{1}{2})^6 + (\frac{1}{2})^7 + (\frac{1}{2})^8$.

Solution:

The first step is to factor $(\frac{1}{2})^3$ out of each term and obtain the expression

$$(\tfrac{1}{2})^3[1 + (\tfrac{1}{2}) + (\tfrac{1}{2})^2 + (\tfrac{1}{2})^3 + (\tfrac{1}{2})^4 + (\tfrac{1}{2})^5].$$

The expression inside the brackets is a geometric series with common multiplier $r = \frac{1}{2}$ and largest exponent 5. Thus the sum is

$$\begin{aligned}(\tfrac{1}{2})^3[\frac{1-(\frac{1}{2})^6}{1-\frac{1}{2}}] &= (\tfrac{1}{2})^3 * 2 * [1-(\tfrac{1}{2})^6] \\ &= (\tfrac{1}{2})^2[1-(\tfrac{1}{2})^6] \\ &= (\tfrac{1}{2})^2 - (\tfrac{1}{2})^8. \quad \blacktriangle\end{aligned}$$

(The reader should explain each of the steps in the preceding computation.)

Example 4.1.7 (Geometric Series).

Express 0.4444444 ... as an infinite geometric series and show that 0.4444444 ... $= \frac{4}{9}$.

Solution:

We start by expressing our infinite decimal (0.4444...) as an infinite sum of fractions in the following manner:

$$\begin{aligned}0.4444444\ldots &= \frac{4}{10} + \frac{4}{10^2} + \frac{4}{10^3} + \frac{4}{10^4} + \frac{4}{10^5} + \ldots + \frac{4}{10^n} + \ldots \\ &= \frac{4}{10}(1 + \frac{1}{10} + \frac{1}{10^2} + \frac{1}{10^3} + \ldots + \frac{1}{10^n} + \ldots)\end{aligned}$$

The right-hand side is an *infinite* geometric series with $r = \frac{1}{10}$ and constant $c = \frac{4}{10}$.

We apply the geometric series summation formula to the first n terms where n is an arbitrarily large integer. Thus

0.4444444 ... is approximately $\frac{4}{10} * \frac{1-(\frac{1}{10})^n}{1-\frac{1}{10}}$ where n is an arbitrary large integer

$$= \frac{4}{10} * \frac{10}{9}[1-(\tfrac{1}{10})^n]. \text{ (Why?)}$$

Because $(\frac{1}{10})^n \to 0$ as n becomes infinite, we have the result

$$0.4444444\ldots = \frac{4}{10} * \frac{10}{9} = \frac{4}{9}. \quad \blacktriangle$$

Query 4.

What fraction is represented by the infinite decimal 0.99999 ... ?

Exercises 4.1

1. (Computational Skill) Simplify the following expressions.

   a. $\dfrac{1-(\frac{1}{5})^8}{1-\frac{1}{5}}$

   b. $\dfrac{1-(\frac{1}{4})^9}{1-\frac{1}{4}}$

   c. $\dfrac{1-(-\frac{1}{3})^{10}}{1-(-\frac{1}{3})}$

2. (Computational Skill) Simplify the following expressions.

   a. $(1+.1)(1+.1)(1+.1)(1+.1)150$

   b. $6+(6)(\frac{1}{2})+(6)(\frac{1}{2})^2+(6)(\frac{1}{2})^3+(6)(\frac{1}{2})^4$

   c. $(\frac{1}{2})^2+(\frac{1}{2})^3+(\frac{1}{2})^4+(\frac{1}{2})^5+(\frac{1}{2})^6+(\frac{1}{2})^7+(\frac{1}{2})^8+(\frac{1}{2})^9$

3. (Calculator Skill) Plot the recursive sequence $L(n) = 500(.8^n)$ for $n = 0,1,2,...10$. Check your plot against the plot in Example 4.1.4.

4. For each of the following recursive sequences, list the first six elements of the sequence.

   a. $b(n) = 2b(n-1),\ b(0) = 3.$

   b. $c(n) = c(n-1)+3,\ c(0) = 2.$

   c. $d(n) = d(n-1)+d(n-2),\ d(0) = 1,\ d(1) = 2.$

5. Rose stands exactly 1 yard away from a door. If she takes 10 steps toward the door and each of her steps takes her halfway to the door, how far from the door is she after the tenth step?

6. Determine a closed form solution to the recursive sequence: $b(n) = 2b(n-1),\ b(0) = 3$, then compute the value of $b(20)$.

7. Assume that a pair of fruit flies produces three pairs of fruit flies every hour. If you start with one pair of fruit flies, how many pairs will you have after one day?

8. Al has two parents and each of his parents had two parents and each of these (grandparents) had two parents and so on. How many ancestors does Al have over the past 30 generations?

9. What fraction is represented by the infinite decimal 0.11111...?

10. What fraction is represented by the infinite decimal 0.33333...?

11. What fraction is represented by the infinite decimal 0.34111...?

12. Develop a model for the length for a single piece of string required to tie up a box, like a shoe box. The string is to make one turn around the long part of the box and one turn around the short side of the box and then the two ends are to be tied. Solve your model.

13. Develop a model for doubling the volume of a cereal box with dimensions 11x8x3 in. if the thickness of the box is to remain 3 in. and the ratio of the height to the width of the box is to

remain the same.

14. Develop a model for the volume of a closed box that can be formed from a rectangular sheet of cardboard. Assume that the lid of the box has 2 sides and 1 front flap, each of which is the depth of the box. You can consider cutting out pieces in order to form the corners or incorporate a folding pattern (as in a pizza box). Let L = length of the cardboard, W = width of the cardboard, and D = the depth of the box. Using your model, determine the maximum volume for a box that can be folded from an $8\frac{1}{2}$ x 11 sheet of cardboard.

15. Develop a recursive sequence model for Chuck's option 2 if interest is compounded quarterly rather than monthly. The quarterly interest rate is $\frac{\text{annual percentage rate}}{4} = \frac{12\%}{4} = 3\%$. Determine the value of this modified option.

16. Develop a recursive sequence model for Chuck's option 2 if interest is compounded semiannually rather than monthly. The semiannual interest rate is $\frac{\text{annual percentage rate}}{2} = \frac{12\%}{2} = 6\%$. Determine the value of this modified option.

17. Develop a recursive sequence model for Chuck's option 2 if interest is compounded daily rather than monthly. (Assume that a year has exactly 360 days.) The daily interest rate is $\frac{\text{annual percentage rate}}{360} = \frac{12\%}{360} = .03\%$. Determine the value of this modified option.

4.2 Modeling (Business)

There are numerous financial situations that can be modeled using our paradigm, such as

(New Situation) = (Old Situation) + (Change).

Savings accounts, car loans, credit card debts, mortgages, annuities, state-sponsored college education accounts, and installment purchases are samples of real-life situations that can be modeled with recursive sequences, based on the preceding paradigm. Many of the financial models that are pre-programed into calculators are based on recursive sequences. In the next example, establishing a college education account, is used to motivate generalizing the recursive sequence model for Chuck's option 2 in Example 4.1.1 to include yearly deposits.

Example 4.2.1 Sue's Plan.

Sue plans to set up a college education account for her new born daughter. A local bank agrees to pay her a annual percentage rate (APR) of 6% for 18 years provided she makes an initial deposit of $100, deposits $50 at the end of each month, and maintains the account for 18 years. How much will Sue have in the account after 18 years?

Solution:

We develop a recursive sequence model for Sue. As always, we begin the process by defining the variables and stating the given conditions. Let

n = number of months
$a(n)$ = balance in the account after n months
$a(0) = 100$ (initial deposit)
$r = \frac{APR}{12} = \frac{.06}{12} = .005$ (monthly interest rate)
$d = 50$ (monthly deposit).

The next step is dependent on the fact that the balance in the account only changes at the end of each month. This means that we can model the situation by expressing next month's beginning balance in terms of this month's beginning balance plus the change that takes place during this month (that is, interest received plus deposit). Thus our model is

(Next month's balance) = (This month's balance) + (Interest + Deposit)

or symbolically

$$a(n) = a(n-1) + ra(n-1) + d \quad \text{with} \quad a(0) = 100$$

or

$$a(n) = (1+r)a(n-1) + d \quad \text{with} \quad a(0) = 100.$$

Substituting $r = \frac{.06}{12} = .005$ and $d = 50$, we have the recursive sequence model for Sue's plan:

$$a(n) = 1.005a(n-1) + 50 \quad \text{with} \quad a(0) = 100.$$

(Compare this result with the recursive sequence model for Chuck's option 2 in Example 4.1.1.)

The next step of the modeling process (see the modeling process diagram, Figure 4.1.1) is to solve the model. We begin by computing $a(n)$ for the first few values of n. When computing

$a(n)$, we must be careful not to disguise any patterns by carrying out simplifying arithmetic operations such as adding or multiplying.

We iterate the values of $a(n)$ in terms of $a(0)$ for the first few values of n as follows:

$$
\begin{aligned}
a(1) &= (1+r)a(0)+d \\
a(2) &= (1+r)a(1)+d \\
&= (1+r)[(1+r)a(0)+d]+d && \text{Substitute for } a(1) \\
&= (1+r)^2a(0)+(1+r)d+d && \text{Simplify} \\
a(3) &= (1+r)a(2)+d \\
&= (1+r)[(1+r)^2a(0)+(1+r)d+d]+d && \text{Substitute for } a(2) \\
&= (1+r)^3a(0)+(1+r)^2d+(1+r)d+d && \text{Simplify} \\
a(4) &= (1+r)a(3)+d \\
&= (1+r)[(1+r)^3a(0)+(1+r)^2d+(1+r)d+d]+d \\
&= (1+r)^4a(0)+(1+r)^3d+(1+r)^2d+(1+r)d+d \\
a(5) &= (1+r)a(4)+d \\
&= (1+r)[(1+r)^4a(0)+(1+r)^3d+(1+r)^2d+(1+r)d+d]+d \\
&= (1+r)^5a(0)+(1+r)^4d+(1+r)^3d+(1+r)^2d+(1+r)d+d \\
a(6) &= (1+r)a(5)+d \\
&= (1+r)[(1+r)^5a(0)+(1+r)^4d+(1+r)^3d+(1+r)^2d+(1+r)d+d]+d \\
&= (1+r)^6a(0)+(1+r)^5d+(1+r)^4d+(1+r)^3d+(1+r)^2d+(1+r)d+d
\end{aligned}
$$

(If the pattern is not clear, the reader should compute $a(n)$ for more values of n.)

Query 1.

Compute $a(7)$ in terms of $a(0)$, r, and d.

We group terms in $a(6)$ and then factor d out of the second term.

$$
\begin{aligned}
a(6) &= (1+r)^6a(0)+(1+r)^5d+(1+r)^4d+(1+r)^3d+(1+r)^2d+(1+r)d+d \\
&= (1+r)^6a(0)+[(1+r)^5d+(1+r)^4d+(1+r)^3d+(1+r)^2d+(1+r)d+d] \\
&= (1+r)^6a(0)+d[(1+r)^5+(1+r)^4+(1+r)^3+(1+r)^2+(1+r)+1]
\end{aligned}
$$

Note that the sum in the bracket, $(1+r)^5+(1+r)^4+(1+r)^3+(1+r)^2+(1+r)+1$, is a geometric series with the ratio $(1+r)$. Summing this geometric series yields

$$a(6) = (1+r)^6a(0)+(50)[\frac{1-(1+r)^6}{1-(1+r)}].$$

We conjecture a solution by generalizing the expression for $a(6)$ to obtain a general solution that includes Sue's situation as a special case. We do this in two steps.

Step 1: Note that 6, the input in $a(6)$, occurs in exactly 3 places, once on the left-hand side and twice on the right-hand side of the equation. Replacing these 3 occurrences of 6 with the variable n gives an expression for $a(n)$, the balance in the account after n months.

Step 2: Simplify the last term on the right-hand side, express it as the difference of 2 terms, combine the first and last terms, and then simplify as follows:

$a(n)$	$=$	$(1+r)^n a(0) + d\frac{1-(1+r)^n}{1-(1+r)}$	Sequence
	$=$	$(1+r)^n a(0) + d\frac{1-(1+r)^n}{-r}$	Simplify the denominator
	$=$	$(1+r)^n a(0) + d[\frac{1}{-r} - \frac{(1+r)^n}{-r}]$	Separate the fraction into two fractions
		$(1+r)^n a(0) + d[\frac{1}{-r}] - d[\frac{(1+r)^n}{-r}]$	Multiply both fractions by d
	$=$	$(1+r)^n a(0) + \frac{d}{-r} - \frac{d}{-r}(1+r)^n$	Simplify
	$=$	$(1+r)^n[a(0) - \frac{d}{-r}] + \frac{d}{-r}$	Group terms containing $(1+r)^n$
	$=$	$(1+r)^n[a(0) + \frac{d}{r}] - \frac{d}{r}$	Simplify

We will follow the common convention and write the general solution in the form

$$a(n) = (1+r)^n c - \frac{d}{r} \text{ where } c = a(0) + \frac{d}{r}.$$

Note that, we can easily solve for c by setting $n = 0$. That is,

$$a(0) = c - \frac{d}{r} \text{ or } c = a(0) + \frac{d}{r}.$$

Because our general solution is based on a conjecture, it needs to be verified. This is done by showing the general solution satisfies the original recursive sequence. That is, when substituted into the recursive sequence an equality is obtained. We ask the reader to trust that our general solution is correct. (In Section 4.6, we will illustrate the method for verifying a conjectured solution.)

To answer Sue's particular question, we substitute

$$n = 18 * 12 = 216 \text{ months}$$
$$r = \frac{.06}{12} = .005 \text{ (monthly interest rate)}$$
$$d = 50 \text{ (monthly deposit)}$$
$$c = 100 + \frac{50}{.005} = 10,100$$

After 18 years, Sue's account will have

$$a(216) = 1.005^{216}(10,100) - \frac{50}{.005} = 19,661.34,$$

a nice "nest egg" with which to start college. ▲

Query 2.

How much would Sue have in her account after 18 years if she deposited \$60 per month rather than \$50?

Our recursive sequence model for Sue's plan is applicable in many different scenarios in which an account balance changes on a regular basis (for example, monthly interest and payment). The next two problems illustrate how this model is adapted to a car loan situation and a credit card situation.

A car loan is like a reverse savings account. You borrow an amount, pay interest each month, and make a monthly payment (that is, a negative deposit). The plan continues until there is a zero balance in the account.

Example 4.2.2.

Bill buys a Ford Ranger with the help of a $15,000 car loan at an annual percentage pate (APR) of 12%. He agrees to pay $400 at the end of each month. Model this financial situation and determine how many months it will take Bill to pay off his loan.

Solution:

We begin the modeling process (as always) by defining the variables and stating the conditions. Let

n = number of months
$a(n)$ = balance due after the nth month
$a(0) = 15,000$
$r = \frac{APR}{12} = \frac{.12}{12} = .01$ (monthly interest rate)
$p = 400$ (monthly payment).

We adapt the model developed for Sue's plan to this car loan situation by considering the monthly payment to be a negative deposit. That is, setting $d = -p$. Thus the balance due after the nth month is given by the expression

$$a(n) = (1+r)^n c - \frac{-p}{r} = (1+r)^n c + \frac{p}{r}.$$

We now solve for the value of c. Because we know the initial value of the loan, $a(0) = 15,000$, we solve the model expression for $n = 0$ and set the result equal to 15,000. (Recall that a number raised to the zero power is defined to be one.)

$$a(0) = (1+r)^0 c + \frac{p}{r} = c + \frac{p}{r} = 15,000$$

and so

$$c = 15,000 - \frac{400}{.01} = -25,000. \text{ (Explain)}$$

Substituting for r, c, and p in the model gives

$$a(n) = 1.01^n(-25,000) + \frac{400}{.01} = 1.01^n(-25,000) + 40,000.$$

The car loan is paid off when $a(n) = 0$. Thus we need to solve for the value of n that makes $a(n) = 0$. We could do this graphically, but instead will illustrate solving using logarithms. Here are the steps.

$$a(n) = 1.01^n(-25,000) + 40,000$$

$a(n) = 0$ implies $1.01^n(-25,000) = -40,000$

which implies $1.01^n = \frac{-40,000}{-25,000}$

which implies $1.01^n = \frac{8}{5}$.

To solve $1.01^n = \frac{8}{5}$ for n, we now take the logarithm of both sides of the equation (recall

$\log(a^b) = b\log(a)$ and $\log(\frac{a}{b}) = \log(a) - \log(b)$).

1.01^n	=	$\frac{8}{5}$	Equation
$\log(1.01^n))$	=	$\log(\frac{8}{5})$	Apply logarithm function
$n\log(1.01)$	=	$\log(8) - \log(5)$	Properties of logarithms
n	=	$\frac{\log(8)-\log(5)}{\log(1.01)}$	Divide both sides by $\log(1.01)$
n	=	47.23	

Because n represents the number of months, the mathematical solution for n must be rounded up to the next full month. Thus 48 months are required for Bill to pay off the loan. Because $a(47)$ is the balance due at the beginning of the 48th month the last payment will be

$$\begin{aligned}\text{48th payment} &= a(47) + ra(47)\\ &= (1+r)a(47)\\ &= (1+.01)93.31\\ &= 94.35.\end{aligned}$$

So the car loan will be paid off in 48 months with the last payment being $94.35. ▲

Query 3.

If Bill paid $500 per month, how many months would it take for him to pay off his car loan?

Query 4.

Solve the equation in the previous example, $1.01^x(-25000) + 40000 = 0$ graphically. (Note that n is replaced by x because calculators plot functions of x.)

Example 4.2.3.

Malcolm has reached the limit on his credit card with an accumulated debt of $5,000. The credit card company demands that Malcolm pay at least the amount to carry his debt in addition to paying for any additional purchases. (Carrying his debt means that Malcolm must pay the monthly interest charge on his debt each month.) The annual percentage rate (APR) is 17.9%. Malcolm decides to first rip up his card to avoid any new expenses and then to pay $100 per month. How long will it take Malcolm to pay off his credit card debt?

Solution:

Paying off a credit card debt is like paying off a car loan, the monthly payment is like a negative deposit in a savings account. Thus we adapt the standard savings account model (the model for Sue's plan), which is

$$a(n) = (1+r)^n c - \frac{d}{r},$$

where n = number of months

$a(n)$ = account balance after the nth month

d = is the monthly deposit

$r = \frac{APR}{12}$ (monthly interest rate)

c = constant = $a(0) + \frac{d}{r}$.

We set $d = -p$, where $p = 100$ is the monthly payment. Thus the balance due the credit card company at the end of the nth month is given by the expression

$$a(n) = (1+r)^n c - \frac{-p}{r} = (1+r)^n c + \frac{p}{r}$$

where

$$r = \frac{APR}{12} = \frac{.179}{12} = .0149$$

and

$$a(0) = c + \frac{p}{r} = 5{,}000,$$

so

$$c = 5{,}000 - \frac{100}{.0149} = -1{,}711.4094.$$

Thus

$$a(n) = 1.0149^n(-1{,}711.4094) + \frac{100}{.0149}$$
$$= 1.01^n(-1{,}711.4094) + 6{,}711.4094.$$

In order to determine when the loan is paid off, we set $a(n) = 0$ and solve for n.

$$a(n) = 1.0149^n(-1{,}711.4094) + 6{,}711.4094 = 0$$

implies

$$1.0149^n(-1{,}711.4094) = -6{,}711.4094.$$

Thus

$$1.0149^n = \frac{-6{,}711.4094}{-1{,}711.4094} = 3.9216.$$

Taking logarithms of both sides of the equation yields

$$n\log(1.0149) = \log(3.9216).$$

So

$$n = \frac{\log(3.9216)}{\log(1.0149)} = 92.4 \text{ months}.$$

Hence it will take Malcolm 93 months to pay off his \$5,000 credit card debt. He will pay \$100 for 92 months, plus $(1.0149)a(92) = \$38.83$ in the 93rd month for a total of \$9,238.83. ▲

Query 5.
How much would Malcolm have to pay each month if he planned to pay off his \$5,000 credit card debt in 5 years?

A retirement annuity is similar to a car loan with the payment going to the person. For example, consider a person who decides to retire having accumulated \$500,000 in her retirement fund. Her bank will pay her an annual percentage rate (APR) of 6% compounded monthly on the balance of her account. If she decides to withdraw \$3,000 at the end of each month for living expenses, how long will her retirement fund last? (Solution: adapt the savings account model as was done in the car loan problem with $a(0) = 500{,}000$, $r = \frac{.06}{12} = .005$, and $p = 3{,}000$.)

Query 6.

Explain why the monthly interest rate is $\frac{\text{Annual Percentage Rate}}{12}$ and the quarterly rate is $\frac{\text{Annual Percentage Rate}}{4}$.

Exponential and Logarithmic Functions

Exponential and logarithmic functions often arise in modeling real-life situations. As illustrated in the problems in this section, exponential functions are used in modeling savings accounts, car loans, and credit card debts. In Section 3.3, we used an exponential function to model the population of senior citizens and in Section 3.4 we used an exponential function to model the temperature of a soda can. Radioactive decay and the spread of a rumor are other examples of situations that can be modeled with exponential functions. These examples have one thing in common:

The amount of change is proportional to the amount of the quantity present.

This is the defining characteristic of exponential functions. (See Section 3.6, Exercise 13.)

Exponential growth is often described in terms of *doubling times*. Recall from Chuck's problem in Section 4.1 that if $a(0)$ dollars is invested in a savings account that pays a monthly interest rate of r percent, then the balance after n months is given by the model $a(n) = (1+r)^n a(0)$. How long will it be before the balance is twice the initial deposit? That is, for what value of n is $a(n) = 2a(0)$? This time period is called the doubling time. The steps in computing the doubling time are as follows:

$$\begin{cases} a(n) = (1+r)^n a(0) \\ a(n) = 2a(0) \end{cases}$$ Solve for n.

$2a(0) = (1+r)^n a(0)$ — Substitute for $a(n)$.

$2 = (1+r)^n$ — Divide by $a(0)$.

$\ln(2) = n\ln(1+r)$ — Apply the logarithm function to both sides of the equation

$\frac{\ln(2)}{\ln(1+r)} = n$ — Divide by $\ln(1+r)$

Thus the doubling time a monthly interest rate of 1 percent is

$$n = \frac{\ln(2)}{\ln(1.01)} = \frac{0.693}{0.010} = 69.3 \text{ months or 5 years, nine months, and 9 days.}$$

Query 7.

If the U.S. population has a 1.3% growth rate, what is its doubling time?
(Answer: 53.67 years)

Query 8.

The following table lists a few situations in which change is proportional to the amount present. Add three more situations to the table.

Situation	Nature of Change
Savings account	Interest earned is proportional to amount invested
Population growth	Births and deaths are proportional to the size of the population
Loss of signal strenth in fiber optics cables	Strength of signal diminishes in proportion to the length of the cable
Temperature change	Proportional to difference between surrounding temperature and temperature of the object
Voltage drop	Proportional to length of wire

Table 4.2.1

Solving exponential models often involves the use of the logarithmic function (the inverse of the exponential function). In Section 3.3, we stated that the exponential and logarithmic operations are inverse operations. That is,

$$\log_b(b^x) = x \text{ and } b^{\log_b(x)} = x \text{ where } b > 0.$$

b is called the base of the exponential or of the logarithm. Scientists and mathematicians usually use the irrational number e~2.17828 as a base. When the base e is used, we refer to the exponential and logarithm functions as the **natural exponential function** and **natural logarithm function**. The natural logarithm is denoted by ln [that is, $\log_e(r) = \ln(r)$]. Often when logarithms are first introduced, 10 is used for the base. Base 10 logarithms are called common logarithms and are usually denoted by log [that is, $\log_{10}(r) = \log(r)$]. Computer scientists usually use a base of 2 (to reflect the on–off nature of electricity). The base 2 logarithm is denoted by the symbol lg [that is, $\log_2(r) = \lg(r)$]. Regardless of the choice of base, as long as it is positive, we have the following fundamental relationship between exponential and logarithmic functions:

$$y = b^x \text{ if and only if } \log_b(y) = x \text{ where } b > 0.$$

The following multiplot of $y = e^x$, $y = x$, and $y = \ln(x)$ illustrates that the logarithm graph and the exponential graph are reflections of one another in the line $y = x$. (This is true for all functions and their inverses.)

Graphs of Exponential and Logarithm Functions

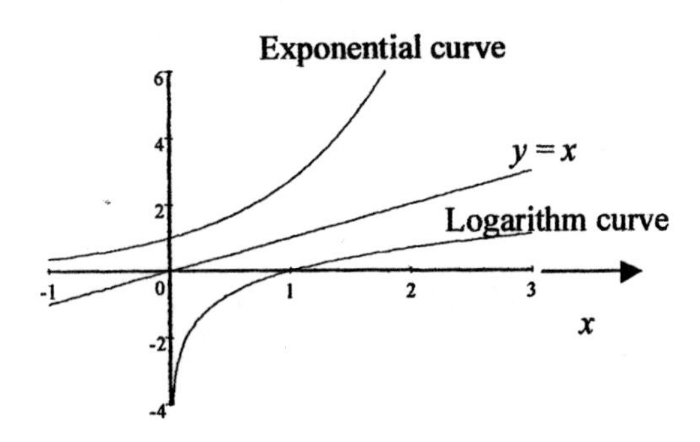

Figure 4.2.1

Query 9.

Verify that the point $(2, 7.39)$ is on the graph of $y = e^x$ and the point $(7.39, 2)$ is on the graph of $y = \ln(x)$.

In order to facilitate computing, we list the algebraic rules for exponents and logarithms.

Rules of Exponents

Let r, s, m, and n be real numbers. The following rules apply.

1. $r^{-n} = \frac{1}{r^n}$ for r not zero
2. $r^0 = 1$ for r not zero
3. $r^1 = r$
4. $r^2 = r * r$, $r^n = r * r * r * r ... r$ $(n\ r's)$ for n a positive integer
5. $r^n r^m = r^{n+m}$ (for example, $2^2 2^3 = 2^{2+3} = 2^5$)
6. $(r^n)^m = r^{nm}$ (for example, $(2^2)^3 = 2^6$)
7. $(\frac{r}{s})^n = \frac{r^n}{s^n}$ (for example, $(\frac{2}{3})^2 = \frac{2^2}{3^2}$)

Rules of Logarithms

Let b, r, s, and n be real numbers with $b > 0$. The following rules apply.

1. If $r = b^s$, then $s = \log_b(r)$, b is called the **base** of the logarithm.
2. $\log_b(r^n) = n \log_b(r)$
3. $\log_b(r^{-1}) = -\log_b(r)$ (set $n = -1$)
4. $\log_b(1) = 0$
5. $\log_b(b) = 1$
6. $\log_b(r * s) = \log_b(r) + \log_b(s)$
7. $\log_b(\frac{r}{s}) = \log_b(r) - \log_b(s)$

Exercises 4.2

1. (Computational Skill) Solve for x in each of the following equations:
   a. $\log(2x) = \log(12) - \log(5)$
   b. $\log(5x) = \log(\frac{9}{4})$
   c. $2^x = \frac{4}{5}$ (use logarithms)
2. (Computational Skill) Solve for x in each of the following equations:
   a. $\log(2^x) = \log(\frac{8}{3})$
   b. $\log(\frac{2x}{5}) = \log(7) + \log(3)$
   c. $3^x = \frac{2^x}{5}$
3. (Computational Skill) Make up and solve three exercises similar to Exercises 1a–c, 2a–c.
4. Given the recursive sequence, $a(n) = 1.05a(n-1)$, $a(0) = 150$, compute the value of $a(17)$.
5. Given the recursive sequence, $a(n) = 0.85a(n-1) + 50$, $a(0) = 150$, compute the value of $a(17)$
6. Make up and solve three exercises similar to Exercise 5.
7. If the annual percentage rate is 12%, determine each of the following.
   a. Semiannual rate
   b. Quarterly rate
   c. Monthly rate
   d. Daily rate (assume 365 days per year)
8. Bulla invests $500 in a savings account that pays an annual percentage rate of 4% with interest compounded quarterly. She asks you to determine her balance after 10 years by developing a recursive sequence model, iterating the model to discover a pattern, formulating the general solution expression, and then specializing the general solution to her particular situation. She then wants you to explore other scenarios by answering the following type questions.
   a. Using your model, determine the balance if $10,000 had been invested rather than $5,000.
   b. Using your model, determine the balance if the annual interest rate was 6% rather than 4%.
   c. Using your model, determine the balance if the 4% interest is compounded monthly rather than quarterly.
9. If the world's population grows by 4% per year, how many years will it take for the population to double? Would it take twice as long for the population to double if the growth rate was 2% per year? Explain.
10. The New Bell Ice Cream Company introduced a new flavor, Blueberry Supreme, in October. For the next 12 months sales grew at a rate of 1% per month. What was the percent increase in sales for the entire 12 months?
11. Derive the model in Example 4.2.2 (car loan) in a manner similar to that which was done for Sue's plan.

12. Derive the model in Example 4.2.3 (credit card) in a manner similar to that which was done for Sue's plan.

13. Assume that you have reached the limit on your credit card with an accumulated debt of $4,000. Assume also that you have ripped up your credit card so as not to use it anymore. Using one of your credit cards or a friend's, convert the APR to a monthly rate, and then determine how much your monthly payments would be for you to pay off your debt the month you intend to graduate from college.

14. Having convinced yourself that you need a car in order to go to school, you have found just the right one. The only hitch is that you will have to borrow $10,000 to purchase it. Because it is a used car, the bank will only give you a maximum of 3 years to pay off the loan at an APR of 10%, payable monthly. What monthly payment would allow you to pay off your loan in 3 years?

15. Assume that you will retire on your 65th birthday. In order to prepare for retirement, you decide to open a savings account and invest a fixed amount at the end of each month until you retire. If the savings account paid interest at an APR of 5%, payable monthly, how much would you have in your account when you retire if your monthly deposit is the following?

    a. $10
    b. $30
    c. $50

16. You decide to open a savings account that pays interest at an APR of 6%, payable monthly, and to make monthly deposits for 4 years.

    a. Compute your balance after 4 years if you originally deposited $500 and made $50 payments at the end of each month.
    b. How much would your payments have to be in order to obtain the balance in part (a) if you made the 48 payments, but made no initial deposit.

17. (Refer to Sue's plan.) How much would Sue have to pay per month for the balance in her savings account after 18 years to be $40,000?

18. Robin has accumulated $500,000 in her retirement fund. Her bank will pay her monthly interest at an APR of 6% on the balance of her account. If she decides to withdraw $3,000 at the end of each month for living expenses, how many years will her retirement fund last?

19. Rosa wants to set up an annuity fund that will pay her 1% each month. What initial investment must she make in order to withdraw the following amounts monthly while maintaining her initial investment?

    a. $2,000
    b. $3,000
    c. $4,000

20. Your parents have just received a nice inheritance and wish to share it with you by providing a retirement investment program for you. To add a bit of challenge to the sharing process, your parents offer you the choice of two plans:

    Plan A: Your parents will place $10,000 in a savings account that guarantees a monthly interest rate of 0.5% with the stipulation that nothing is withdrawn from the account until you retire.

    Plan B: Your parents will place $10,000 in an account and add $100 to it each month, but there will be no interest and you cannot withdraw any amount until you retire.

Model each of the two plans and then determine which plan would provide more retirement money if you retired in

a. 5 years

b. 10 years

c. 15 years

d. 20 years

21. Bankers use the Rule of 70 to approximate the doubling time for interest-bearing accounts. The Rule of 70 says to determine the number of years for an interest-bearing account to double its amount, divide 70 by 100 times the yearly interest rate. For example, an account receiving 10% annual interest would double in $\frac{70}{10} = 7$ years, and an account receiving 6% interest would double in $\frac{70}{6} = 11.67$ years.

a. Lee owes $2000 on a credit card account that charges 18% annual interest. Use the Rule of 70 to approximate how long it will take for his debt to double if he does not pay nor charge anything to his account.

b. On Sanda's seventeenth birthday, her grandparents invested $1000 in a retirement fund for her. If the fund pays 5% annual interest, use the Rule of 70 to approximate the value of the fund on Sanda's sixty-fifth birthday.

22. Compare the growths of linear, quadratic, and exponential functions for positive values of x by forming a multiplot of the following three functions:

$$f(x) = 3x + 1, \qquad g(x) = x^2 - 2, \qquad h(x) = 2^x$$

Note that all three functions become unbounded as $x -> \infty$. Which function grows the fastest? Apply the what-if technique by changing coefficients and/or exponents. Write a paragraph explaining why an exponential function will eventually become larger than any polynomial function as $x -> \infty$.

23. Verify the following rules for (natural) logarithms by applying the exponential function and reducing the resulting equation to an identity. Let r, s, and n be real numbers.

a. If $r = e^s$, then $s = \ln(r)$.

b. $\ln(r^{-1}) = -\ln(r)$

c. $\ln(1) = 0$

d. $\ln(e) = 1$

e. $\ln(r^n) = n\ln(r)$

f. $\ln(r * s) = \ln(r) + \ln(s)$

g. $\ln(\frac{r}{s}) = \ln(r) - \ln(s)$

24. Write an essay, including illustrative examples, reflecting on the effect exponential and logarithm functions have on the representation of data. For example, the logarithm function has the property that large changes in the input values yield small changes in the output values. In contrast, the exponential function has the property that small changes in the input yield large changes in the output.

4.3 Modeling (Motion Problems)

Sir Isaac Newton's (1642-1727) study of motion led to his being one of the two principal inventors of the calculus. The other was Gottfried Wilhelm Leibniz (1646-1716). In this section, we model the trajectory of a moving object. We begin with vertical motion and then introduce parametric functions to generalize to non-vertical motion such as the trajectory of a thrown javelin. The basis for our analysis is Newton's Second Law of Motion.

Newton's Second Law of Motion

The net force acting on an object is the product of its mass and its acceleration.

Force = (mass)(acceleration)

Because acceleration is the rate of change of velocity and velocity is the rate of change of position with respect to time, Newton's Second Law of Motion gives a second-order rate of change, that is, a second-order differential equation model for motion. When you take calculus, you will solve this differential equation and obtain the following position model for vertical motion near the earth's surface (ignoring air resistance):

$$s(t) = -16t^2 + vt + c$$

where

t = time in seconds
$s(t)$ = position of object at time t
v = initial velocity (feet per second)
c = initial position

(The coefficient of t^2 is $\frac{1}{2}$ the acceleration due to gravity which we approximate to be 32 feet per second^2 and is negative because gravity exerts a downward (negative) force on an object.)

Note that when both the initial velocity and the initial position are zero, Newton's Second Law of Motion reduces to $s(t) = -16t^2$, which was the result of Example 3.6.9.

Example 4.3.1.

A ball is thrown upward with an initial velocity of 10 feet per second off the top of a 100-foot cliff. How long does it take for the ball to hit the ground at the base of the cliff?

Solution:

We choose the positive direction to be upward, and we place the origin at the base of the cliff. Based on these conventions, $v = 10$ (ball is thrown in the positive direction) and $c = 100$. (Why is $c = 100$?)

Substituting for v and c in the position model for vertical motion, we have

$$s(t) = -16t^2 + 10t + 100.$$

To determine the time it takes before the ball hits the ground, we set $s(t) = 0$ and solve for t. We first do this graphically and then analytically using the **quadratic formula.** We only consider positive values for t. (Why?)

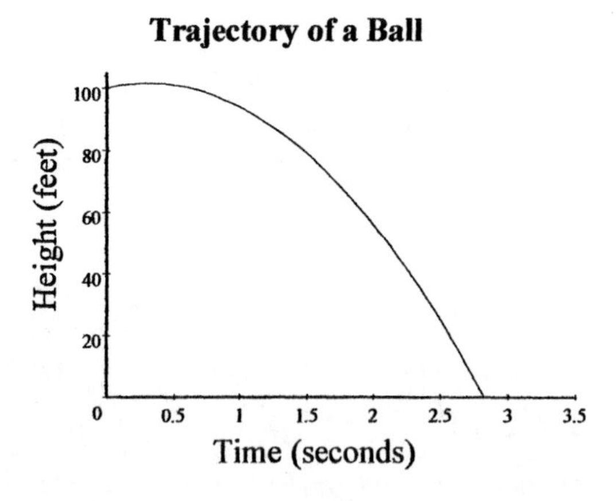

Figure 4.3.1

The ball appears to hit the ground after approximately 2.8 seconds. In order to obtain an exact answer, we apply the quadratic formula. This formula states

$$\text{if } at^2 + bt + c = 0, \text{ then } t = \frac{-b \pm \sqrt{b^2 - 4ac}}{2a}.$$

Substituting the coefficients from the position model,

$a = -16$
$b = 10$
$c = 100,$

into the quadratic formula yields

$$t = \frac{-10 \pm \sqrt{100 + 6400}}{32} = \frac{-10 \pm 80.6226}{-32}.$$

We thus have two solutions,

$$t = -2.207 \text{ and } t = 2.832.$$

Although both solutions are correct mathematically, only the positive solution makes sense in the real-world setting of the problem. Thus the answer is that it will take 2.832 seconds before the ball hits the ground. ▲

Query 1.
If the ball in Example 4.3.1 had been thrown downward with a velocity of 10 feet per second, how long would it be before the ball hit the ground?

The importance of the quadratic formula is due, in part, to the fact that the solutions to many models for motion contain a quadratic function (thanks to Newton's Second Law of Motion). We will derive the quadratic formula after first making an observation on the terms in the equation

$$(x + b)^2 = x^2 + 2bx + b^2.$$

Observe that the middle term, $2bx$, of the expanded expression, $x^2 + 2bx + b^2$, is two times the product of the first and last terms of the factored expression, $(x + b)^2$. In the derivation, we use this observation to rewrite $ax^2 + bx + c$ as a squared expression equal to a sum of constants. We then take the square root and solve for the variable.

Derivation of the Quadratic Formula:

$$\text{If } ax^2 + bx + c = 0, \text{ then } x = \frac{-b \pm \sqrt{b^2 - 4ac}}{2a}.$$

$ax^2 + bx + c$	$=$	0	Quadratic equation
$a(x^2 + \frac{b}{a}x + \frac{c}{a})$	$=$	0	Factor out the coefficient of x^2
$x^2 + \frac{b}{a}x + \frac{c}{a}$	$=$	$\frac{0}{a} = 0$	Divide each term by a
$x^2 + \frac{b}{a}x + (\frac{b}{2a})^2 - (\frac{b}{2a})^2 + \frac{c}{a}$	$=$	0	Add and subtract $(\frac{b}{2a})^2$
$[x^2 + \frac{b}{a}x + (\frac{b}{2a})^2] - [(\frac{b}{2a})^2 - \frac{c}{a}]$	$=$	0	Group terms
$(x + \frac{b}{2a})^2 - [(\frac{b}{2a})^2 - \frac{c}{a}]$	$=$	0	Factor left-hand term
$(x + \frac{b}{2a})^2$	$=$	$[(\frac{b}{2a})^2 - \frac{c}{a}]$	Transpose constant terms
$x + \frac{b}{2a}$	$=$	$\pm\sqrt{(\frac{b}{2a})^2 - \frac{c}{a}}$	Take square roots
x	$=$	$-\frac{b}{2a} \pm \sqrt{(\frac{b}{2a})^2 - \frac{c}{a}}$	Transpose $\frac{b}{2a}$
x	$=$	$\frac{-b \pm \sqrt{b^2 - 4ac}}{2a}$	Simplify right-hand side

(The reason for adding and subtracting $(\frac{b}{2a})^2$ in the fourth step was to express the left-hand side as a squared term plus a constant term.) Explain the algebra in the last step.

Query 2.

What are the mathematical solutions of the quadratic equation, $x^2 + 5x - 10 = 0$?

Parametric Functions

In order to apply Newton's Second Law to nonvertical motion, we need to develop the concept of horizontal and vertical components of motion. We begin our development by returning to the definition of slope of a straight line as rise over run (Section 2.6). We speak of the run as the horizontal component and the rise as the vertical component as illustrated in the following diagram.

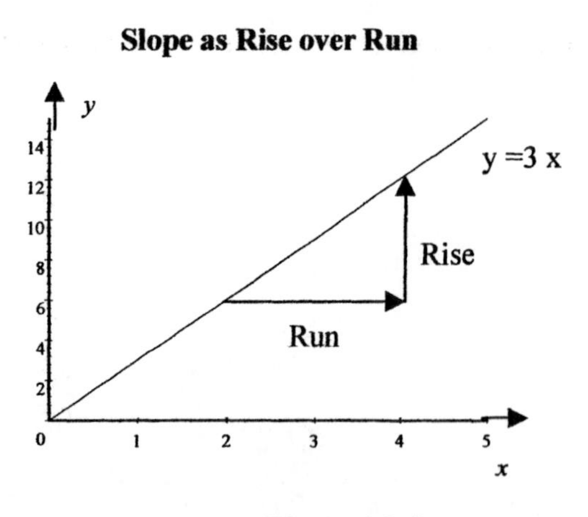

run = horizontal component

rise = vertical component

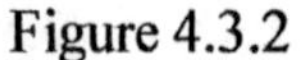

Figure 4.3.2

In order to analytically describe the horizontal and vertical components separately, we define x and y as functions of a third variable, say t, called a **parameter**. It is easy to do this when y is a known function of x. The procedure is to set x equal to t and then use the relation between x and y to define y as a function of t. In this situation of the straight line, $y = 3x$, we define

$$x(t) = t \text{ and } y(t) = 3t \quad \text{(substitute } t \text{ for } x \text{ in the relation } y = 3x\text{)).}$$

The functions $x(t) = t$ and $y(t) = 3t$ are called **parametric functions** for the line $y = 3x$. Each value of t, determines values for both $x(t)$ and $y(t)$. Thus each value of t determines the ordered pair $(x(t), y(t))$ which represents a point on the line. We can therefore define our line function as $l(t) = (x(t), y(t)) = (t, 3t)$. Note that l maps a value of the parameter t into an ordered pair. Thus l maps a number into a point in the plane.

Example 4.3.2.

Plot and identify the curve described by the parametric functions

$$x(t) = t - 1 \qquad y(t) = t^2 - 2$$

and then undo the parametrization to express the curve as the plot of a function of x.

Solution:

We begin by constructing a table of values for the variables t, x, and y.

t	x	y
−3	−4	7
−2	−3	2
−1	−2	−1
0	−1	−2
1	0	−1
2	1	2
3	2	7

Table 4.3.1

Plot these points $(x(t), y(t))$ and then sketch a smooth curve through the points to obtain the following plot.

Curve Defined by Parametric Functions

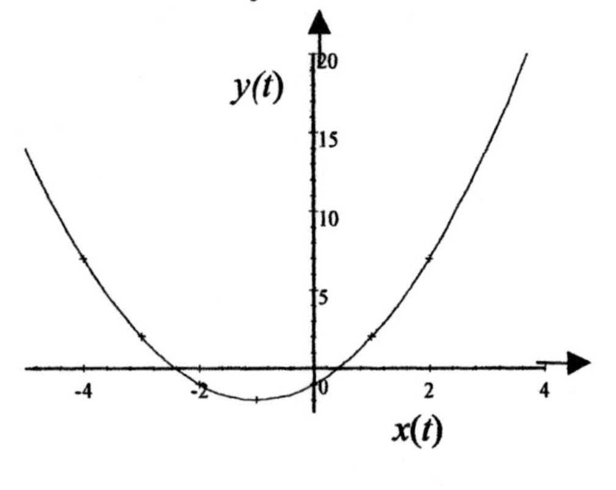

Figure 4.3.3

The curve looks like a parabola, which suggests that it should be described by a quadratic function. In order to verify this, we undo the parametrization by eliminating t from the parametric equations,

$$\begin{cases} x = t - 1 \\ y = t^2 - 2 \end{cases}.$$

Solve for t in the first equation ($t = x + 1$) and substitute it into the second equation ($y = (x+1)^2 - 2 = x^2 + 2x - 1$). Thus the curve represented by the parametric functions is the parabola defined by

$$y(x) = x^2 + 2x - 1. \quad \blacktriangle$$

The preceding discussion and example may have seemed just an exercise in changing notation; however, it will lead to much more, including modeling nonvertical motion.

Example 4.3.3.

Define a circle as the graph of a parametric function, something that we were unable to do previously (Query 2, Section 2.2).

Solution:

As usual, we begin with a diagram. Consider a circle of radius r.

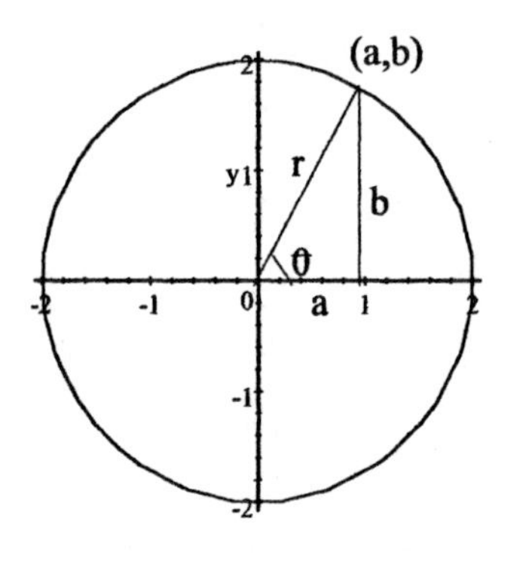

Figure 4.3.3

Our task is to express the a and b components of the ordered pair (a,b) as functions of a new parameter. Because the radius of the circle, r, is fixed, the only variable that changes as the point (a,b) moves around the circle is the central angle θ. The components a and b are related to r and θ by the sine (sin) and cosine (cos) functions as defined by $\cos(\theta) = \frac{a}{r}$ and $\sin(\theta) = \frac{b}{r}$.

Therefore, we can define components a and b as functions of θ:

$$\text{horizontal component: } a(\theta) = r\cos(\theta)$$
$$\text{vertical component: } b(\theta) = r\sin(\theta).$$

This allows us to define a parametric function for the circle of radius r as $c(\theta) = (r\cos(\theta), r\sin(\theta))$. This is a step forward as previously we could not describe a circle as the graph of a function. We list a few of the input–output ordered pairs for this parametric function.

θ	$c(\theta)$	=	$(2\cos(\theta), 2\sin(\theta))$
0	$c(0)$	=	$(2,0)$
$\frac{\pi}{2}$	$c(\frac{\pi}{2})$	=	$(0,2)$
π	$c(\pi)$	=	$(-2,0)$
$\frac{3\pi}{2}$	$c(\frac{3\pi}{2})$	=	$(0,-2)$

Table 4.3.2 ▲

In the previous examples of the line, parabola, and circle, we were considering a situation in which there was a known relation between the x and y components of a point on the curve. The next step in developing the concept of horizontal and vertical components of motion is to consider a curve for which we do not know the relation between the x and y components of a point on the curve. We consider the trajectory of a rocket. Suppose a miniature rocket is blasted off the ground with a velocity of 50 feet per second at an angle of 60 degrees with the horizontal.

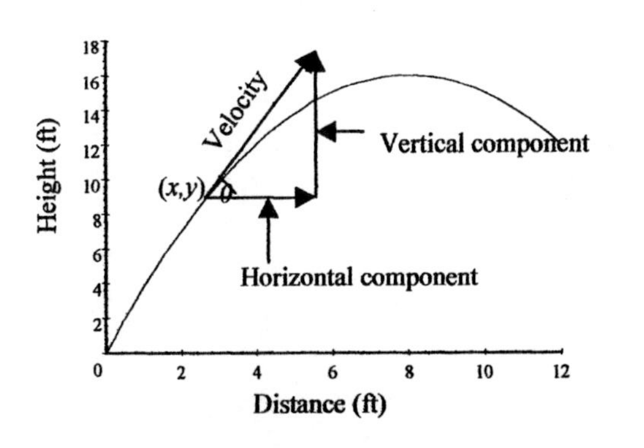

Figure 4.3.5

The velocity t at the point (x,y) is directed in the direction of the tangent to the curve. The triangle formed by the velocity, horizontal component, and vertical component corresponds to the triangle in the previous circle diagram. This suggests we define the horizontal and vertical components as we did in the circle situation.

horizontal component $= (velocity)\cos(\theta)$
vertical component $= (velocity)\sin(\theta)$.

We are now ready to develop a parametric model for projectile motion.

Example 4.3.4 Chuck's Javelin Throw.
Chuck, an aspiring javelin thrower, releases his javelin at a 45 degree angle with initial velocity of 50 feet per second from a height of 7 feet. Where does the javelin land?

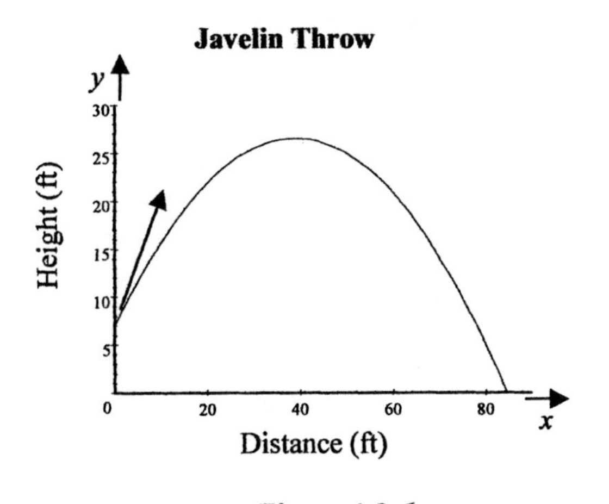

Figure 4.3.6

Solution:

We develop a model by recalling the solution expression for a vertical motion model is $s(t) = -16t^2 + vt + c$ where

t = time in seconds
$s(t)$ = position of object at time t
v = initial velocity (feet per second)
c = initial position.

In order to generalize this model to apply to our javelin throw (non-vertical motion), we express the position of the javelin's center by the ordered pair $(x(t), y(t))$, the initial position by the ordered pair (a, b), and decompose the initial velocity into horizontal and vertical components.

The only two forces on the javelin are the initial velocity force and the force due to gravity (we ignore air resistance). The initial velocity force has both a horizontal and vertical component. However, the force of gravity is a vertical force, and thus it has a 0 horizontal component. We therefore have

$$\begin{aligned} x(t) &= (\text{horizontal component of initial velocity}) * t + (\text{initial horizontal position}) \\ y(t) &= (\text{vertical component of the gravity force}) \\ &\quad + (\text{vertical component of initial velocity}) * t + (\text{initial vertical position}). \end{aligned}$$

The vertical component of the gravity force is $-16t^2$ (Newton's Second Law of Motion), and the initial position is (a, b). Thus

$$\begin{aligned} x(t) &= (\text{horizontal component of initial velocity}) * t + a \\ y(t) &= -16t^2 + (\text{vertical component of initial velocity}) * t + b. \end{aligned}$$

Chuck releases his javelin at a height of 7 feet. Therefore, we set the initial position at the point $(0, 7)$. That is, $a = 0$ and $b = 7$.

The horizontal component of the initial velocity $= 50\cos(45°) = 50(\frac{1}{\sqrt{2}}) = \frac{50}{\sqrt{2}}$.

The vertical component of the initial velocity $= 50\sin(45°) = 50(\frac{1}{\sqrt{2}}) = \frac{50}{\sqrt{2}}$.

Thus a parametric function for the position of the javelin is

$$s(t) = (x(t), y(t)) = (\tfrac{50}{\sqrt{2}}t, -16t^2 + \tfrac{50}{\sqrt{2}}t + 7).$$

To determine the time the javelin is in flight, we solve the equation $y(t) = 0$. (Why?) That is,

$$-16t^2 + \tfrac{50}{\sqrt{2}}t + 7 = 0$$

Applying the quadratic formula, we have

$$\begin{aligned} t &= \frac{\frac{-50}{\sqrt{2}} \pm \sqrt{(\frac{50}{\sqrt{2}})^2 - 4(-16)(7)}}{-32} \\ &= \frac{\frac{-50}{\sqrt{2}} \pm \sqrt{1250 + 448}}{-32} \\ &= \frac{-35.36 \pm 41.21}{-32} \\ &= 2.395, -0.18. \end{aligned}$$

Because a negative value for t does not make sense in our setting, we know that the javelin hits the ground 2.395 seconds after being thrown. The domain of the parametric function is therefore the interval $[0, 2.395]$. (Explain.)

To determine how far the javelin is thrown, we evaluate $x(t)$ at $t = 2.395$

$x(2.395) = \frac{50}{\sqrt{2}}(2.395) = 84.68$ feet. (Not a world record, but a good throw.) ▲

Parametric functions are useful in describing non-vertical motion (Example 4.3.4) and curves that do not satisfy the vertical line test, such as a circle. Describing a parabola opening to the right is another example of the use of a parametric function to represent a curve that does not satisfy the vertical line test. The equation of the parabola with vertex at the point $(-1,2)$ and opening to the right is $x+1 = (y-2)^2$ or $x = y^2 - 4y + 3$. We obtain a parametric function for this curve by setting $y = t$ and solving for x in terms of t. This gives the parametric function $p(t) = (t^2 - 4t + 3, t)$

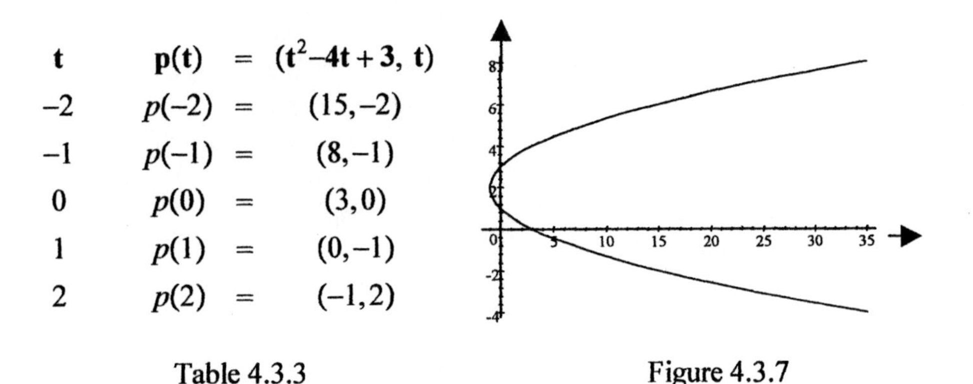

t	p(t)	=	$(t^2-4t+3,\ t)$
-2	$p(-2)$	=	$(15,-2)$
-1	$p(-1)$	=	$(8,-1)$
0	$p(0)$	=	$(3,0)$
1	$p(1)$	=	$(0,-1)$
2	$p(2)$	=	$(-1,2)$

Table 4.3.3

Figure 4.3.7

Example 4.3.5 Leon's Rocket Shoot.

Leon is preparing for a rocket contest that requires all rockets to be fired from the ground with a velocity of 100 kilometers per hour. The winner is the one whose rocket travels the furthest horizontal distance before touching down. The contest rules also require that distance be measured in meters. Because the only variable Leon controls is the angle of elevation, he decides to experiment to determine the angle that yields the greatest horizontal distance. Rather than actually firing his rocket several times and measuring the distances, he decides to model the situation with angles of 30, 45, and 60 degrees and use his graphing calculator to determine the corresponding distances.

Solution:

The problem is similar to Example 4.3.2. Because the motion is not vertical, we will use parametric functions to model the trajectories of Leon's rockets. We begin by defining our variables and stating the given conditions.

t = time in seconds

$p(t)$ = position of the rocket at time t

$p(t) = (x(t), y(t))$ measured in meters

$x(t)$ = (horizontal component of initial velocity) t + (initial horizontal position)

$y(t) = -16t^2$ + (vertical component of initial velocity) t + (initial vertical position)

v = initial velocity $= 100$ = kilometers per hour $= \frac{(100)(1{,}000)}{(60)(60)} = \frac{250}{9}$ meters per second

initial position $= (0,0)$

As stated in Example 4.3.4,

$$\text{horizontal component of initial velocity} = v\cos(\theta) = \tfrac{250}{9}\cos(\theta)$$
$$\text{vertical component of initial velocity} = v\sin(\theta) = \tfrac{250}{9}\sin(\theta).$$

We measure θ in radians. Thus the parametric function is

$$p(t) = \left(\tfrac{250}{9}\cos(\theta)t,\ -16t^2 + \tfrac{250}{9}\sin(\theta)t\right)$$

where θ is the launch angle of elevation.

Leon needs to convert the angle measure from degrees to radians and then plot the resulting three parametric functions. Recall the conversion between degree and radian measure is based on the central angle of a circle. That is, 360 degrees = 2π radians. Thus 1 degree is $\frac{2\pi}{360} = \frac{\pi}{180}$ radians. For example, 30 degrees = $(30)\frac{\pi}{180} = \frac{\pi}{6}$ radians. The conversion into radian measure for the three angles (30, 45, 60 degrees) and the corresponding parametric functions are displayed in the following table.

Trial	Degrees	Radian	Parametric Function
#1	30	$\pi/6$	$p(t) = \left(\frac{500}{9}\cos(\frac{\pi}{6}),\ -16t^2 + \frac{250}{9}\sin(\frac{\pi}{6})\right)$
#2	45	$\pi/4$	$p(t) = \left(\frac{250}{9}\cos(\frac{\pi}{4}),\ -16t^2 + \frac{250}{9}\sin(\frac{\pi}{4})\right)$
#3	60	$\pi/3$	$p(t) = \left(\frac{250}{9}\cos(\frac{\pi}{3}),\ -16t^2 + \frac{250}{9}\sin(\frac{\pi}{3})\right)$

Table 4.3.4

The plots for the three parametric functions are displayed in the following multiplot.

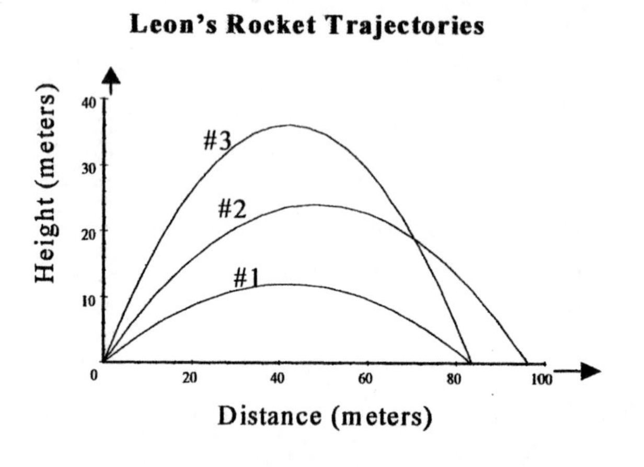

Figure 4.3.8

For the three angles tested, the trajectory angle of 45 degrees gives the greatest distance, but is 45 degrees the best choice for obtaining the maximum distance? Does the distance increase as the trajectory angle increases from 0 to 45 degrees and then decrease as the trajectory angle increases from 45 to 90 degrees? More exploration is needed to answer these questions. See Exercise 10 in this section. ▲

The preceding multiplot of Leon's rocket trajectories suggests that the first and third rockets (30-

and 60-degree trajectory angles) land at the same place. Do they land at the same time? That is, do they collide on landing? A little thought suggests that they would not collide because the lengths of their trajectories are different, and thus their flight times would be different.

Query 3.

In Leon's rocket experiment, what are the flight times for Trial 1 (30 degree trajectory angle) and for Trial 3 (60 degree trajectory angle)? Hint: set the y component equal to zero and solve for t.

Intersection Points or Collision Points?

Two vehicles traveling on intersecting roads can safely pass through a common intersection provided they arrive at the intersection at different times. However, if they arrive at the common intersection at the same time, the result is a collision. The parametric plots of the paths of two moving objects show possible intersection points, but do not indicate if the points are collision points. An analysis of the time variable is required to determine if an intersection point is also a collision point. See Exercise 14 in this section.

Collision-type problems pervade athletics, particularly team sports. Consider a soccer player passing the ball to a charging teammate, a quarterback throwing a pass, or a ballplayer running to catch a fly ball. Each involves determining two trajectories on a collision course. A more lethal example is a missile defense system. Determining collision trajectories is a frequent and often times a difficult problem.

Trigonometric Functions

Trigonometric functions are periodic functions and thus are often used to model wave motion and cyclic phenomena. A function f is periodic if there exists a number p such that $f(x+p) = f(x)$. The smallest positive number p for which this is true is called the period of the function. The six trigonometric functions are defined in terms of ratios of sides of a triangle formed from the center point of a circle of radius r and a point, (a,b), on a circle circumference, as shown in the following diagram.

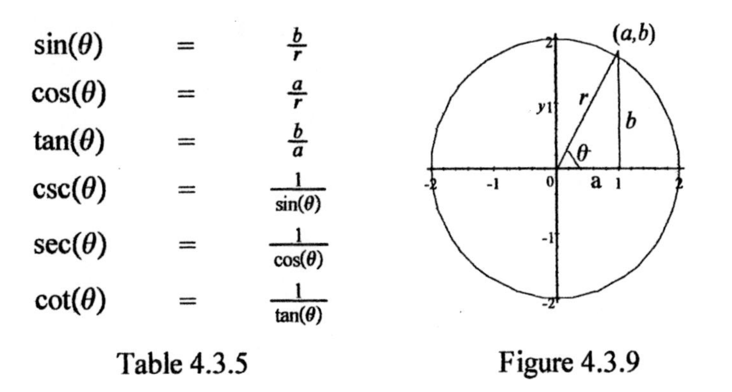

$\sin(\theta)$	=	$\frac{b}{r}$
$\cos(\theta)$	=	$\frac{a}{r}$
$\tan(\theta)$	=	$\frac{b}{a}$
$\csc(\theta)$	=	$\frac{1}{\sin(\theta)}$
$\sec(\theta)$	=	$\frac{1}{\cos(\theta)}$
$\cot(\theta)$	=	$\frac{1}{\tan(\theta)}$

Table 4.3.5

Figure 4.3.9

Notice the following patterns:

$\theta = 0$	$(a,b) = (r,0)$	and	$\sin(0) = 0$
$\theta = \frac{\pi}{4}$	$(a,b) = (\frac{r}{\sqrt{2}}, \frac{r}{\sqrt{2}})$	and	$\sin(\frac{\pi}{4}) = \frac{1}{\sqrt{2}}$
$\theta = \frac{\pi}{2}$	$(a,b) = (0,r)$	and	$\sin(\frac{\pi}{2}) = 1$
$\theta = \frac{3\pi}{4}$	$(a,b) = (\frac{-r}{\sqrt{2}}, \frac{r}{\sqrt{2}})$	and	$\sin(\frac{3\pi}{4}) = \frac{1}{\sqrt{2}}$
$\theta = \pi$	$(a,b) = (-r,0)$	and	$\sin(\pi) = 0$
$\theta = \frac{5\pi}{4}$	$(a,b) = (\frac{-r}{\sqrt{2}}, \frac{-r}{\sqrt{2}})$	and	$\sin(\frac{5\pi}{4}) = \frac{-1}{\sqrt{2}}$
$\theta = \frac{3\pi}{2}$	$(a,b) = (0,-r)$	and	$\sin(\frac{3\pi}{2}) = -1$
$\theta = \frac{7\pi}{4}$	$(a,b) = (\frac{r}{\sqrt{2}}, \frac{-r}{\sqrt{2}})$	and	$\sin(\frac{7\pi}{4}) = \frac{-1}{\sqrt{2}}$

Table 4.3.6

Small-Group Activity

Draw a circle on a piece of graph paper. One member of the group moves her index finger counterclockwise around the circumference of the circle three times, pausing at several different points. At each pause point, another member of the group measures the vertical distance of the point from the horizontal axis. Then another member of the group constructs a bar chart in which the columns correspond to the pause points and the height of a column is the vertical distance at that pause point. Assume that the radius of the circle is one. Then he height of a column is equal to the value of $\sin(\theta)$ at that pause point. Let the horizontal label of the column be the angle at that pause point. Because you are tracing the circle three times, the pause angles are in the interval $[0, 6\pi]$. Does your bar chart suggest the behavior of a periodic function? Modify this activity so that the height of a column in the bar chart is $\cos(\theta)$.

Query 4.

Explain why the periods of the $\sin(\theta)$ and $\cos(\theta)$ are both 2π.

Query 5.

Complete the following table.

$\theta = 0$	$(a,b) = (r,0)$	and	$\cos(0) =$
$\theta = \frac{\pi}{4}$	$(a,b) = (\frac{r}{\sqrt{2}}, \frac{r}{\sqrt{2}})$	and	$\cos(\frac{\pi}{4}) =$
$\theta = \frac{\pi}{2}$	$(a,b) = (0,r)$	and	$\cos(\frac{\pi}{2}) =$
$\theta = \frac{3\pi}{4}$	$(a,b) = (\frac{-r}{\sqrt{2}}, \frac{r}{\sqrt{2}})$	and	$\cos(\frac{3\pi}{4}) =$
$\theta = \pi$	$(a,b) = (-r,0)$	and	$\cos(\pi) =$
$\theta = \frac{5\pi}{4}$	$(a,b) = (\frac{-r}{\sqrt{2}}, \frac{-r}{\sqrt{2}})$	and	$\cos(\frac{5\pi}{4}) =$
$\theta = \frac{3\pi}{2}$	$(a,b) = (0,-r)$	and	$\cos(\frac{3\pi}{2}) =$
$\theta = \frac{7\pi}{4}$	$(a,b) = (\frac{r}{\sqrt{2}}, \frac{-r}{\sqrt{2}})$	and	$\cos(\frac{7\pi}{4}) =$

Table 4.3.7

Query 6.

Explain why the following definitions of horizontal and vertical components of force make sense when considering a force directed along the hypotenuse of a right triangle of angle θ:

The horizontal component of the force is defined to be (magnitude of the force)*cos(θ).

The vertical component of the force is defined to be (magnitude of the force)*sin(θ).

We close this section by recalling the graphs of sin(θ) and cos(θ).

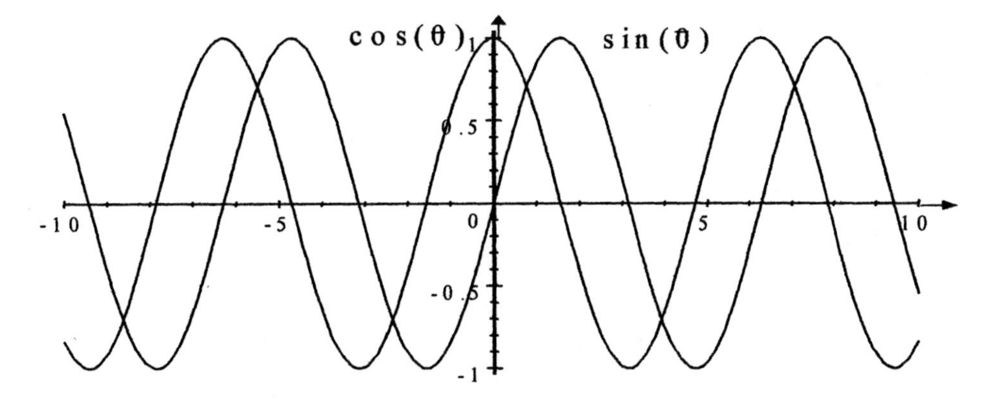

Figure 4.3.10

Exercises 4.3

1. (Computational Skill) Solve for x in the following:

   a. $x^2 + 3x - 1 = 0$

   b. $2x^2 - x - 2 = 0$

   c. $x^2 = 4x + 1$

2. (Computational Skill) Compute the following conversions:

   a. 44 degrees to radian measure

   b. $\frac{3}{5}$ radians to degrees

   c. 60 miles per hour to feet per second.

3. (Computational Skill) Make up and solve three exercises similar to Exercises 1a–c, 2a–c

4. On July 7, 1994, the Associated Press reported the following data for men's 100-meter dash. The third column represents the date in terms of years rounded to 2 decimal places. Graphically determine a model for this data. Suggestion: plot the first entry as $(12.43, 10.6)$, the second entry as $(23.31, 10.4)$, and so on. What is the fastest time permitted by your model?

Record Holder	Date (month/day/year)	Date (yr)	Time (sec)
Donald Lippincott	July 6, 1912	1912.43	10.6
Charles Paddock	April 23,1921	1923.31	10.4
Percy Williams	August 9, 1930	1930.61	10.3
Jesse Owens	June 20, 1936	1936.47	10.2
Willie Williams	August 3, 1956	1956.59	10.1
Armin Hary	June 21, 1960	1960.47	10
Jim Hines	June 20, 1968	1968.47	9.99
Jim Hines	October 14, 1968	1968.79	9.95
Calvin Smith	July 3, 1983	1983.50	9.93
Carl Lewis	September 24, 1988	1988.73	9.92
Leroy Burrell	June 14, 1991	1991.45	9.9
Carl Lewis	August 25, 1991	1991.65	9.86
Leroy Burrell	July 6, 1994	1994.51	9.85

Table 4.3.8

5. A ball is thrown downward toward the ground from the top of a 1,000-foot building with a velocity of 25 feet per second. Determine how long it takes for the ball to reach the ground.

6. A fireworks rocket is blasted upward from the ground with a velocity of 75 feet per second. Determine the following:

   a. How long before the rocket returns to the ground

   b. How high the rocket soared

7. What would be the length of Chuck's javelin throw if he released it with a velocity of 60 feet per second rather than 40 feet per second?

8. What would be the length of Chuck's javelin throw if he released it at 60° rather than at 45°?

9. Complete the following table.

$\theta = 0$	$(a,b) = (r,0)$	and	$\tan(0) =$
$\theta = \frac{\pi}{4}$	$(a,b) = (\frac{r}{\sqrt{2}}, \frac{r}{\sqrt{2}})$	and	$\tan(\frac{\pi}{4}) =$
$\theta = \frac{\pi}{2}$	$(a,b) = (0,r)$	and	$\tan(\frac{\pi}{2}) =$
$\theta = \frac{3\pi}{4}$	$(a,b) = (\frac{-r}{\sqrt{2}}, \frac{r}{\sqrt{2}})$	and	$\tan(\frac{3\pi}{4}) =$
$\theta = \pi$	$(a,b) = (-r,0)$	and	$\tan(\pi) =$
$\theta = \frac{5\pi}{4}$	$(a,b) = (\frac{-r}{\sqrt{2}}, \frac{-r}{\sqrt{2}})$	and	$\tan(\frac{5\pi}{4}) =$
$\theta = \frac{3\pi}{2}$	$(a,b) = (0,-r)$	and	$\tan(\frac{3\pi}{2}) =$
$\theta = \frac{7\pi}{4}$	$(a,b) = (\frac{r}{\sqrt{2}}, \frac{-r}{\sqrt{2}})$	and	$\tan(\frac{7\pi}{4}) =$

Table 4.3.9

10. Continue Leon's exploration of the relation between the rocket's trajectory angle and the horizontal distance it travels. Construct a table showing the conversion from degree to radian measure and the corresponding parametric functions for five different trajectory angles: 45 degrees, two angles less than 45 degrees, and two angles greater than 45 degrees. Construct a multiplot of the plots of the five parametric functions. Interpret the results.

11. Write a two-paragraph paper arguing that the 45-degree trajectory angle yields the maximum distance when air resistance is ignored (see: Example 4.3.5 and exercise 10).

12. Bertha can throw a softball at 60 miles per hour. If she releases the ball at a point 6 feet off of the ground, how far can she throw the ball?

13. Determine how hard (initial velocity) a soccer ball should be kicked if it has a trajectory angle of 45 degrees and is to land 100 feet away.

14. Otis and Oscar are practicing throwing and catching a football. Otis can throw the ball 75 feet per second at an angle of $\frac{\pi}{4}$ radians with a release point 7 feet off the ground. Oscar, the promising end, can run 30 feet per second. If Oscar runs directly away from Otis, how long should Otis wait before throwing the ball if Otis is to catch it three feet off the ground?

4.4 Modeling (Physical Sciences)

In this section, we analyze a cooling problem in physics and two chemistry problems. We model the cooling problem in three ways. First graphically, second approximating the situation by assuming that change only takes place at the end of 10-minute periods, and third assuming that change takes place continuously. The second and third methods depend on Newton's Law of Cooling. The chemistry problems involve balancing chemical reactions and determining concentrations at chemical equilibrium.

Example 4.4.1 (Don's Cooling Cup of Coffee).

Don conducted an experiment to learn something about the rate of cooling of a cup of coffee. He placed a thermometer in a cup of coffee and took the following readings over a sixty-minute time span. The room temperature was 70°F.

Time (minutes)	Degrees (F)
0	120
10	111
20	103
30	97
40	93
50	89
60	86

Table 4.4.1

He observed that the rate of cooling was greatest when the coffee temperature was the highest. For example, in the first ten minutes the coffee cooled 9° from 120° to 111° whereas during the last ten minutes the coffee cooled only 3°, from 89° to 86°. This seems like a reasonable phenomenon because the difference between the room temperature and the coffee temperature is greatest when the coffee is hottest. Two other observations are (1) the coffee temperature is decreasing and (2) the temperature of the coffee will approach but never fall below the room temperature (70°). When will the temperature of the coffee be 80°?

We answer this question in three ways, with a graphical model, a recursive sequence model, and a continuous model, and then compare the results.

Solution (Graphical Model):

The preceding observations suggest that the temperature curve starts at the point (0,120), falls rather steeply, and then gradually flattens out to be asymptotic to the line $y = 70$. These are characteristics of the graph of an exponential function with a negative exponent. (Recall the soda temperature problem, Example 3.4.2.) Because physical scientists typically use the natural exponential function, the exponential function with base e~2.71828, we will consider the function

$$f(t) = ae^{-bt} + 70$$

where $f(t)$ is the temperature of the coffee t minutes after the initial reading. (What is the reason for including the constant 70?) Because the initial temperature is 120 degrees and $f(0) = ae^0 + 70 = a + 70$, we have $a + 70 = 120$ and so $a = 50$. Thus we start with the function

$$f(t) = 50e^{-bt} + 70$$

and scale it to fit the data. The following plot shows the data points and the (graphical) scaling process using the graph of f for $b = 0.1,\ 0.05,$ and 0.02.

Don's Cooling Coffee

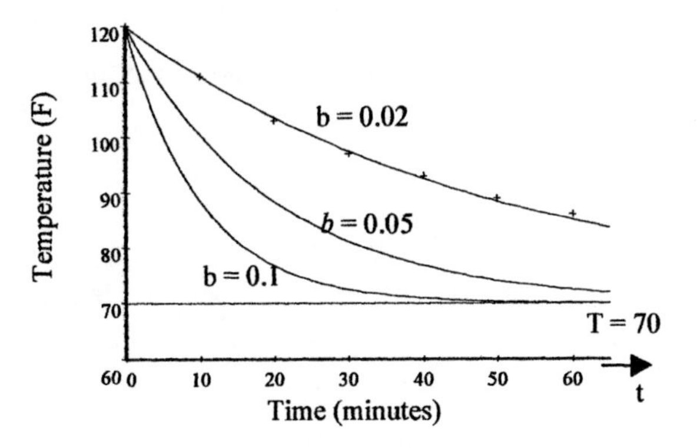

Figure 4.4.1

$f(t) = 50e^{-0.02t} + 70$ appears to be a reasonable model. Using this model, we compute the time when the temperature of the coffee is 80°F by setting $f(t) = 80$ and solving for t.

$$f(t) = 50e^{-0.02t} + 70 = 80$$

implies

$$50e^{-0.02t} = 10$$

or

$$e^{-0.02t} = \tfrac{1}{5}.$$

Taking the natural logarithm (base e) of both sides of the equation yields

$$-0.02t = \ln(\tfrac{1}{5})$$

or

$$t = \frac{\ln(1) - \ln(5)}{-0.02} = 80.47.$$

Thus it will take almost 81 minutes for the temperature of the coffee to cool from 120° to 80°. ▲

Solution (Recursive Sequence Model):

Sir Isaac Newton, one of the world's greatest scientists, formulated what is known today as **Newton's Law of Cooling**, which states that the rate of change in temperature of a cooling or warming body is proportional to the difference between the temperature of the body and the surrounding temperature.

We derive a recursive sequence model in a manner similar to that used to derive a model for Sue's plan in Section 4.2. This requires that we divide the time domain into a sequence of uniform periods. We consider ten-minute periods and assume that the temperature at the beginning of time period $n + 1$ is equal to the temperature at the beginning of time period n plus the change in temperature during the nth time period.

As always, we begin by defining our variables and stating the given conditions. Let

n = number of ten-minute periods
$T(n)$ = temperature of the coffee after n ten-minute periods
$T(0) = 120$
$T(1) = 111$
p = proportionality constant (Newton's Law of Cooling)
RM = Room temperature: $RM = 70$.

Our general recursive sequence model based on Newton's Law of Cooling is

$$T(n) - T(n-1) = p[T(n-1) - RM]$$

or

$$T(n) = T(n-1) + p[T(n-1) - RM].$$

We now solve for the proportionality constant, p. Because we know the temperature when $n = 0$, when $n = 1$, and the room temperature, we can substitute into the preceding equation and obtain

$$T(1) = T(0) + p[T(0) - 70].$$

Substituting for $T(0)$ and $T(1)$, we solve for p as follows:

$$\begin{aligned} 111 &= 120 + p(120 - 70) \\ -9 &= p * 50 \\ -\tfrac{9}{50} &= p \end{aligned}$$

Therefore our model for this problem is

$$T(n) = T(n-1) - \tfrac{9}{50}[T(n-1) - 70] \text{ with } T(0) = 120$$

or

$$T(n) = \tfrac{41}{50}T(n-1) + \tfrac{630}{50} \text{ with } T(0) = 120.$$

The next step in the modeling process (see the modeling process diagram in Section 4.1) is to solve the model. We begin by computing $T(n)$ in terms of $T(0)$ for the first few values of n. Because our purpose is to discover a general expression for $T(n)$, we avoid disguising the pattern by not carrying out any arithmetic simplification.

$$
\begin{aligned}
T(1) &= \tfrac{41}{50}T(0) + \tfrac{630}{50} \\
T(2) &= \tfrac{41}{50}T(1) + \tfrac{630}{50} \\
&= \tfrac{41}{50}[\tfrac{41}{50}T(0) + \tfrac{630}{50}] + \tfrac{630}{50} && \text{Substitute for } T(1) \\
&= (\tfrac{41}{50})^2T(0) + (\tfrac{41}{50})(\tfrac{630}{50}) + \tfrac{630}{50} && \text{Expand} \\
T(3) &= \tfrac{41}{50}T(2) + \tfrac{630}{50} \\
&= \tfrac{41}{50}[(\tfrac{41}{50})^2T(0) + (\tfrac{41}{50})(\tfrac{630}{50}) + \tfrac{630}{50}] + \tfrac{630}{50} && \text{Substitute for } T(2) \\
&= (\tfrac{41}{50})^3T(0) + (\tfrac{41}{50})^2(\tfrac{630}{50}) + (\tfrac{41}{50})(\tfrac{630}{50}) + \tfrac{630}{50} && \text{Expand} \\
T(4) &= \tfrac{41}{50}T(3) + \tfrac{630}{50} \\
&= \tfrac{41}{50}[(\tfrac{41}{50})^3T(0) + (\tfrac{41}{50})^2(\tfrac{630}{50}) + (\tfrac{41}{50})(\tfrac{630}{50}) + \tfrac{630}{50}] + \tfrac{630}{50} && \text{Substitute for } T(3) \\
&= (\tfrac{41}{50})^4T(0) + (\tfrac{41}{50})^3(\tfrac{630}{50}) + (\tfrac{41}{50})^2(\tfrac{630}{50}) + (\tfrac{41}{50})(\tfrac{630}{50}) + \tfrac{630}{50} && \text{Expand}
\end{aligned}
$$

(If the pattern is not clear, the reader should compute $T(n)$ for more values of n.)

Query 1.

Compute $T(5)$ in terms of $T(0)$.

We obtain a general expression for $T(n)$ by following the method illustrated in solving Sue's plan, Section 4.2. We first rewrite the expression for $T(4)$, factoring the $\frac{630}{50}$ out of the last four terms. This gives the following:

$$
\begin{aligned}
T(4) &= (\tfrac{41}{50})^4T(0) + (\tfrac{41}{50})^3(\tfrac{630}{50}) + (\tfrac{41}{50})^2(\tfrac{630}{50}) + (\tfrac{41}{50})(\tfrac{630}{50}) + \tfrac{630}{50} \\
&= (\tfrac{41}{50})^4T(0) + \tfrac{630}{50}[(\tfrac{41}{50})^3 + (\tfrac{41}{50})^2 + (\tfrac{41}{50}) + 1].
\end{aligned}
$$

Note that the sum in the bracket, $(\frac{41}{50})^3 + (\frac{41}{50})^2 + (\frac{41}{50}) + 1$, is a geometric series with the ratio $r = \frac{41}{50}$. Summing this geometric series (see Section 4.2) yields

$$
\begin{aligned}
T(4) &= (\tfrac{41}{50})^4T(0) + \tfrac{630}{50}\frac{1-(\frac{41}{50})^{3+1}}{1-\frac{41}{50}} \\
&= (\tfrac{41}{50})^4T(0) + \tfrac{630}{50}\frac{1-(\frac{41}{50})^4}{\frac{9}{50}} \\
&= (\tfrac{41}{50})^4T(0) + \tfrac{630}{9}[1-(\tfrac{41}{50})^4] \\
& (\tfrac{41}{50})^4T(0) + 70[1-(\tfrac{41}{50})^4] \\
&= (\tfrac{41}{50})^4T(0) + 70 - 70(\tfrac{41}{50})^4 \\
&= (\tfrac{41}{50})^4[T(0) - 70] + 70 \\
T(4) &= (\tfrac{41}{50})^4[T(0) - 70] + 70.
\end{aligned}
$$

Having rewritten the expression for $T(4)$, we generalize it to an expression for $T(n)$ by noting where 4, the input value, occurs in the expression for $T(4)$. The 4 occurs as the input value in $T(4)$ and as the exponent on $(\frac{41}{50})$. Replacing each occurrence of the input value 4 by n yields the expression for $T(n)$ that we have been seeking. We now have a symbolic solution, based on Newton's Law of Cooling, for the temperature after n ten-minute periods, which is

$$T(n) = \left(\tfrac{41}{50}\right)^n[T(0) - 70] + 70.$$

Now to compute the number of minutes for the coffee to cool to 80°, we set $T(n) = 80$, solve for n, and then multiply by 10. (Why multiply by 10?)

$$\begin{aligned}
T(n) &= 0 \\
\Rightarrow \left(\tfrac{41}{50}\right)^n[T(0) - 70] + 70 &= 80 \\
\left(\tfrac{41}{50}\right)^n[120 - 70] + 70 &= 80 \\
\left(\tfrac{41}{50}\right)^n[50] &= 10 \\
\left(\tfrac{41}{50}\right)^n &= \tfrac{1}{5} \\
n\ln\left(\tfrac{41}{50}\right) &= \ln\left(\tfrac{1}{5}\right) \\
n &= \frac{\ln(.2)}{\ln(.82)} \\
n &= 8.11 \\
10n &= 81.10
\end{aligned}$$

(You should pause at this point to explain each of the algebraic and/or arithmetic manipulations in the preceding calculations.)

According to our model, it will take 81.10 minutes for the coffee to cool to 80°. ▲

Solution (Continuous Model):

The recursive sequence model is called a **discrete** model because it measures the temperature at discrete periods of time (10-minute intervals). In calculus, you will study the continuous model of Newton's Law of Cooling that is obtained by solving the differential equation $\frac{dT}{dt} = p[T(t) - RM]$. This model is

$$T(t) = ce^{-pt} + RM$$

where

t	=	time in minutes
$T(t)$	=	temperature of the body at time t
p	=	proportionality constant
RM	=	surrounding temperature
e	=	base of the natural logarithm
c	=	coefficient to be determined

For our problem, the body is the coffee and the surrounding temperature is the room temperature. Furthermore, we know that the initial body (coffee) temperature is 120 and the room temperature is 70. Because the initial temperature of the body is $T(0) = c + RM$, we can solve for c as $c = T(0) - RM = 120 - 70 = 50$. This reduces the model to $T(t) = 50e^{-pt} + 70$. We now solve for the value of p by evaluating this equation at a known data point, say (10, 111).

$$
\begin{array}{lcll}
T(10) & = & 111 & \text{Implies} \\
50e^{-10p}+70 & = & 111 & \\
50e^{-10p} & = & 41 & \text{Subtract 70 from both sides} \\
e^{-10p} & = & \frac{41}{50} & \text{Divide both sides by 50} \\
-10p & = & \ln(\frac{41}{50}) & \text{Apply logarithm function to both sides, } \log(e) = 1 \\
p & = & \frac{\ln(0.82)}{-10} & \text{Divide both sides by } -10 \\
p & = & 0.0198 & \text{Computation.}
\end{array}
$$

The continuous model for our problem is thus.

$$T(t) = 50e^{-0.0198t} + 70$$

Now in order to compute the number of minutes for the coffee to cool to 80°, we set $T(t) = 80$ and solve for t.

$$T(t) = 50e^{-0.0198t} + 70 = 80 \Rightarrow e^{-0.0198t} = \frac{10}{50}$$

$$\Rightarrow t = 81.285 \text{ minutes. (Using logarithms.)} \quad \blacktriangle$$

The results of our three models for the time it takes for the coffee to cool to 80° are as follows:

graph model:	80.47 minutes
discrete model	81.10 minutes
continuous model	81.29 minutes

Example 4.4.2.

Determine a natural exponential function whose graph passes through the 2 points $(0, 100)$ and $(10, 30)$, lies above the t-axis, and is asymptotic to the positive t-axis.

Solution:

The conditions could describe a cooling function, and so we use the solution of the continuous model for Newton's Law of Cooling, $T(t) = ce^{bt} + RM$. The parameter RM represents the surrounding temperature and therefore RM represents the asymptote in the graph. (Why?) In this example, the asymptote is the t-axis and thus $RM = 0$. Therefore our solution function reduces to $T(t) = ce^{pt}$.

Substituting the two data points into our solution expression gives two equations in the parameters a and b.

$$\begin{cases} \text{Substituting } (0,100) \text{ yields } T(0) = ce^0 = c = 100. \\ \text{Substituting } (10,30) \text{ yields } T(10) = ce^{10b} = 30. \end{cases}$$

Thus
$$\begin{array}{c} c = 100 \\ ce^{10p} = 30 \end{array}.$$

Substituting for c in the second equation yields

$$100e^{10p} = 30 \text{ or } e^{10p} = \tfrac{3}{10}.$$

To solve for p, we apply the natural logarithm to both sides of this equation, obtaining

$$10p = \ln(0.3) \text{ or } p = \tfrac{1}{10}\ln(0.3) = -0.1204.$$

Thus $c = 100$, $p = -0.1204$, and therefore the desired function is $T(t) = 100e^{-0.1204t}$. ▲

Query 2.

If in the preceding example, the curve was asymptotic to the line $T = 10$, what would be the resulting function?

Small-Group Activity

Consider how the temperature of a potato changes when placed in a 350°F oven and left to bake. Sketch a temperature graph of the potato. Is the temperature graph concave up like the right-hand side of a parabola, or is it concave down like the arc of the cosine function from 0 to $\frac{\pi}{2}$ or is it an *S*-shaped logistic type curve illustrated in Figure 4.4.4? Explain your reasoning to your class.

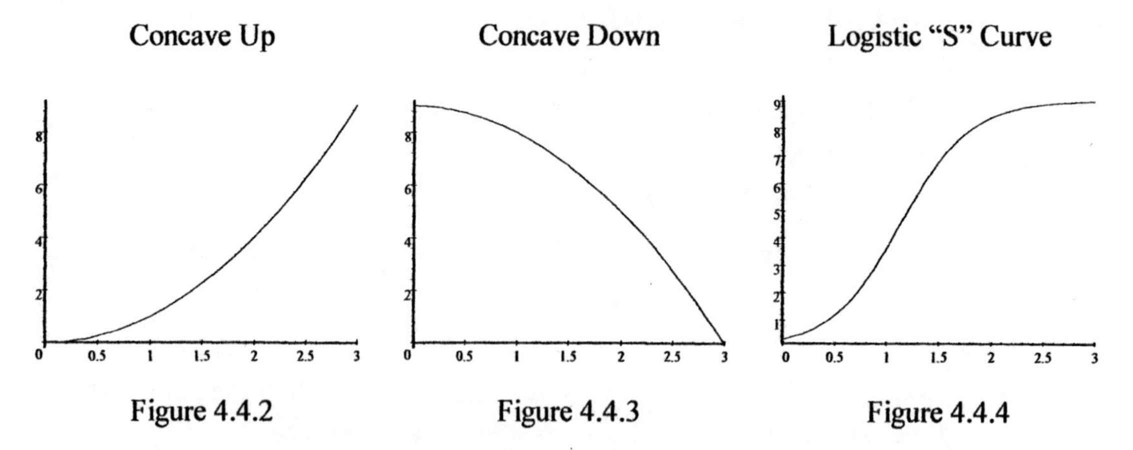

Figure 4.4.2 Figure 4.4.3 Figure 4.4.4

Is there a maximum temperature that the potato can obtain? If so, what is it?

Query 3.

If the graph of a function is classified as increasing and concave down, how is the graph of the negative of the function described?

Chemistry—Balancing Chemical Reactions

The law of conservation of mass states that there is no detectable change in mass during the course of an ordinary (nonnuclear) chemical reaction. This law was first stated by Antoine Lavoisier in his work *Traite Elementaire de Chemie* (1789). This means that there must be as many atoms of each element, combined or uncombined, after a chemical reaction as before the chemical reaction. *Balancing a chemical reaction* means to determine the number of atoms of each element involved in the reaction. (See Section 2.8.)

Bottled gas, such as is found in outside gas grills, contains butane gas, which burns with oxygen to form carbon dioxide and water. The chemical reaction is

$$C_4H_{10} + O_2 \rightarrow CO_2 + H_2O.$$

Example 4.4.3

Balance the chemical reaction: $C_4H_{10} + O_2 \rightarrow CO_2 + H_2O$. That is, determine the number of atoms of each chemical (carbon, hydrogen, oxygen) in order to satisfy the law of conservation of mass.

Solution:

We begin the solution process by using variables to denote the number of molecues in each chemical combination involved in the reaction. We let

x_1 = number of molecues of C_4H_{10} (butane gas) each of which contains 4 atoms of C and 10 atoms of H

x_2 = number of molecues of O_2 (oxygen) each of which contains 2 atoms of O

x_3 = number of atoms of CO_2 (carbon dioxide) each of which contains 1 atom of C and 2 atoms of O

x_4 = number of atoms of H_2O (water) each of which contains 2 atoms of H and 1 atom of O

Applying the law of conservation of mass, gives the following equations.

The number of atoms of carbon (C) before the reaction is equal to the number of atoms of carbon after the reaction. Thus

$$4x_1 = x_3.$$

The number of atoms of hydrogen (H) before the reaction is equal to the number of atoms of hydrogen after the reaction. Thus

$$10x_2 = 2x_4.$$

The number of atoms of oxygen (O) before the reaction is equal to the number of atoms of oxygen after the reaction. Thus

$$2x_2 = 2x_3 + x_4.$$

To balance the chemical reaction, we need to solve these three equations for positive integer values of $x_1,\ x_2,\ x_3,\ x_4$.

Because there are three equations and four variables, we expect there are infinitely many solutions. We write down the system of equations, form the associated augmented matrix, use our calculators to find the rref matrix, and then write down the equivalent system of equations.

System:
$$\begin{cases} x_1 \quad -x_3 \quad = 0 \\ 10x_1 \quad -2x_4 = 0 \\ 2x_2 \ -2x_3 \ -x_4 = 0 \end{cases}$$

Augmented matrix:
$$\begin{bmatrix} 1 & 0 & -1 & -0 & 0 \\ 10 & 0 & 0 & -2 & 0 \\ 0 & 2 & -2 & -1 & 0 \end{bmatrix}$$

rref matrix: $\begin{bmatrix} 1 & 0 & 0 & -.2 & 0 \\ 0 & 1 & 0 & -1.3 & 0 \\ 0 & 0 & 1 & -.8 & 0 \end{bmatrix}$

Equivalent system: $\begin{cases} x_1 - .2x_4 = 0 \\ x_2 - 1.3x_4 = 0 \\ x_3 - .8x_4 = 0 \end{cases}$.

Thus $x_1 = 0.2x_4$, $x_2 = 1.3x_4, x_3 = 0.8x_4$. Because our solution needs to be in positive integer values and because x_1, x_2, x_3 are expressed in terms of x_4, we chose an integer value for x_4 that will make the other three variables integers. Let $x_4 = 10$. Our solution is

$$\begin{cases} x_1 = 2 \\ x_2 = 13 \\ x_3 = 8 \\ x_4 = 10 \end{cases}$$

and the balanced equation is: $2C_4H_{10} + 13O_2 = 8CO_2 + 10H_2O$. ▲

The study of chemical equilibrium often presents difficult challenges. The study is based on the observation that after a sufficiently long time following a chemical reaction, the amount of reactants and products are independent of time. Consider a generic acid HA, which dissociates as follows:

$$HA <-> H^+ + A^-.$$

After a sufficiently long time, we say this reaction is in equilibrium, meaning that the concentrations (amounts) of stuff on the left, reactants and products, and the concentrations of stuff on the right are constant. This information, obtained from experiment with a specific ratio of the concentrations, is expressed as

$$K = \frac{[H^+][A^-]}{[HA]}$$

where K is known as the equilibrium constant. The bracket notation $[HA], [A^-], [H^+]$ is the standard notation for indicating the concentrations of each species.

Example 4.4.4.

The generic weak acid HA dissociates in water according to the reaction $HA <-> H^+ + A^-$. Because HA is a weak acid, the dissociation is not complete and only a small amount will dissociate. At equilibrium, the equilibrium constant K satisfies the equation

$$K = \frac{[H^+][A^-]}{[HA]}.$$

We assume that initially we have HA in water at an initial concentration of c. What are the concentrations of all species at equilibrium given that $c = 1$ and $K = 1.8 * 10^{-5}$? (The values for c and K are reasonable for acetic acid.)

Solution:

We begin by defining our variables and stating the given conditions. Let

x be the number of molecules of HA that dissociate
$K = 1.8 * 10^{-5}$
$c = 1$

We note that for every molecule of HA that dissociates, we obtain one H^+ and one A^- molecule. Thus at equilibrium the concentration of HA is $c - x$, the concentration of H^+ is x, and the concentration of A^- is x. We determine the value of x by substituting into the equilibrium equation and solving the resulting quadratic equation.

$$K = \frac{x * x}{c - x} \text{ or } x^2 + Kx - cK = 0$$

Applying the quadratic formula (Section 4.3), we have

$$x = \frac{-K \pm \sqrt{K^2 + 4ck}}{2}$$

and substituting the values for K and c, gives

$$x = 4.234 * 10^{-3}.$$

At equilibrium the concentrations are

$$[HA] = 9.96 * 10^{-1}, [H+] = [A^-] = 4.234 * 10^{-3}.$$ ▲

Query 4.

Rework Example 4.4.3 using the values $c = 0.8$ and $K = 3.5 * 10^{-4}$, which are appropriate for hydrofluoric acid.

Exercises 4.4

1. (Computation Skill) Simplify the following expressions.
   a. $\frac{4^2}{2^8}$
   b. $1 + 2 + 2^2 + 2^3 + \ldots + 2^{10}$
   c. $1 - 2 + 2^2 - 2^3 + \ldots + (-2)^{10}$

2. (Computational Skill) Solve for x using logarithms.
   a. $4^x = 17$
   b. $12 * 2^{3x} = 30$
   c. $(2^{3x} + 8^x) = 47$

3. (Computational Skill) Make up and solve three exercises involving computations with exponents and logarithms.

4. Derive a symbolic formula for the kth term of the recursive sequence $a(0) = 5$ and $a(n + 1) = 0.2a(n) + 3$. What is the long-term value of the sequence? That is, what is the value of $a(k)$ for large values of k?

5. Rework Example 4.4.2 with the first point is changed from $(0, 100)$ to $(0, 75)$.

6. Determine a natural exponential function whose graph passes through the points $(1,-100),(10,-30)$, lies below the *t*-axis and is asymptotic to the positive *t*-axis. Hint: Review Example 4.4.2.

7. A potato is placed in an 350°F oven and left to bake for one-hour. Assume the temperature of the potato is 70°F when placed in the oven and the proportionality constant in Newton's Law of Cooling is 1.7. Determine both the temperature of the potato after one hour and after several hours in the oven using the following.

   a. Continuous model

   b. Discrete model

8. For each of the following conditions, sketch a graph illustrating the condition and then determine a scaled and/or shifted plot of $f(x) = ax^2 + b$ that approximates your graph.

   a. An increasing function whose graph is concave up

   b. An increasing function whose graph is concave down

   c. A decreasing function whose graph is concave up

   d. A decreasing function whose graph is concave down

9. Repeat Exercise 8 using the exponential function, $g(x) = ae^{bx}$.

10. Balance these chemical reactions.

    a. Ammonia reacts with oxygen to produce nitrogen and water: $NH_3 + O_2 \rightarrow N_2 + H_2O$.

    b. Potassium nitrate reacts with oxygen to produce potassium nitrate: $KNO_2 + O_2 \rightarrow KNO_3$.

11. Balance these chemical reactions:

    a. Sodium chloride reacts with hydrogen sulfate to produce sodium sulfate and chlorine: $NaCL + H_2SO_4 \rightarrow Na_2SO_4 + HCL$.

    b. Sodium Carbonate reacts with calcium hydroxide to produce sodium hydroxide and sodium carbonate: $Na_2CO_3 + Ca(OH)_2 = NaOH + CaCO_3$

12. Balance these chemical reactions:

    a. Nitropropane produces Water plus Carbon Monoxide plus Nitrogen gas plus Oxygen gas: $C_3H_5(NO_3)_3 \rightarrow H_2O + CO + N_2 + O_2$.

    b. Sodium carbonate (Na_2CO_3) plus bromine gas (Br_2) produces sodium bromide (NaBr) plus sodium bromate ($NaBrO_3$) plus carbon dioxide (CO_2): $Na_2CO_3 + Br_2 \rightarrow NaBr + NaBrO_3 + CO_2$.

13. Silver chloride (AgCl), a solid, dissociates in water with the process $AgCl = Ag^+ + Cl^-$. The equilibrium concentrations satisfy the following equation $K = [Ag^+][Cl^-]$, where [] denotes concentration, the amount of matter per unit volume. The measured value of K is $1.6 * 10^{-10}$.

    a. Assume no silver or chloride ion is present initially. Determine the equilibrium concentrations of silver ion and chloride ion.

    b. Assume the solution initially contains chloride ion at a concentration of 0.00100 moles per liter. Determine the equilibrium concentrations of silver ion and chloride ion.

14. Repeat Exercise 10a for silver sulfide (Ag_2S), whose dissociation process is $Ag_2S = 2Ag^+ + S^{-2}$ and equilibrium equation is $K = [Ag^+]^2[S^{-2}]$.

4.5 Modeling (Blend Problems)

Blend problems occur in a wide spectrum of settings from the home kitchen (eggs into a batter) to public swimming pools (chlorine into the water) to sewage treatment plants to oil refineries. A key to solving blend problems is understanding the definition of concentration as is illustrated in the following problems.

Example 4.5.1 (Washing Clothes).

Larry decides to save money on laundry detergent by mixing his own solution of bleach and water. The label on the bleach container recommends a laundry mixture of 1 cup of bleach to 1 quart of water. What is the recommended concentration? How much bleach is in a 10-quart premixed solution? How much water is in a 10-quart premixed solution?

Solution:

The concentration is the amount of bleach divided by the total volume (bleach plus water). Thus the recommended laundry concentration is $\frac{1 \text{ cup}}{5 \text{ cups}} = \frac{1}{5} = 0.20.$

(1 quart = 4 cups). Why is the denominator 5 rather than 4?

The amount of bleach in a 10-quart premixed solution is the concentration times the volume. That is, $(0.20)(10) = 2$ quarts or 8 cups. Thus there is 8 quarts or 32 cups of water in the 10-quart premixed solution. ▲

Example 4.5.2 (Mopping the Floor).

Larry decides to dilute his premixed bleach laundry solution, which has a concentration of 0.20 to make a floor washing solution that has 1 cup of bleach per gallon of water. How much water should Larry add to 1 gallon of the premixed laundry solution to get the correct concentration for the floor-washing solution?

Solution:

We first compute the concentration for the floor-washing solution. This is the amount of bleach divided by the total volume (bleach plus water). Thus the concentration is

$$\frac{1 \text{ cup}}{1 \text{ cup } + 1 \text{ gallon}} = \frac{1 \text{ cup}}{17 \text{ cups}} = \frac{1}{17} \text{ (1 gallon = 4 quarts, 1 quart = 4 cups).}$$

We now define the unknown. Let

x = number of gallons of water added to the premixed solution.

The amount of bleach in 1 gallon of the premixed solution is the concentration times the volume. Thus there is (0.20)(1 gallon) = 0.2 gallons of bleach in the premixed solution. The concentration after adding x gallons of water should be $\frac{1}{17}$. That is,

$$\frac{0.2 \text{ gallons}}{1 + x \text{ gallons}} = \frac{0.2}{1+x} = \frac{1}{17}.$$

Hence $3.4 = 1 + x$ and so $x = 2.4$. Larry should add 2.4 gallons of water to the premixed solution. ▲

Example 4.5.3 (Coffee—Regular, Decaffeinated, Lite, or a Mixture?).

There was a time when ordering coffee, the server asked, "Cream and sugar?" A few years ago, decaffeinated coffee appeared on the market and then the server asked, "Regular or decaffeinated? Cream and sugar?" Now with lite brands (50% decaffeinated) available, the server asks, "Regular, lite, or decaffeinated? Cream and sugar?" In the future, as blends proliferate, the server will probably ask, "Caffeine rating? Cream and sugar?"

Margaret really enjoys the full, rich flavor of regular coffee; however, her doctor has advised her to reduce the amount of caffeine she ingests. Thus she decides to compromise and mix the decaffeinated coffee with a lite coffee, mixing 1/3 of the lite with 2/3 decaffeinated. The caffeine rating for decaffeinated coffee is 3%, and the caffeine rating for the lite coffee is 50%. (The caffeine rating for regular coffee is 100%.) What is the caffeine rating for Margaret's mixture?

Solution:

The caffeine rating is proportional to the caffeine concentration. Thus Margaret's mixture has a caffeine rating of

$$\tfrac{1}{3}(0.50) + \tfrac{2}{3}(.03) = \tfrac{0.56}{3} = 0.1867 \text{ or } 18.67\%. \quad ▲$$

This next example was submitted by Wayne Sadik, an environmental chemical engineer for the EXXON Petroleum Corporation.

Chemical plants and refineries have many pipes that need to be heated or cooled at various times during the refining process. The cooling is usually done with heat exchangers in which cool water flows over hot pipes. (A car radiator is a heat exchanger in which the cooling fluid is the air rushing over the warm tubes in the radiator.) The warm water from the heat exchangers then needs to be cooled. This is done in cooling towers where the warm water is run over a lattice of wood, plastic, or ceramic, and air is blown through it. This action causes some of the water to be evaporated, which must then be replaced. Over time the residual chemicals and minerals in the water accumulate and sometimes become so concentrated as to cause corrosion. To counter this accumulation of chemicals and minerals, the water is treated (very expensive) or a small amount of water containing the chemicals is removed (continuously or at intervals). The water removed is called "blow-down."

Example 4.5.4.

EXXON's Baton Rouge plastics plant makes various grades of polyethylene. The plant has several small cooling towers and one large one called the base plant cooling tower. It is too large to have its own blow-down line, and so chemicals and minerals tend to concentrate in it. Water is fed into this cooling tower from three sources. The rate of flow is measured in gallons per minute (gpm), and the concentration of silica in the water is measured in parts per million (ppm). The flow rates and silica concentrations for these three sources are as follows.

Well Water: 550 gpm and 24 ppm SiO_2
Reverse Osmosis (RO) treated boiler water: 90 gpm and 128 ppm SiO_2
Blow-down from E-Line cooling tower: 25 gpm and 109 ppm SiO_2

A new source of water is being evaluated to offset a portion of the well water and reduce pumping costs. The new source of water is from treated oily water sewer permeate. If this stream is used, the flow rates and silica concentration for the four sources would be as follows:

Well water: 460 gpm and 24 ppm SiO_2
RO treated boiler water: 90 gpm and 128 ppm SiO_2
Blow-down from E-Line cooling tower: 25 gpm and 109 ppm SiO_2
Oily water sewer permeate: 90 gpm and 63 ppm SiO_2

Will the concentration of silica using the oily water sewer permeate be greatly different from the present concentration? An outside vender saw this as an opportunity and offered to provide a special chemical at $500 per month to alleviate the silica buildup. Wayne Sadik did the following computations and showed that the increase in silica did not require the special chemical.

Solution:

The silica concentration is the number of parts per million of silica in solution divided by the total number of gallons.

In the present situation, the amount of silica from the three sources is

Well water: 550 gpm x 24 ppm =	13,200 ppm
RO unit: 90 gpm x 128 ppm =	11,520 ppm
Blow-down: 25 gpm x 109 ppm =	2,725 ppm
Total	27,445 ppm

and the total number of gallons per minute is $550 + 90 + 25 = 665$ gpm.

Thus the silica concentration is $\frac{27{,}445}{665} = 41.3$ ppm.

In the proposed situation, the amount of silica from the four sources is

Well water: 460 gpm x 24 ppm =	11,040 ppm
RO unit: 90 gpm x 128 ppm =	11,520 ppm
Blow-down: 25 gpm x 109 ppm =	2,725 ppm
Oily water: 90 gpm x 63 ppm =	5,670 ppm
Total	30,955 ppm

and the total number of gallons per minute is $460 + 90 + 25 + 90 = 665$ gpm.

Thus the silica concentration is $\frac{30{,}955}{665} = 46.5$ ppm.

The small increase in silica concentration is acceptable, and therefore Wayne Sadik was able to save the company money by utilizing the treated oily water sewer permeate without having to pay an outside vendor to treat the solution. ▲

Query 1.

Compute the concentration of silica if the proposed situation had been to use 465 gallons of well water, 90 gallons of the RO unit, 25 gallons from the blow-down, and 85 gallons from the oily water.

There are numerous situations in which mixing takes place on a continuous basis, such as evaporating water in the process of making maple syrup, pollution leaching into groundwater, chlorinating swimming pools, and the mixing of hot and cold water. We will approximate a continuous mixing with a discrete model by assuming that the change takes place at periodic intervals. The analysis is based on our paradigm,

(New Situation) = (Old Situation) + (Change).

Example 4.5.5 (The Brine Tank).

A stream of water flows into a 50 gallon brine tank at the rate of 1 gallon per minute. A second stream flows out of the tank at the same rate. Every 10 minutes, 5 pounds of salt is added to the tank. If the 50 gallons of brine initially contained 10 pounds of salt, how much salt is in the tank after 2 hours? What is the long-term concentration in the tank? (Assume that the solution is being constantly mixed to produce a uniform concentration.)

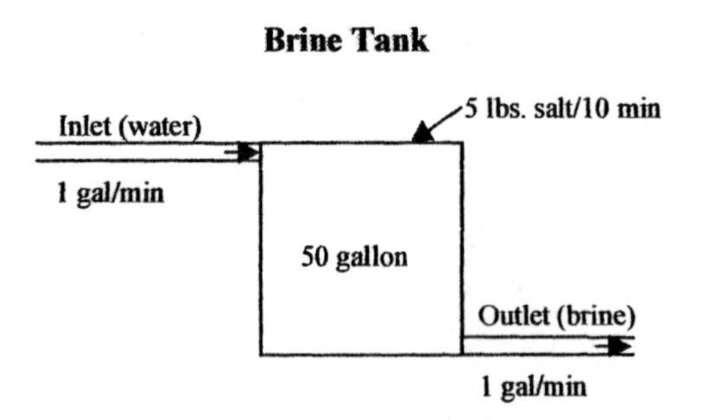

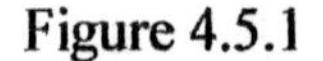

Figure 4.5.1

Solution:

We shall assume that all changes take place at the end of 10-minute time periods. Thus at the end of each 10-minute time period, 10 gallons of water is added to the tank, 10 gallons of brine is removed from the tank, and 5 pounds of salt is added to the tank. We now define the variables and state the initial conditions.

t = number of 10-minute periods
$s(t)$ = number of pounds of salt in the tank after t 10-minute periods
$s(0)$ = 10 lbs, initial amount of salt in the tank

The concentration of salt at time t is

$$\frac{\text{lbs. of salt}}{\text{volume}} = \frac{s(t)}{50}.$$

Therefore the amount of salt removed from the tank during the tth 10-minute time period is the concentration times the volume of solution removed, $\frac{s(t-1)}{50}(10) = \frac{s(t-1)}{5}$.

Thus the change that takes place during a 10-minute time period is the amount of salt added minus the amount removed or $5 - \frac{s(t-1)}{50}(10)$.

Now using our paradigm, **(New Situation)** = **(Old Situation)** + **(Change)**, we form the following model:

$$s(t) = s(t-1) + (5 - 10\tfrac{s(t-1)}{50}) = \tfrac{4}{5}s(t-1) + 5 \text{ with } s(0) = 10.$$

This recursive sequence is similar to the model for Sue's plan (Section 4.2). In that problem, the solution of the sequence

$$a(n) = (1+r)a(n-1) + d \text{ was } a(k) = (1+r)^k c + \tfrac{d}{1-(1-r)}$$
$$\text{where } c = a(0) - \tfrac{d}{1-(1-r)}.$$

Applying this solution format to the brine tank problem gives

$$s(k) = (\tfrac{4}{5})^k c + \tfrac{5}{1-\frac{4}{5}} = (\tfrac{4}{5})^k c + 25$$
$$\text{where } c = 10 - \tfrac{5}{1-\frac{4}{5}} = 10 - 25 = -15.$$

The number of pounds of salt in the brine tank after n 10-minute time periods is

$$s(n) = -15(\tfrac{4}{5})^n + 25.$$

Thus after 2 hours, that is 12 10-minute time periods, we have

$$s(12) = -15(\tfrac{4}{5})^{12} + 25 = 23.97 \text{ lbs. of salt.}$$

In order to determine the long-term concentration, we need to first determine the long-term amount of salt in the brine tank. Because $(\frac{4}{5})^k \to 0$ as k becomes very large, the long-term amount of salt in the brine tank is 25 lbs. Thus the long term concentration of salt in the brine tank is $\frac{25}{50} = \frac{1}{2}$ pound of salt per gallon. ▲

Query 2.

In Example 4.5.5, how many 10-minute time periods does it take before the concentration level is greater than 0.45? Explain your reasoning.

Exercises 4.5

1. (Computation Skill)
   a. Determine the concentration after 3 cups of milk is added to 2 gallons of water.
   b. How much salt is in 5 gallons of brine with a concentration of 0.32?
   c. How much water needs to be added to 10 gallons of brine to reduce the concentration from 0.3 to 0.15?
2. (Computational Skill) Solve the following recursive sequences (refer to Sue's plan, Section 3.2).
   a. $a(n) = 0.7a(n-1); a(0) = 10$
   b. $a(n) = 0.7a(n-1) + 5; a(0) = 10$
   c. $a(n) = 2a(n-1) - 4; a(0) = 10$

3. (Computational Skill) Make up and solve three exercises similar to Exercise 1.

4. The instructions on the label of a can of insect dust says in order to form a spray mixture, combine 2 tablespoons of the dust to a quart of water. How many pounds of dust should Uncle Will add to the 50-gallon tank on his spraying machine if there are 64 tablespoons of dust in 1 pound?

5. If 1 gallon of 90°F water is mixed with 5 quarts of 60°F water, what is the resulting water temperature?

6. (Small-Group Activity) Solve the following problem by developing a recursive sequence model (refer to the solution of Sue's plan in Section 4.2 and Example 4.4.1). 90°*F* water flows at the rate of 1 quart per minute into a 3-gallon sink filled with 60°F water. Water flows out through the overflow at the same rate of 1 quart per minute. Assume the water in the sink is being constantly mixed in order to maintain a uniform temperature. Determine how long it takes for the water in the sink to reach 70°F.

7. (Small Group Activity) The label on a gallon of antifreeze says that "This antifreeze protects against winter freeze up and summer boil over" and provides the following chart. Extend the chart to include percentages from 0 to 100 in jumps of 10. Write a paragraph explaining your reasoning.

Zerex %	Water %	Freeze Point (F)	Boil Point (F)
70	30	–84	276
60	40	–62	270
50	50	–34	265

8. Will wants to flush the cooling system in his dump truck that holds about 3 gallons of coolant. He attaches a tap-in on the heater hose to run water into the system at the rate of 1 half gallon per minute and opens the radiator's petcock to release the coolant at the same rate. Will runs the motor to keep the liquid thoroughly mixed during the flushing process. If the coolant is 60% antifreeze at the start, how long will it take to reduce the antifreeze percentage to 10%? Develop a recursive sequence model for the situation, obtain a general solution of the model, and then answer Will's question. Hint: let $a(n)$ = the amount of antifreeze in the coolant after n minutes.

9. (Small-Group Activity) Linda and her son prepare 2.5 gallons of punch according to the recipe: "Stir 8 cups of sugar and 2 cups of strawberry jam into 1.5 gallons of water. Add 3 cups grape juice, 3 cups pineapple juice, 8 sliced bananas. Just before serving add 2 quarts of cherry Koolaid frozen into cubes and 6 ounces of frozen grapefruit juice concentrate." Unfortunately, her son dissolved 12 ounces of grapefruit concentrate rather than the 6 ounces called for in the recipe. Linda decides to serve the punch and after each cup of punch is served, she will add a cup of water. How many cups of punch need to be served before the grapefruit concentration agrees with the recipe concentration, that is, 6 ounces per 40 cups?

10. (Small-Group Activity) Leroy prepares two gallons of shrimp creole for a dinner party. Unfortunately, he adds 2 tablespoons of Hot Cajun sauce per gallon rather than the 2 teaspoons called for in the recipe. (There are 3 teaspoons in a tablespoon. Thus Leroy added 12 teaspoons of Hot Cajun sauce rather than 4 teaspoons!) What to do? He doesn't want to throw out the 2 gallons of shrimp creole and besides some people might like it hot. He decides to serve the shrimp creole over rice, two cups to a serving. After each serving, he adds 2 cups of white sauce to the shrimp creole bowl. After how many servings of shrimp creole will the concentration of Hot Cajun sauce be reduced to that specified in the recipe?

4.6 Modeling (Life Sciences)

In the previous sections of this chapter, we have modeled a broad spectrum of situations with recursive sequences based on our paradigm (**New Situation**) = (**Old Situation**) + (**Change**). These recursive sequences have the form $a(n) = r^n c + b$ where $a(n)$ represents the quantity present at the end of the nth period, r is the rate of change, and c and b are constants. In some situations, the rate of change, r, was greater than 1, and in other situations the rate of change was less than 1. In Example 4.2.1, Sue's plan for a college education account modeled by $a(n) = 1.005^n(10,100) - 10,000$, the rate of change is 1.005. Whereas in Example 4.5.5, the amount of salt in a brine tank after n hours modeled by $s(n) = 0.8^n(-15) + 25$, the rate of change is 0.8. In these examples, as well as in the others that we have analyzed, we were concerned with a fixed value of n. In this section, we will also focus on long-term effects. That is, what happens when n becomes very large?

The long-term behavior can generally be described as belonging to one of the following three categories.

1. **Equilibrium**, the quantity converges to a constant amount
2. **Unbounded**, the quantity grows or decays without bound
3. **Oscillation**, the quantity oscillates

For example, the amount of salt in the brine tank (Example 4.5.5) converges to the equilibrium state of 25 lbs., Sue's college education account (Example 4.2.1) would grow without bound if the account were allowed to continue forever, and a model of the hours of sunlight in Beddington, Maine, would show oscillatory behavior.

Query 1.
Explain how the rate of change, r, determines the category of long-term behavior of the recursive sequence $a(n) = r^n c + b$.

This section concludes with an analysis of a **logistic model** that is frequently used to model population growth. The graph of a logistic model is an S-shaped curve.

In this next example, we develop a recursive sequence model by a method other than iteration.

Example 4.6.1 (Litton's Medical Dilemma).

After an examination, Litton's doctor prescribed 600 mg of medication (2 pills) every 4 hours for 2 days to treat his medical condition. Being a curious person, Litton wondered if at the end of 2 days the amount of medication in his system would be different if he took 1 pill every 2 hours for 2 days rather than what was prescribed.

Solution:

In order to answer Litton's question, we will develop and solve a model for each of the two scenarios based on the following two assumptions.

1. The medication is instantaneously ingested into his body's system.
2. 20% of the medication in his system is expelled from his system each hour.

Model A: (2 pills every 4 hours)

We begin by defining the variables and stating the given conditions. Let

n = number of 4 hour time periods
$m(n)$ = amount (mg.) of medication in Litton's system after the nth time period.
$m(0) = 600$, the initial amount of medication
600 mg. = 4 hour dosage (2 pills).

Now to develop a recursive sequence model. Recall our paradigm is

(New Situation) = (Old Situation) + (Change).

The new situation is $m(n)$, the old situation is $m(n-1)$, and the change is what happens during the nth 4-hour time period. We now determine the rate of change from one period to the next.

The period begins with Litton having $m(n)$ mg of medication in his system.

At the end of the first hour, Litton retains $0.8m(n)$ of the medication. (He loses 20% per hour.)
At the end of the second hour, Litton retains $0.8[0.8m(n)] = 0.8^2m(n)$ of the medication.
At the end of the third hour, Litton retains $0.8[0.8^2m(n)] = 0.8^3m(n)$ of the medication.
At the end of the fourth hour, Litton retains $0.8[0.8^3m(n)] = 0.8^4m(n)$ of the medication and then takes a new dosage of 600 mg.

Thus the rate of change is $.8^4$, and the model (with numbers rounded to 2 decimal places) is

$$m(n) = 0.8^4m(n-1) + 600 = 0.41m(n-1) + 600 \text{ with } m(0) = 600.$$

This model is very similar to the model developed in Example 4.2.1 for Sue's plan for a college education account. We could follow the same iteration process for developing a closed form solution to our model as we did for Sue's plan. However, we will follow an alternative approach based on our experience iterating a recursive sequence and the assumption that the amount of medication in Litton's system will eventually stabilize at an equilibrium level.

By inspection, we know that one aspect of iterating the recursive sequence, $m(n) = 0.41m(n-1) + 600$, involves multiplying each iterate by the factor 0.41. Thus the nth iterate will contain the term 0.41^nC, where C is some constant. The other terms will involve sums and products of the numbers 0.41 and 600. We therefore conjecture that the format of a closed form solution is

$$m(n) = 0.41^nC + B, \quad \text{where } C \text{ and } B \text{ are constants.}$$

We let E denote the equilibrium level at which the amount of medication in Litton's system stabilizes. When the equilibrium level has been obtained, there will be no change in amount of medication in Litton's system in successive time periods. Thus, because $m(n)$ = amount (mg) of medication in Litton's system after the nth time period, at equilibrium, $m(n) = m(n-1) = E$.

Another important observation with respect to our conjectured solution format is that 0.41^n grows smaller as n grows larger. In fact, for very large values of n, 0.41^n is very close to zero and so 0.41^nC is very close to zero. Thus for very large values of n, $m(n)$ is approximately equal to B. However, for very large values of n, $m(n)$ is approximately equal to the equilibrium value E. Hence $B = E$ and our conjectured closed form solution is $m(n) = 0.41^n C + E$, where C is a constant and E is the equilibrium value.

In order to determine the value of E, we consider n to be very large and therefore we may substitute E for both $m(n)$ and $m(n-1)$ in the recursive sequence and then solve for E.

$m(n)$	$= 0.41m(n-1)+600$	Recursive sequence
E	$= 0.41E+600$	Substitute E for $m(n)$ and $m(n-1)$
$E-0.41E$	$= 600$	Subtract $0.41E$ from both sides
$0.59E$	$= 600$	Simplify
E	$= \frac{600}{0.59}$	Divide both sides by 0.59
E	$= 1{,}016.95$	Simplify

Our conjectured closed form solution is now $m(n) = 0.41^n C + 1016.95$, where C is a constant.

Because this is a conjectured solution, we need to verify that it is in fact a solution of the recursive sequence by substituting the solution into the original recursive sequence to see if the result is an equality. If it is, then we will have verified that the conjectured solution is, in fact, a solution.

So, we substitute our conjectured solution into the recursive sequence

$m(n)$	$= 0.41m(n-1)+600$	Recursive sequence
$0.41^n C + 1{,}016.95$	$= 0.41(0.41^{n-1}C + 1{,}016.95) + 600$	Substitute, note the exponents
	$= 0.41^n C + 0.41(1{,}016.95) + 600$	Simplify
	$= 0.41^n C + 416.95 + 600$	Simplify
	$= 0.41^n C + 1{,}016.95$	Equality

We have verified that our conjectured solution is a solution. The final step is to determine the value of the constant C. We do this by evaluating the solution at $n = 0$. Because $m(0) = 600$ (initial value) and $m(0) = 0.41^0 C + 1{,}016.95 = C + 1{,}016.95$,

$$C + 1{,}016.95 = 600$$

or

$$C = 416.95.$$

Therefore our closed form solution is

$$m(n) = -416.95(0.41^n) + 1{,}016.95.$$

The equilibrium level is shown in the following plot.

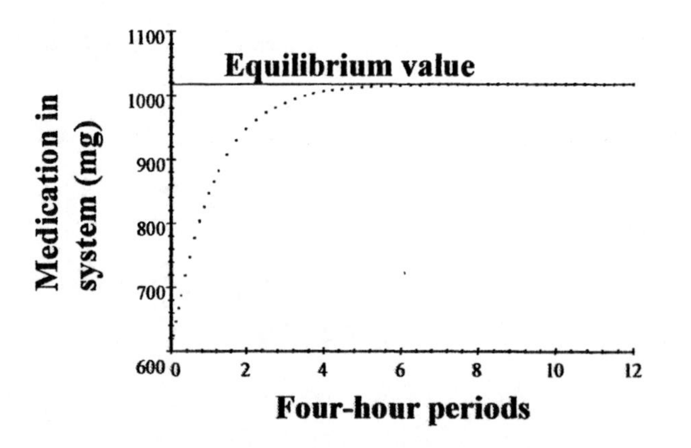

Figure 4.6.1

Because there are 12 (4-hour) periods in 2 days and $m(0) = 600$, the amount of medication in Litton's system after 2 days is

$$m(12) = -416.95(0.41^{12}) + 1{,}016.95 = 1{,}016.94 \text{ mg (rounded to two decimal places).}$$

Model B (1 pill every 2 hours)

We begin by defining the variables and stating the given conditions. Let

n = number of 2-hour time periods
$w(n)$ = amount (mg.) of medication in Litton's system after the nth time period
$w(0) = 300$, the initial amount of medication
300 mg. = two hour dosage (1 pill).

Query 2.
Develop a model for Model B following the process used to develop Model A.
(Answer: $w(n) = 0.8^2 w(n-1) + 300 = 0.64w(n-1) + 300,\ w(0) = 300$)

Query 3.
Develop a closed form solution to Model B following the process used for Model A.

The closed form solution for Model B is

$$w(n) = -533.33(0.64^n) + 833.33.$$

Because there are 24 (2-hour) periods in 2 days and $w(0) = 300$, the amount of medication in Litton's system after 2 days is

$$w(24) = -533.33(0.64^{24}) + 833.33 = 833.32 \text{ mg.} \quad \blacktriangle$$

Interpretation of the Results

Following the doctor's dosage rate of 2 pills every 4 hours, Litton will have 1,016.94 grams of medication in his system at the end of the second day. However, if he modifies the prescription and takes 1 pill every 2 hours, he will end the second day with only 833.32 grams of medication in his system, even though he has taken the prescribed number of pills. This is quite different than what the doctor had anticipated. As always, the results are dependent on the assumptions made. Changing our two assumptions will yield different results. However, our analysis has shown that changing the dosage rate can have substantial effects.

We iterate and then plot to obtain numerical and graphical representations of the amount of medication in Litton's system.

Model A

Hours	**Medication (mg)**
0	600
4	846
8	946.86
12	988.21
16	1,005.17
20	1,012.12
24	1,014.97
28	1,016.14
32	1,016.62
36	1,016.81
40	1,016.89
44	1,016.93
48	1,016.94

Table 4.6.1

Model B

Hours	**Medication (mg)**	**Hours**	**Medication (mg)**
0	300	26	831.72
2	492	28	832.3
4	614.88	30	832.67
6	693.52	32	832.91
8	743.85	34	833.06
10	776.06	36	833.16
12	796.68	38	833.22
14	809.87	40	833.26
16	818.31	42	833.28
18	823.72	44	833.30
20	827.18	46	833.31
22	829.39	48	833.32
24	830.81		

Table 4.6.2

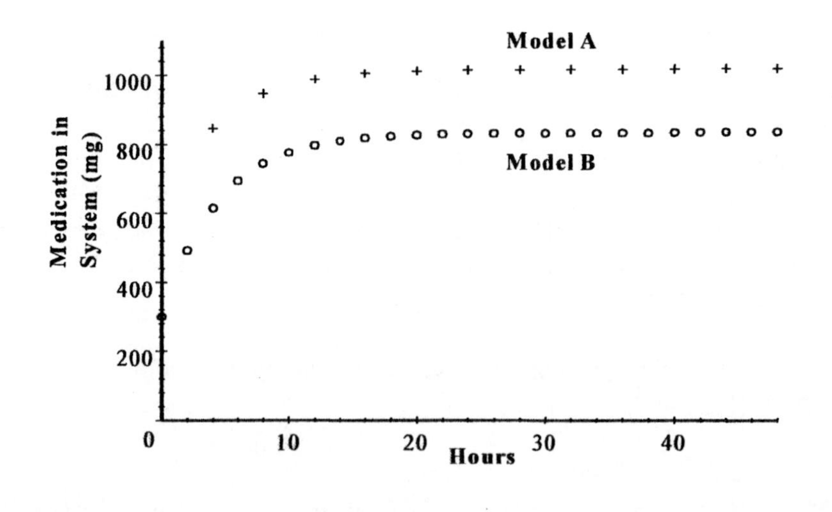

Figure 4.6.2

Both the numerical and graphical models indicate that the level of medication in Litton's system is close to the equilibrium level after the first 12 hours (1,016.94 for Model A and 833.33 for Model B). We now look at the symbolic models to see how we could have determined these relatively constant levels of medication.

Model A: $m(n) = 0.41^n C + 1{,}016.95,\ m(0) = 600$ (C is a constant)

Model B: $w(n) = 0.64^n D + 833.33,\ w(0) = 300$ (D is a constant)

Note that the rate of change in both models is a positive number less than 1 raised to the nth power and thus these terms will approach zero as n becomes very large. (Why?) This means that the equilibrium levels of medication are determined by the second terms. The important point to understand is how these second terms were formulated. At equilibrium there is no change and thus in Model A $m(n) = m(n-1) = E$ and similarly for Model B. Substituting E for $m(n)$ and $m(n-1)$ into the recursive sequence $m(n) = 0.8^4 m(n-1) + 600$ and solving gives

$$E = \frac{600}{1-0.8^4} \quad \text{(equilibrium value for Model B} = \frac{300}{1-0.8^2}\text{)}.$$

Thus the equilibrium value (or long-term value) is the quotient of the dosage divided by one minus the retention rate of the medication per period. That is,

$$\text{Amount of medication in system in the long term} = \frac{\text{dossage}}{1 - \text{retention rate}}.$$

Query 4.

What dosage in milligrams should a doctor prescribe a patient to take every 6 hours if 60% of the drug is eliminated every 6 hours and the doctor wants the patient to have 1,000 grams of the medication in her system after 4 days of taking the medication?

Logistic Model

Problems involving growing populations need to take into account environmental restrictions on the maximum sustainable size of the population. Ecologists call this maximum the **carrying capacity** of the environment. Biologists usually describe carrying capacity of a lake in terms of fish mass. A less accurate but more readily understandable description heard in a tropical fish store is "one inch of fish per gallon of water." Thus the approximate carrying capacity of a 20-gallon aquarium would be 20 inches of fish.

Example 4.6.2 (Growth of a Yeast Culture).

An experiment in the growth of a yeast culture in a restricted volume produced the data in the following table where

n = number of hours
$b(n)$ = volume of biomass at the end of n hours
$b(0)$ = 9.6, initial amount.

Time (hr)	Yeast Biomass (gram) b(n)	Change in Biomass (gram) b(n) – b(n – 1)
0	9.6	
1	18.3	8.7
2	29.0	10.7
3	47.2	18.2
4	71.1	23.9
5	119.1	48.0
6	174.6	55.5
7	257.3	82.7
8	350.7	93.4
9	441.0	90.3
10	513.3	72.3
11	559.7	46.4
12	594.8	35.1
14	640.8	11.4
15	651.1	10.3
16	655.9	4.8
17	659.6	3.7
18	661.8	2.2

Table 4.6.3

The third column in the Table 4.6.3 indicates changes in the growth rate. The rate is slow at first, picks up and becomes large, and then decreases to almost nothing. The greatest growth appears to occur near the eighth hour. The following plot provides a visual picture of the growth rate. From the plot, we could estimate that the carrying capacity in this experiment is approximately 675.

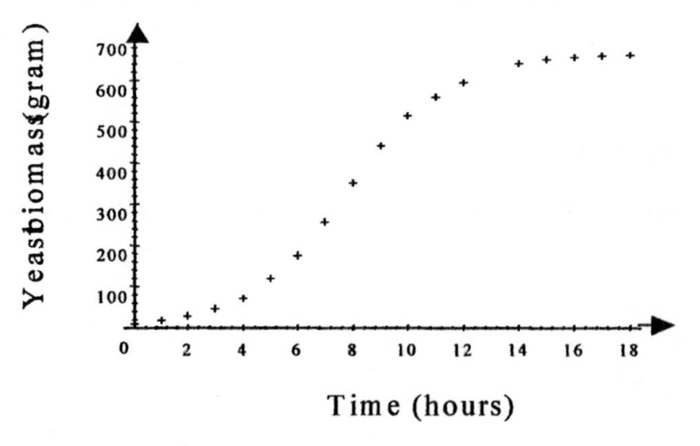

Figure 4.6.3

Query 5.

Is the concept of carrying capacity value the same as the concept of equilibrium value? (Explain.)

We now develop a recursive sequence model for this growth of a yeast culture. This will be a more complicated model than we have previously considered. We shall depend on Sections 3.5 (Shifting and Scaling Graphs) and 3.7 (Graphical Approximations) as well as the work on exponential functions in Section 4.2 to suggest and guide the development of a model. We begin, as always, by defining our variables and stating the given conditions. Let

n = number of hours
$b(n)$ = weight (gram) of biomass at the end of n hours
$b(0)$ = 9.6, the initial amount
L = the carrying capacity (~ 675).

We observe the following from the scatter plot.

1. For the first few hours, the shape of the plot suggests that the data is exponential as confirmed by the following plot. (Curve A is the graph of $f(x) = e^x$ and Curve B is the graph of a scaled version of A, $g(x) = 9.6e^{.5x}$.) Thus the change in the volume of biomass is proportional to the amount present. (Recall in Section 4.2, we said that the defining characteristic of an exponential function is that the amount of change is proportional to the quantity present.) So for the first few hours, $b(n) - b(n-1) = mb(n-1)$ for some proportionality constant m.

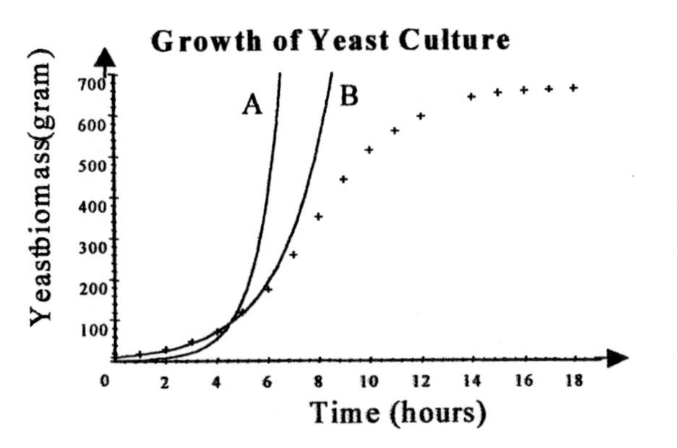

Curve A: $f(x) = e^x$

Curve B: $g(x) = 9.6e^{.5x}$

Figure 4.6.4

2. When n becomes large, the data points converge to the carrying capacity and the growth approaches zero. Also the last few data points in the plot appear to have the same shape as would the first few points if they were pivoted about the data point (8,350.7). An approximation to this pivoting can be obtained through four scaling transformations of $g(x) = 9.6e^{.5x}$ to yield $h(x) = -9.6e^{-.5(t-16)} + 665$. This is confirmed in Figure 4.6.5. Thus the shape of the last few points is an exponential with negative exponent and therefore the last few data points have a relationship to the carrying capacity (L) that is similar to the relationship that the first few data points have to the horizontal axis. That is, the amount of change in the weight of the biomass is proportional to the difference between the carrying capacity and the amount of biomass present. So for the last few hours, $b(n) - b(n-1) = j[L - b(n-1)]$ for some proportionality constant j.

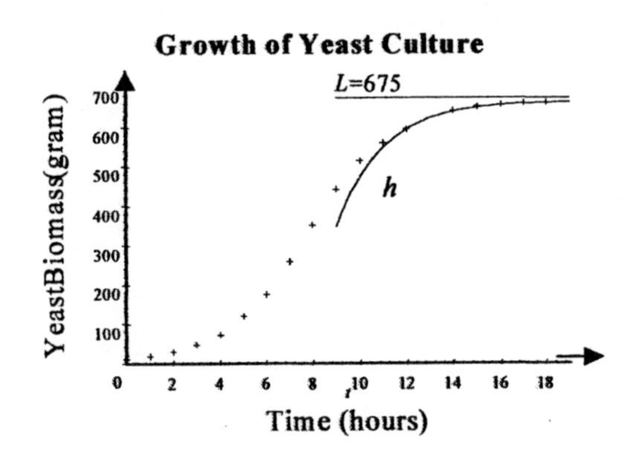

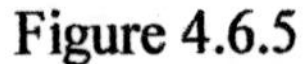

Figure 4.6.5

The challenge is how to capture the partial results of the two observations into one change equation. One possibility is to combine them in the following manner.

$$b(n) - b(n-1) = kb(n-1)[L - b(n-1)]$$

where $L = 675$ is the carrying capacity and k is a proportionality constant. The reasoning is that when $b(n-1)$ is small, $[L - b(n-1)] \sim L$ and so $b(n) - b(n-1) \sim kLb(n-1)$. Thus the change in the growth of the biomass is proportional to the amount present, which is Observation 1. When $b(n-1)$ is close to the carrying capacity, $b(n-1) \sim L$ and thus $b(n) - b(n-1) \sim kL[L - b(n-1)]$. Thus the change in the growth of the biomass is proportional to the difference between the carrying capacity and the amount present, which is Observation 2.

Solving for k, we have $k = \dfrac{b(n) - b(n-1)}{b(n-1)[675 - b(n-1)]}$. We focused on the two ends of the data pattern to formulate the basic model. Let us now focus on the middle of the data pattern to determine a value for k. We compute the k values for the four hours ($n = 7, 8, 9, 10$) that give the large changes in the volume of the biomass and then average the results. This gives $k = 0.00083$.

We are now ready to formulate a model. Recall the paradigm

(New Situation) = (Old Situation) + (Change).

The new situation is $b(n)$, the old situation is $b(n-1)$, and the change is $kb(n-1)[L - b(n-1)]$. Thus our model is

$$b(n) = b(n-1) + 0.00083b(n-1)[675 - b(n-1)], \text{ with } b(0) = 9.6.$$

We check the reasonableness of our model by numerically and graphically comparing the observed values with the predicted values for $b(n)$.

Hours	Observed Values	Predicted Values
0	9.6	9.6
1	18.3	14.9
2	29.0	23.1
3	47.2	35.6
4	71.1	54.5
5	119.1	82.6
6	174.6	123.2
7	257.3	179.6
8	350.7	253.4
9	441.0	342.1
10	513.3	436.6
11	559.7	523.0
12	594.8	589.0
13	629.4	631.0
14	640.8	654.0
15	651.1	665.4
16	655.9	670.7
17	659.6	673.1
18	661.8	674.2

Table 4.6.4

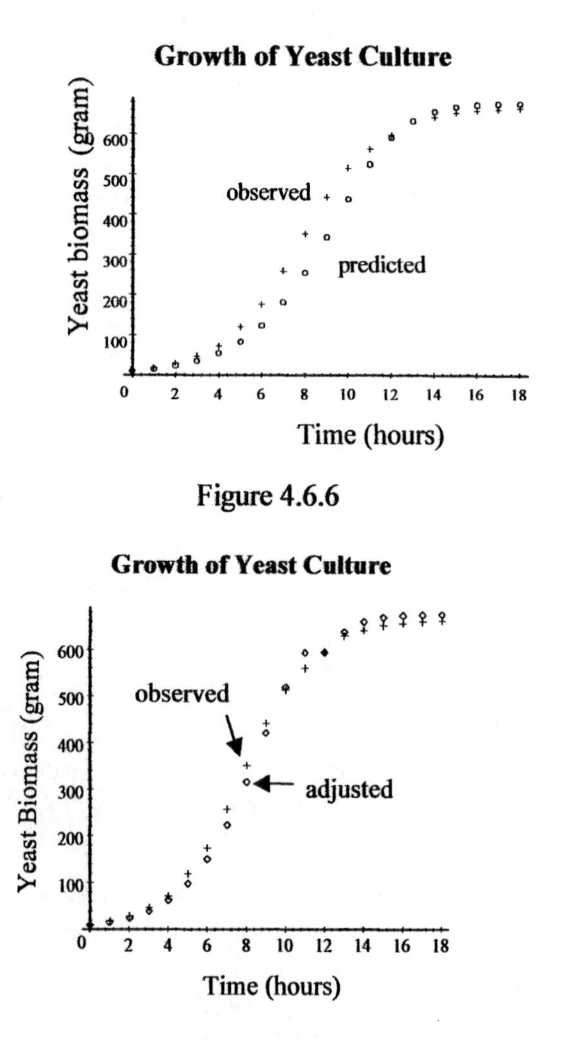

Figure 4.6.6

Figure 4.6.7

Although the fit is good on the ends, it is not in the middle. However, the model successfully captures the shape of the data, which is the most important consideration in constructing a model. (The coefficients in the solution can be adjusted to obtain a better fit as illustrated in Figure 4.6.7 with $k = .00092$.) This general model,

$$b(n+1) = b(n) + kb(n)[L - b(n)]$$

representing the S shaped data curve is called the **logistic model.** This is the general discrete model for population growth in a constrained environment. In calculus, you will formulate the differential equation $\frac{db}{dt} = kb(t)[L - p(t)]$ as a continuous logistic model and obtain the following solution expression,

$$b(t) = \frac{L}{1 + ce^{-Lkt}}$$

where L is the carrying capacity, k is the proportionality constant, and c is a parameter dependent on the initial condition.

Exercises 4.6

1. (Computational Skills) Verify the truth of the following equations.
   a. $(0.41)^4 + (0.41)^3 + (0.41)^2 + 0.41 + 1 = \frac{1-(0.41)^5}{1-0.41}$
   b. $1 + 2 + 2^2 + 2^3 + \dots + 2^{10} = \frac{1-2^{11}}{1-2}$
   c. $\frac{1}{2} + (\frac{1}{2})^2 + (\frac{1}{2})^3 + (\frac{1}{2})^4 + (\frac{1}{2})^5 + \dots = 1$
2. (Computational Skills)
   a. Compute $b(2)$ given $b(n+1) = b(n) + 0.00083b(n)[675 - b(n)]$ and $b(0) = 9.6$.
   b. Compute the value of c given $b(t) = \frac{L}{1+ce^{-Lkt}}$ and $L = 675, k = 0.00083$, and $b(0) = 9.6$.
   c. Compute $b(12)$ given $b(t) = \frac{L}{1+ce^{-Lkt}}$ where $L = 675, k = 0.00083$, and $c = 69.31250$.
3. (Computational Skills) Make up and solve two geometric series exercises similar to Exercise 1 and one iteration problem similar to Exercise 2a.
4. (Reference to Litton's Medical Dilemma) Determine the amount of medication in Litton's system after 2 days if he took 4 pills every 8 hours.
5. Rework Litton's medical dilemma with the assumption that 40% rather than 20% of the medication in his system is expelled from his system each hour.
6. Answer Queries 2 and 3.
7. If Lili takes 900 gms of medication and if 15% of the medication in her system is expelled from her system each hour, how many hours will it be before she has only 5 grams of medication in her system?
8. Don's doctor has advised him to take 1 baby aspirin (81 mg) per day in order to thin his blood and prevent heart attacks. Determine the long-term level of aspirin in Don's system using the assumption that his body expels 20% of the aspirin in his system every day.

9. Don's doctor has advised him to take 1 baby aspirin (81 mg) per day in order to thin his blood and prevent heart attacks. While camping in a remote area, Don ran out of baby aspirin. He considered taking regular aspirin (324 mg), but did not know what dosage would produce the same aspirin level in his body as his daily dose of baby aspirin. Develop a model and determine how frequently Don should take a regular aspirin in order for him to maintain the doctor's recommended aspirin level in his body. Assume that Don's body expels 20% of the aspirin in his system every day.

10. Describe each of the four scaling transformations in the development of the model for the growth of the yeast biomass that transformed $g(x) = 9.6e^{.5x}$ to $h(x) = -9.6e^{-.5(t-16)} + 665$.

11. Jed starts a rumor that the college president has cancelled all classes next Friday in order to give the students an opportunity to catch up. Assume that there are 600 students in the college. Sketch a graph of a rumor function where the input (independent variable) is time and the output (dependent variable) is the number of people who have heard Jed's rumor. Describe your reasoning. What effect, if any, would there be if the college were a commuter college rather than a residential college?

12. Consider the logistic equation $f(t) = \frac{L}{1+ce^{-kt}}$. Explore how the shape of the S curve changes as you change the parameters l, c, and k. Begin with $L = 50$, $c = 5$, $k = .2$, and domain of $f = [0, 100]$. Write a paragraph describing your discoveries.

13. Let $p(t) = \frac{100}{1+4e^{-t}}$ be the solution of a logistic model for the growth of a population. Approximate the following.
    a. The carrying capacity of the environment
    b. The time it takes for the population to reach half the size of the carrying capacity

14. Form a multiplot of the observed data in the growth of a yeast culture problem and the solution to the continuous logistic model using the values $L = 675$, $c = 70$, and $k = 0.00083$. Can you adjust the k value to get a better fit? How? (Recall that we approximated the k value using only some of the observed data values.)

15. Plot the following data and then graphically determine a logistic model that approximates the data. Estimate the carrying capacity. Hint: Recall the solution expression for the continuous logistic model.

x	$f(x)$
0	1
50	2.890
100	6.306
150	9.534
200	11.172
250	11.750
300	11.927

Figure 4.6.5

16. In the 1840s, the Belgian mathematician P. F. Verhulst predicted that there was a carrying capacity for the world's population. He based his prediction, in part, on the following United States census data. Plot this data and then experimentally determine a logistic model for the data.

The logistic model has the form $p(t) = \frac{L}{1+ce^{-Lkt}}$ where $p(t)$ represents the United States population at year t, L is the carrying capacity, k is a proportionality constant, and c is a constant dependent on the initial condition. Conjecture a value for the carrying capacity and present your answer in graphical form.

U.S. Census Data

Date	**Population (millions)**
1790	3.2929
1800	5.308
1810	7.240
1820	9.638
1830	12.866
1840	17.069

Figure 4.6.6

4.7 Modeling (Economics)

In this section, we look at two important aspects of economics related to decision making. The first is to emphasize that committing resources (time, money, influence, equipment, etc.) to one option means not committing these resources to other options. Hence a cost of selecting an option is the value of the other options that could have been chosen. This is called the opportunity cost of the option. The second is to understand the multiplier effects associated with consumption, investment, spending, and taxes when modeling an economy.

Opportunity Cost

Economics is the study of making choices under constraint of scarcity. In most situations, desires exceed resources and thus create a scarcity environment in which choices need to be made. Each choice involves a cost. The cost associated with choosing one option over other available options is called opportunity cost. The **opportunity cost** of an option is the value of what was given up in choosing that option. Thus the opportunity cost of accepting a free lunch is the value of the best alternative for the use of the lunch time. The opportunity cost of selecting apple pie over ice cream for dessert is the value associated with a serving of ice cream. The opportunity cost of taking a day off from work is the pay that could have been earned by working. Identifying and measuring opportunity costs are important aspects of decision making.

Every choice has an opportunity cost.

Query 1.

What is the opportunity cost of extending the evening serving time in your college dining hall by one-half hour?

Query 2.

The role of government-mandated fuel-economy standards for cars and trucks is a contentious issue in global warming debates. In the United States, power plants are the largest source of carbon dioxide emissions, and gasoline burning cars and trucks are the second-largest contributor. What choices does the government have with respect to mandating fuel-economy standards? What is the opportunity cost for each of these choices?

The following example illustrates how opportunity cost can be used to assign workers to specific tasks.

Example 4.7.1.

Bianca and Jim own and operate Central Pizza. They have one station for making pizzas and one station for making sandwiches. Bianca can make 8 pizzas per hour and Jim can make 5. Also Bianca can make 16 sandwiches per hour compared to Jim's 12. Who should be making pizzas and who should be making sandwiches during rush times?

Solution:

We compare the opportunity cost of Bianca shifting from making sandwiches to making pizzas to the opportunity cost of Jim shifting from making sandwiches to making pizzas. Bianca's opportunity cost of making 8 pizzas is 16 sandwiches. Thus her opportunity cost for one pizza is 2 sandwiches. On the other hand, the opportunity cost for Jim making 5 pizzas is 12 sandwiches. Thus his opportunity cost for one pizza is $\frac{12}{5} = 2.4$ sandwiches. Since Bianca's opportunity cost is lower than Jim's, she should make the pizzas and Jim should make the sandwiches. ▲

The preceding example illustrates how the concept of opportunity cost is utilized in international trade. When trading countries focus on doing what they do best, then both the quantity of trade and the level of profits increase. An illustration of this is the growth in trade between the United States and Mexico during the past few years.

Multipliers in the Economy

Budgets are planning documents. Some are simple and straightforward, whereas others are complicated. In particular, the federal budget is a very complicated document designed to guide the economic system to meet political, social, and economic goals. It contains thousands of pages that detail proposed government spending and expected revenues. Because budgets are developed with only partial understanding of how political, social, and economic forces will evolve, modifications are often necessary. In order to develop an understanding of the dynamics of modifying an economic system, in particular the role of multipliers, we begin with an oversimplified economy—one that contains no government and no imports or exports. Our work illustrates the basic modeling process of making simplifying assumptions to enable one to traverse the triangular modeling diagram as presented in Example 4.1.1 and then relaxing the assumptions to increase the realism of the model.

Part I. A macroeconomic analysis of a simple economy, one in which there is no government and no imports or exports.

We begin with the following basic assumptions and definitions.

1. Macroeconomics looks at the total output and total expenditure in the economy in contrast to microeconomics which looks at individual units such as an industry.
2. **Gross Domestic Product** (*GDP*) is the total national income or the total output of the economy. That is, *GDP* is the sum of the goods and services produced by the economy.
3. Consumption (C) is a linear function of GDP: $C = a + m\ GDP$. The slope, m, is called the Marginal Propensity to Consume (*MPC*). Thus $MPC = \dfrac{\text{change in consumption}}{\text{change in } GDP}$. The constant a represents the level of consumption if *GDP* were zero. On a personal basis, *MPC* is the percent of additional income you would spend rather than save. For example if you received a \$100 bonus, the *MPC* is the amount of the bonus you would spend.
4. Investment (I) is the creation of capital stock. Buildings, equipment, and inventories are all part of a firm's investment.
5. The total expenditure of the economy is the sum of household consumption (C) and business

investment (I).

6. In macroeconomics, equilibrium is achieved when the total output is equal to the total expenditure. In symbols, equilibrium of the economic system is represented by the equation $GDP = C + I$ where GDP = total output of goods and services, C = total consumption by households, and I = total investment by firms.

7. Households dispose of income in only two ways—consume (C) or save (S). Thus for households, $GDP = C + S$.

8. Equilibrium requires investment to equal savings, $I = S$. (From assumption 7, $GDP = C + S$ and from assumption 6, $GDP = C + I$. Thus $I = S$.)

9. In order to simplify the modeling process, we assume that changes in the economy take place at the end of regular time periods.

The relations among income, consumption, savings, and investments are illustrated in the following diagram.

Economic Flow Diagram

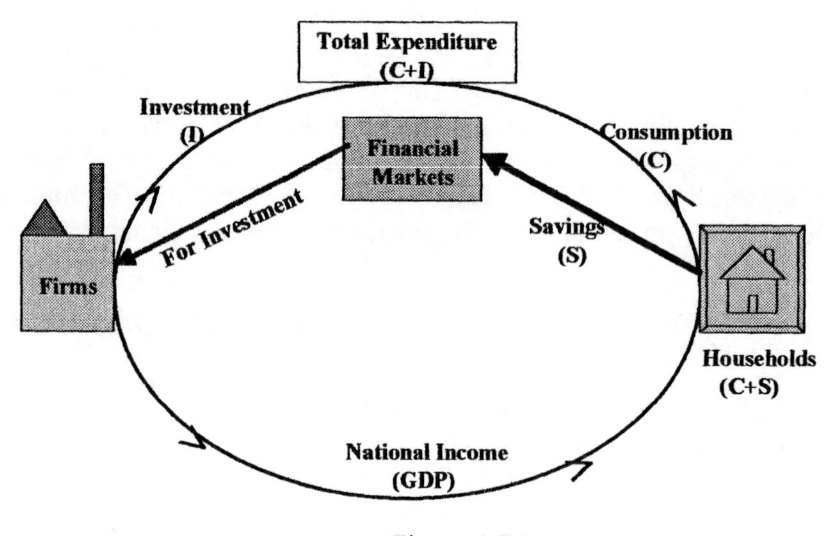

Figure 4.7.1

Households control the level of consumption and firms control the level of investment except for inventories. Inventories may change as a result of changes in consumer interest. Consider, for instance, the impact of increasing fuel prices on the inventory of cars and trucks. An unplanned change in inventory disrupts the equilibrium state of the economy in the following manner.

1. a. If the unplanned change increases the inventory, then the actual income (that is, GDP) is greater than the planned expenditure ($GDP > C + I$). In this case, firms have an incentive to reduce output (and thus income).

   b. If the unplanned change decreases the inventory, then the actual income (that is, GDP) is less than the planned expenditure ($GDP < C + I$). In this case, firms have an incentive to increase output (and thus income).

The following examples illustrate how to determine equilibrium levels.

Example 4.7.2.

If the consumption function is $C = 50 + 0.7\ GDP$ and the investment level is $I = 70$, what level of income creates equilibrium in the economy? (All amounts are in billions.)

Solution:

We answer this question in two different ways, first using a graphical analysis and secondly using an algebraic analysis.

Graphical Analysis

We plot total expenditure, $C + I = (50 + 0.7\ GDP) + 70 = 120 + 0.7\ GDP$, as a function of GDP and then superimpose the equilibrium line $C + I = GDP$. The income value of the point where the two lines intersect is the level of income that creates equilibrium in the system. We see from the following multiplot that the income equilibrium is 400.

Expenditure as a Function of Output

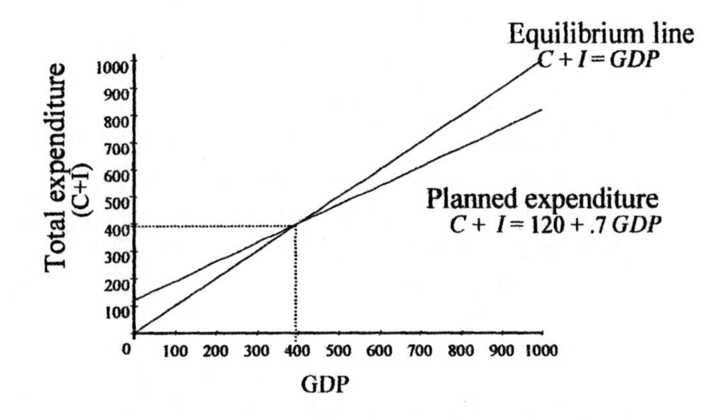

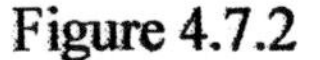
Figure 4.7.2

Algebraic Analysis

Equilibrium exists when $GDP = C + I$. We substitute for C and I and then solve for GDP.

GDP	$=$	$C + I$	Equilibrium condition
	$=$	$(50 + 0.7\ GDP) + 70$	Substitute $C = 50 + 0.7\ GDP$ and $I = 70$
	$=$	$120 + 0.7\ GDP$	Simplify
$0.3\ GDP$	$=$	120	Subtract $0.7\ GDP$ from both sides
GDP	$=$	400	Divide both sides by .3

The income equilibrium is thus 400 billion. ▲

Query 3.

Would the income equilibrium in Example 4.7.2 have been larger or smaller if the $MPC = 0.8$? Explain.

We now apply the what-if technique to Example 4.7.2 by considering the effect of an unplanned increase of 90 billion in the total investment of the economy.

Example 4.7.3.

Rework Example 4.7.2 modified to include an unplanned increase of 90 billion in the total investment of the economy. This new level of investment will be maintained in future years. Thus the initial conditions are

$GDP = 400$ (result of Example 4.7.2)
$C = 50 + 0.7\,GDP$ (as in Example 4.7.2)
$I = 70 + 90 = 160$ (unplanned increase of 90)

Determine the new income equilibrium value.
(In addition to the graphical and algebraic analyses, we also give a recursive sequence analysis.)

Solution:

Graphical analysis. Graphically, the unplanned addition of 90 billion to investment has the effect of shifting the expenditure line 90 units upwards as shown in Figure 4.7.3. As can be seen in this multiplot, the equilibrium line intersects the new expenditure line at the point (700,700). Thus equilibrium is achieved when income is increased to 700 billion. Note an increase of 90 billion in investments requires an increase of 300 billion in output in order to achieve equilibrium.

Expenditure as a Function of Output

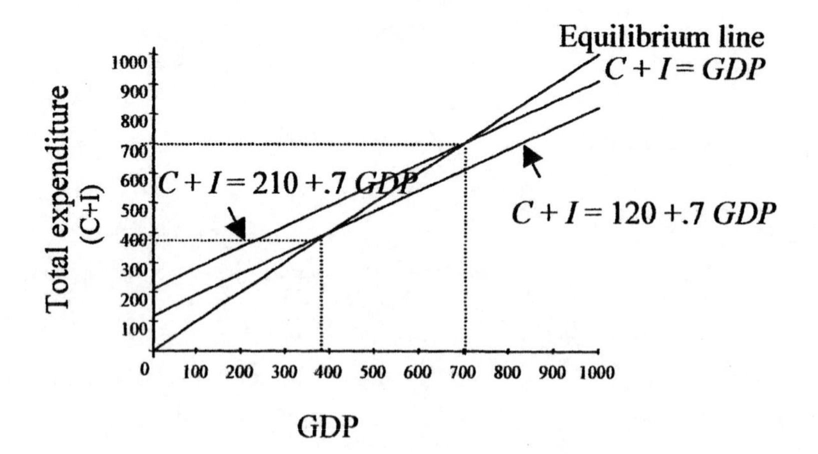

Figure 4.7.3

Algebraic analysis.

This is similar to the algebraic analysis in Example 4.7.2 and thus will be presented without comment.

$GDP = C + I$		Equilibrium condition
$= (50 + 0.7\ GDP) + (70 + 90)$		Substitute for C and I
$= 210 + 0.7\ GDP$		Simplify
$0.3\ GDP = 210$		Subtract .7 from both sides
$GDP = 700$		Divide both sides by .3

The income equilibrium is thus 700 billion.

Recursive sequence. (Although this method is more involved than the previous two, the additional insights gained make the extra work worthwhile.)

Because of the unplanned addition of 90 billion to investments, the economy is not in equilibrium with expenditures ($C + I$) being 90 billion more than GDP. This causes firms to increase production in the next period. This increased production results in more hours of labor, giving households more income, and thus consumption rises according the consumption function. This additional consumption results in an unplanned reduction of inventories causing firms to increase production again in the next period. This leads to more household income, more consumption, and lower inventories. Let us model this cyclic behavior of the economy with a recursive sequence. We begin by defining variables.

n = number of periods
$GDP(n)$ = income level at end of period n
$AE(n)$ = actual expenditure during period n
$MPC = 0.7$ (Marginal Propensity to Consume, slope of the consumption line)
$GDP(0) = 400$ (original planned output)

Because of the cyclic behavior of increased production -> increased income -> increased consumption -> increased production, etc., we consider a recursive sequence model for actual expenditures. In the paradigm

(New Situation) = (Old Situation) + (Change)

the

New Situation = Actual expenditure in period n, $AE(n)$
Old Situation = Actual expenditure in period $n-1$, $AE(n-1)$
Change = MPC *(change in expenditure during period $n-1$)
= $MPC * [AE(n-1) - AE(n-2)]$

Thus the model is

$$AE(n) = AE(n-1) + 0.7 * [AE(n-1) - AE(n-2)]$$

with

$$AE(0) = GDP(0) \text{ and } AE(1) = GDP(0) + 90.$$

We iterate this recursive sequence to discover a pattern.

$$\begin{aligned}
AE(0) &= GDP(0) && \text{Initial condition}\\
AE(1) &= GDP(0) + 90 && \text{Initial condition}\\
AE(2) &= AE(1) + 0.7\,[AE(1) - AE(0)] &&\\
&= AE(0) + 90 + 0.7(90) && \text{Substitute}\\
AE(3) &= AE(2) + 0.7\,[AE(2) - AE(1)] &&\\
&= AE(0) + 90 + .7(90) + 0.7[0.7(90)] && \text{Substitute}\\
&= AE(0) + 90 + 0.7(90) + 0.7^2(90) &&\\
AE(4) &= AE(3) + 0.7\,[AE(3) - AE(2)] &&\\
&= AE(0) + 90 + 0.7(90) + 0.7^2(90) + 0.7[0.7^2(90)] && \text{Substitute}\\
&= AE(0) + 90 + 0.7(90) + 0.7^2(90) + 0.7^3(90) &&
\end{aligned}$$

The reader should continue iterating until the pattern is recognized. The results for the next two iterations are

$$\begin{aligned}
AE(5) &= AE(0) + 90 + 0.7(90) + 0.7^2(90) + 0.7^3(90) + 0.7^4(90)\\
AE(6) &= AE(0) + 90 + 0.7(90) + 0.7^2(90) + 0.7^3(90) + 0.7^4(90) + 0.7^5(90).
\end{aligned}$$

The reader should explain how to determine that

$$AE(n) = AE(0) + 90 + 0.7(90) + 0.7^2(90) + 0.7^3(90) + \ \dots\ + 0.7^{n-1}(90).$$

Note the development of the geometric series, with ratio = .7, on the right-hand side of the equations. Applying the geometric series summation formula, we rewrite $AE(n)$ as

$$AE(n) = AE(0) + 90 * \frac{1 - 0.7^n}{1 - 0.7}$$

As n becomes large, 0.7^n converges to zero, and so

$$AE(n) \to AE(0) + 90 * \tfrac{1}{0.3} = 400 + 300 = 700.$$

Thus the-long term effect of the cycling process is to increase income by 300 billion producing a new equilibrium level of 700 billion. ▲

Note that the fraction $\frac{1}{0.3}$ in the preceding equation is $\dfrac{1}{1-0.7} = \dfrac{1}{1-MPC}$. This fraction, $\dfrac{1}{1-MPC}$ is called the **multiplier**.

*** Three important insights that have been illustrated are

1. An economy out of equilibrium cycles toward equilibrium.
2. The multiplier is $\dfrac{1}{1-MPC}$.
3. (new income equilibrium) = (old income equilibrium)
   + (multiplier)∗(change in expenditure).

In symbols, the third insight is

$$(\text{new } GDP) = (\text{old } GDP) + \frac{1}{1 - MPC} * (\text{change in expenditure})).$$

Query 4.

In the equation for $AE(5)$, explain the meaning of the terms 0.7(90) and $0.7^2(90)$ in terms of the cycling process.

We now generalize our macroeconomic analysis by including government spending and taxation in our simple economy.

Part II. A macroeconomic analysis of a simple economy containing government spending and taxation, but no imports or exports.

New terms and modifications to augment the Basic Assumptions in Part I.

1. G = amount of government spending on goods and services
2. T = net taxes = tax payments received – transfer payments to households (for example, unemployment insurance)
3. D = disposable income = GDP – net taxes, that is $D = GDP - T$
4. Consumption function: $C = a + MPC * D$
5. Total expenditure = $C + I + G$
6. Equilibrium condition: $D = C + I + G$

The relations among income, consumption, savings, and investments are illustrated in the following diagram.

Economic Flow Diagram

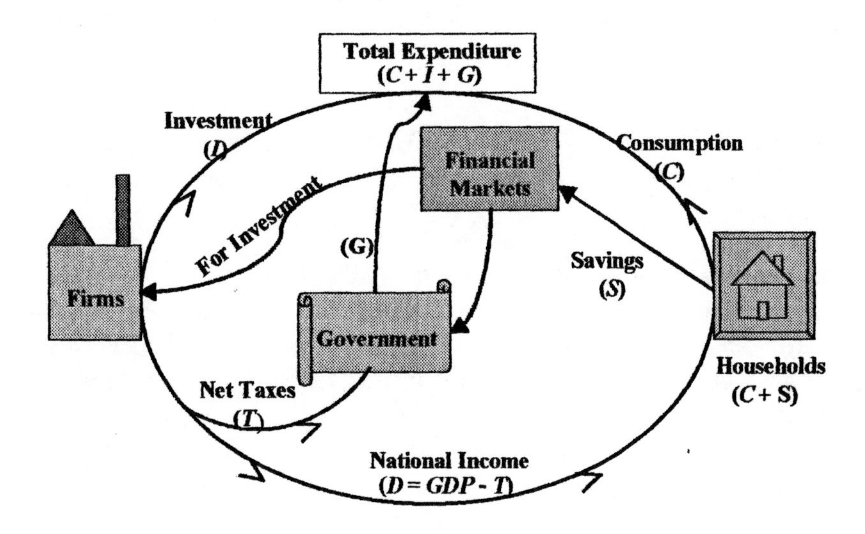

Figure 4.7.4

Government influences the economy through its fiscal policy, that is, its policy on spending and taxing. The next two examples illustrate the effect of multipliers in the government's fiscal policy. Example 4.7.4 illustrates the effect of the spending multiplier and Example 4.7.5 illustrates the effect of the taxing multiplier. The scenario in both examples is to start in a state of equilibrium ($D = C + I + G$). Assume that

$$\begin{aligned} \text{Disposable Income:} \quad D &= 10{,}000 \\ \text{Consumption:} \quad C &= 50 + 0.7D \qquad \text{Note that } MPC = 0.7 \\ \text{Investment:} \quad I &= 1{,}500 \\ \text{Government Spending:} \quad G &= 1{,}000 \end{aligned}$$

(Amounts are measured in billions.)

Example 4.7.4.

The government decides to stimulate the economy by increasing spending by 60 billion. (Increasing government spending, causes disequilibrium with expenditure exceeding income. This initiates a spiraling effect resulting in a new income equilibrium that is greater than the original income equilibrium.) What is the new income equilibrium?

Situation:

The analysis—graphic, algebraic, and using a recursive sequence—is exactly the same as in Example 4.7.2 because an increase in government spending causes the actual expenditures to exceed D, just as an increase in investment caused the total expenditure to exceed the GDP. Thus the government spending multiplier is $\frac{1}{1 - MPC}$. The new income equilibrium is

$$\begin{aligned} &D + \tfrac{1}{1-MPC} * \text{(increased amount of government spending)} \\ &\quad = 10{,}000 + \tfrac{1}{0.3} * 60 = 10{,}000 + 2{,}000 = 10{,}200. \end{aligned}$$

Increasing government spending by 60 billion results in a 200 billion increase in income. ▲

Query 5.

Verify the results of Example 4.7.4 through an algebra analysis. (See Example 4.7.2.)

Example 4.7.5.

The government decides to stimulate the economy by reducing taxes by 60 billion. (Decreasing taxes, increases household income and thus consumption. This results in disequilibrium with expenditures exceeding income. This, in turn, initiates a spiraling effect resulting in a new income equilibrium that is greater than the original income equilibrium.) What is the new income equilibrium?

Solution:

A decrease in taxes increases total expenditure by increasing consumption. The amount of increase is therefore equal to

$$MPC * \text{(amount of tax decrease)}.$$

Because the change in output is equal to the multiplier times the change in the total expenditure, we have

$$\text{(change in } D) = \frac{1}{1-MPC} * [MPC * \text{(change in taxes)}]$$
$$= \frac{MPC}{1-MPC} * \text{(change in taxes)}$$

Because a reduction in taxes increases total expenditures through increased consumption, it will cause an increase in the *GDP*. Thus a negative change in taxes results in a positive change in *GDP*. Similarly a tax increase (positive change) will result in a negative change in *GDP*. Thus we consider the tax multiplier to be a negative multiplier and define

$$\text{Tax Multiplier} = -\frac{MPC}{1-MPC}.$$

The new output equilibrium level is now

$$D - \frac{MPC}{1-MPC} * \text{(change in tax receipts)}$$
$$= 10{,}000 - \tfrac{0.7}{0.3} * (-60) = 10{,}000 + 1{,}40 = 10{,}140.$$

Thus a tax cut of 60 billion results in a 140 billion increase in *GDP*. ▲

Query 6.

Why doesn't a tax cut of 60 billion increase the *GDP* as much as increasing spending by 60 billion?

Small-Group Activity.

Generalize Part II to include imports and exports in the economy.

Exercises 4.7

1. Approximate the solutions to the following pairs of equations by plotting.
   a. $y = 3x - 5$ and $y = -2x + 7$
   b. $y = \sin(x)$ and $y = x^2$
   c. $y = x^2$ and $y = 2^x$
2. Solve the following pairs of equations algebraically.
   a. $y = 3x - 5$ and $y = -2x + 7$
   b. $y = x - 6$ and $y = x^2$
   c. $y = x^3 - 1$ and $y = 3x^2 - 3x$
3. Sum the following series.
   a. $1 + .2 + .2^2 + .2^3 + .2^4 + \ldots$
   b. $1 + (\tfrac{2}{5}) + (\tfrac{2}{5})^2 + (\tfrac{2}{5})^3 + (\tfrac{2}{5})^4 + \ldots$
   c. $6 + .2 + .2^2 + .2^3 + .2^4 + \ldots$

4. Tomorrow you are scheduled to take hour tests in both math and history. Assume you have 1 hour of study time to prepare for the tests. List your study choices, their opportunity costs, and an approximate measure for each opportunity cost.

5. Rework Example 4.7.1 by considering the opportunity costs for Bianca and Jim to shift from making pizzas to making sandwiches.

6. Estimate your own *MPC*. For example if you received a $100 gift from your grandparents, how much of it would you spend and how much would you save? Do you think your personal *MPC* is the same as your parents? Grandparents? Write a half-page essay explaining your answers.

7. Determine the equilibrium level of *GDP* for an economy with no government and no imports or exports whose consumption function is $C = 75 + 0.6\ GDP$ and whose investment is $I = 200$. (Amounts are measured in billions.)

8. Redo Exercise 7 with $MPC = 0.5$.

9. Determine the equilibrium level of investment for an economy with no government and no imports or exports whose consumption function is $C = 100 + 0.6\ GDP$ and whose income equilibrium is 600. (Amounts are measured in billions.)

10. Given $D = 9{,}000, C = 60 + 0.8D,$ and $I = 1{,}000$ in an economy with government and no imports or exports, determine the level of government spending needed to achieve equilibrium. (Answer: 740 billion)

11. Based on the assumption that households only have two choices to distribute their income—consume (*C*) or save (*S*)—define the Marginal *MPS* in terms of *MPC*. Then derive a savings multiplier from the (spending) multiplier.

12. Assume country H receives one billion dollars in foreign aid. Will its GDP increase faster with a high MPC or a low MPC? Write a one-page essay explaining the reasoning leading to your answer.

4.8 Modeling (Music and Art)

Musicians and artists as well as mathematicians are engaged in showing the beauty and harmony in nature.

> A mathematician, like a painter or poet, is a maker of patterns. If his patterns are more permanent then theirs, it is because they are made with ideas . . . . The mathematician's patterns, like the painter's or poet's, must be beautiful; the ideas, like the colors or words, must fit together in a harmonious way. Beauty is the first test; there is no permanent place in the world for ugly mathematics.
>
> G. H. Hardy

In musical composition, the dominant factor was determined by a sound that was pleasing to the ear rather than by a mathematical relation between the frequencies of the notes. This dominant factor is the ***fifth*** in an octave. That is, the span from the ***tonic*** or ***root*** or ***home key*** of an octave to the fifth key, called the dominant key of the octave. In art, the ratio of the lengths of the sides of the rectangle that is considered the "most pleasing to the eye" is called the Golden Mean or Golden Ratio. In mathematics, instances of congruence and symmetry, the basis of geometry, are exhibited in the earliest examples of pottery, weaving, and drawings dating back to the Neolithic era. The musician, artist, and mathematician share a common goal of exhibiting beauty, harmony, and order within nature.

In this section, we focus on piano tuning in music and on the role of the Golden Mean or Golden Ratio in art. We conclude this section by showing how the Fibonacci sequence relates mathematics, music, and art.

Music

Pythagoras (c. 500 *B.C.*), for whom the famous theorem relating the lengths of the sides of a right triangle is named, knew that the frequency of a vibrating string is proportional to its length. For example, halving the length of a string doubles the frequency. No one knows who first realized that plucked strings vibrate and that different vibrations produce different musical sounds, although by the time of Pythagoras high notes were associated with vibration at high speeds and low notes with vibrations at slow speeds. The major contribution of Pythagoras and his school to the field of music was the discovery of the beautiful numerical relation between the lengths of strings and musical intervals.

Plucking or striking a taut string causes a transverse wave, a **sine wave**, to travel along the string and be reflected at the ends. Thus each vibration of a stretched string causes a transverse wave to travel the length of the string twice, once in each direction, as the following diagrams illustrate.

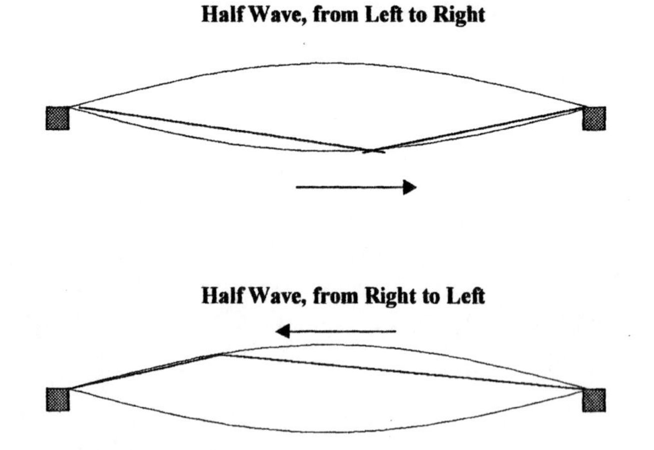

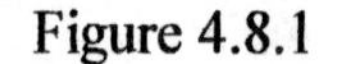
Figure 4.8.1

Hence the frequency, the number of vibrations per second, is given by

$$f = \frac{v}{2L}$$

where v is the velocity of the wave and L is the length of the string. The velocity of the wave is a function of both the tension and the mass of the string, $v = \sqrt{\frac{T}{M}}$, and therefore the frequency is a function of length, tension, and mass of the string. We shall assume that the tension and mass are constant and focus on just the length of the string. When struck or plucked a taught string vibrates with several frequencies. These frequencies, called *harmonics*, are integer multiples of the lowest frequency, f_0, which is called the *fundamental* or *first harmonic*. The second harmonic has frequency 2 f_0, the third harmonic has frequency 3 f_0, and so on.

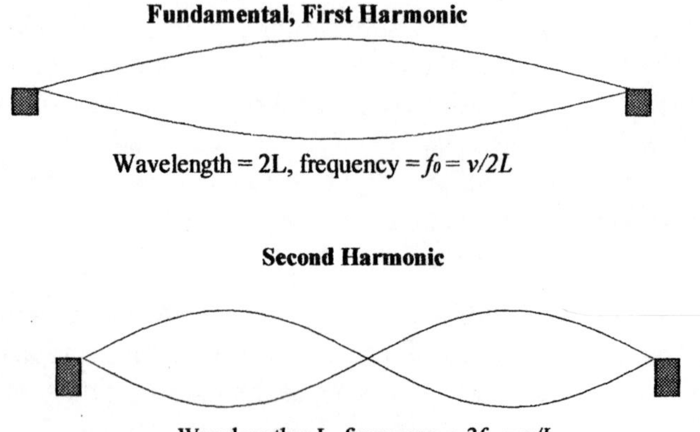

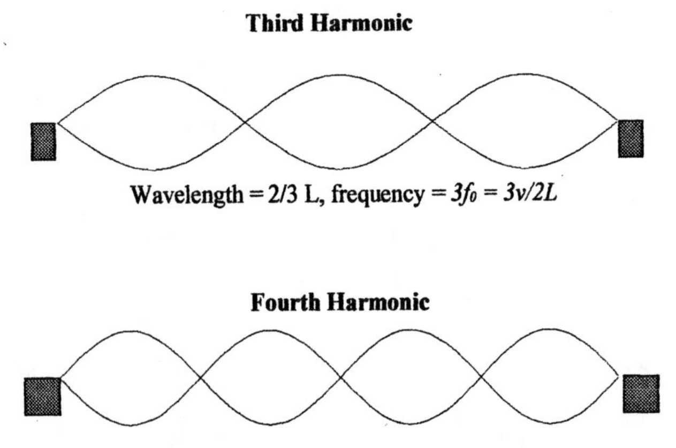

Figure 4.8.2

The frequency range between the first and second harmonics is called an *octave*. The octave range is divided into 12 notes with the ratio of the frequencies of two successive notes, higher to lower, being the same. If r denotes the ratio, then the frequency of the second note is rf_0, the frequency of the third note is $r(rf_0) = r^2f_0$, the frequency of the fourth note is $r(r^2f_0) = r^3f_0$ and so on. Thus the sequence of frequencies is a geometric sequence with ratio r.

The most widely accepted naming convention for notes, in place since the mid-nineteenth century, assigns note names to specific frequencies. The note with frequency of 440 hertz is called *A*. (One hertz is one cycle per second.) On the piano, this note is the 40th key from the right-hand end of the keyboard and is referred to as "*A* above middle *C*." Including both end notes, the notes in the octave based on *A* are labeled in order of increasing frequency

$$A,\ A\#,\ B,\ C,\ C\#,\ D,\ D\#,\ E,\ F,\ F\#,\ G,\ G\#,\ A.$$

This sequence of note labels repeats, labeling all 88 keys on the piano keyboard. The symbol # is read "sharp." On a piano, the black keys sound the sharp notes.

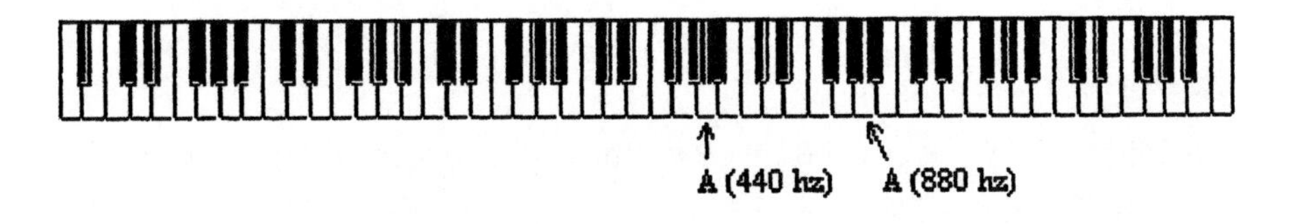

Figure 4.8.3

Query 1.
How many *A* keys are there on a piano keyboard? What are the frequencies of the *A* keys?

Query 2.
Consider the sequence of thirteen notes listed earlier: *A*, *A#*, *B*, ..., *A*. Express the frequency of the second *A* in terms of the frequency of the first *A* and r, the ratio of the frequencies of successive notes.

A piano tuner tunes a piano by octaves. This means that the tuner plays two keys that are an octave apart, say the A above middle C and the next higher A. If no beats are heard, the notes are in relative tune, meaning that the higher note has twice the frequency of the lower note. However if beats are heard, the keys are not in relative tune. In this case, the tuner adjusts the tension on one of the strings until no beats are sounded when the two keys are played. The following two plots illustrate the beat phenomenon. The plot on the left is of $f(x) = \sin(40x) - \sin(20x)$ representing the change in doubling the frequency ("in tune"). The plot on the right is of $f(x) = \sin(41x) - \sin(20x)$ representing the change when the frequency is not exactly doubled ("out of tune"). The resulting undulating nature of the waves produces the beat sound.

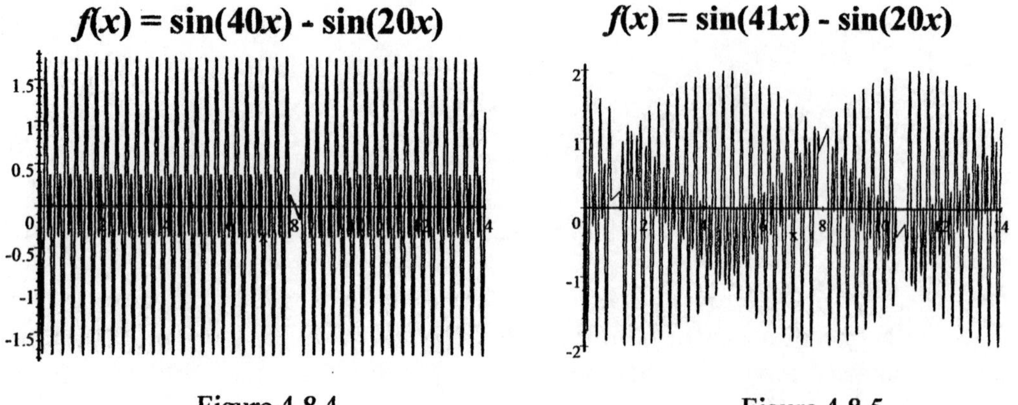

Figure 4.8.4

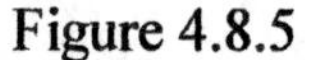

Figure 4.8.5

Art

> Geometry has two great treasures: one is the theorem of Pythagoras; the other, the division of a line into extreme and mean ratios. The first we may compare to a measure of gold; the second we may name a precious jewel.
>
> Johannes Kepler (1571–1630)

Golden Mean or Golden Ratio

Divide a line segment into two pieces such that the ratio of the length of the longer piece to the length of the shorter piece is equal to the ratio of the length of the segment to the length of the longer piece. For example, if the length of the shorter piece is 1 and the length of the longer piece is x, then $\frac{x}{1} = \frac{1+x}{x}$. This ratio is called the Golden Mean or Golden Ratio.

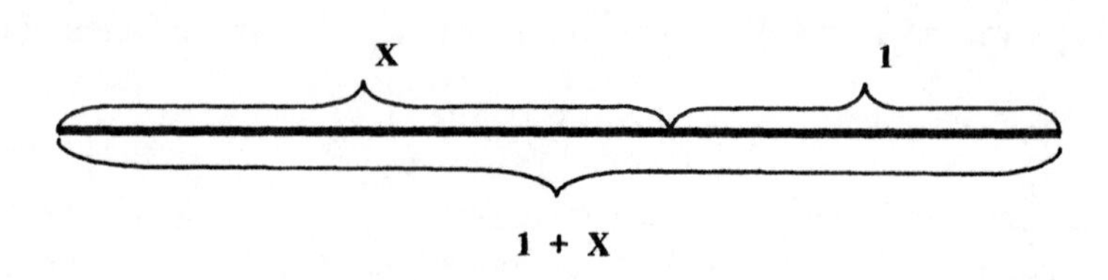

Figure 4.8.6

We solve for x in the Golden Ratio by first multiplying and then applying the quadratic formula. Here are the steps.

$\frac{x}{1}$	$=$	$\frac{1+x}{x}$	Golden Ratio
x^2	$=$	$1+x$	Cross-multiply
x^2-x-1	$=$	0	Group terms on left-hand side
x	$=$	$\frac{1\pm\sqrt{1+4}}{2}$	Quadratic formula
x	$=$	$\frac{1+\sqrt{5}}{2}$	x is a positive number
x	$=$	1.6180	Approximate (4 decimal places)

Thus the Golden Mean or Golden Ratio is $\frac{x}{1} = 1.6180$ (4 decimal places). In some books, the reciprocal or $\frac{1}{x} = 0.6180$ is referred to as the Golden Mean or Golden Ratio.

Example 4.8.1.

Consider any nonzero number, x, invert it and add 1, $1+\frac{1}{x}$. Invert this result and add 1, $1+\frac{1}{1+\frac{1}{x}}$. Continue this process. What is the long-term result?

Solution:

We model this recursive process with a recursive sequence. We begin, as always, by defining the variables. Let

n = number of iterations
$a(n)$ = the numerical value after the first n iterations.

Model:

$$a(n) = 1 + \frac{1}{a(n-1)}.$$

To gain insight of the long-term behavior, we iterate this sequence three times using different initial conditions.

n	a_n	a_n	a_n
0	0.5	2.5	1
1	3	1.4	2
2	1.333	1.714	1.5
3	1.750	1.583	1.667
4	1.571	1.632	1.600
5	1.636	1.613	1.625
6	1.611	1.620	1.615
7	1.621	1.617	1.619
8	1.617	1.618	1.618
9	1.618	1.618	1.618
100	1.618	1.618	1.618

Table 4.8.1

These iterations strongly suggest that the long-term result is approximately 1.618, the Golden Mean or Golden Ratio! ▲

Query 3.

Start with any number $x > -1$, add one to it and then take the square root. Add one to this result and then take the square root. Continue this process. What is the long-term result? Answer by first modeling this process with a recursive sequence and then plotting the sequence for several different initial values.

VASE

Figure 4.8.7

VIOLIN

Figure 4.8.8

Beauty is truth, truth is beauty - that is all
Ye know on earth and all ye need to know.

John Keats, "Ode to a Grecian Urn"

The origin of the Golden Mean or Golden Ratio dates to Pythagoras and the ancient Greek civilization. Pythagoras was a teacher, philosopher, mystic, and, to his followers, almost a god. Numerology based on patterns and number relations, particularly those occurring in nature, permeated his thinking about mathematics and life and formed the basis for the Pythagorean school he established. Pythagoras assigned importance to patterns and number relations based on their special aesthetic, geometric, or algebraic properties. For rectangles, the most important esthetic property was the shape that was most pleasing to the eye. This was considered to be one in which the ratio of the lengths of the sides formed the Golden Ratio. Note the 3-by-5 inch card, which is so common today. The ratio $\frac{5}{3} = 1.6667$ is a good approximation of the Golden Ratio. The page size of the *New York Times* is 13.5 by 22 inches, giving a ratio of $\frac{22}{13.5} = 1.6296$. Similarly dividing a segment according to the Golden Ratio was considered to be the most pleasing division. Curl your knuckles into a loose fist as illustrated in Figure 4.89. The ratio of the lengths of the sides of the pictured rectangle approximate the Golden Mean or Golden Ratio. Your navel divides your body length according to the Golden Ratio. Likewise the ratio of the length of your arm to the width of your shoulders approximates the Golden

Ratio.

Figure 4.8.9

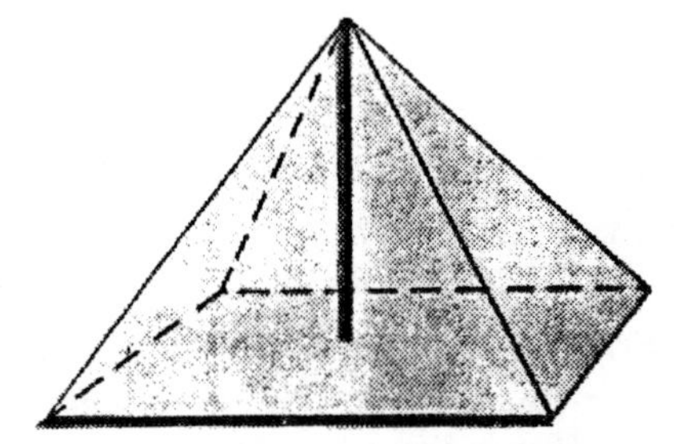

Figure 4.8.10

Numerous examples of the Golden Mean or Golden Ratio appear in design. How many examples can you find in the picture of the Pyramid of Gizeh? In the picture of the Parthenon?

Figure 4.8.11

Fibonacci Sequence

Leonardo of Pisa was the greatest Italian mathematician and possibly the greatest European mathematician in the middle ages. He lived in the sixth century *B.C.* on the island of Samos in the Aegean Sea, in Egypt, in Babylon, and in southern Italy. Leonardo called himself Fibonacci, a shortened version of Filius Bonacci, which means son of Bonacci. His book, *Liber abbaci* (Book of Calculating), published in the early 1200s, is credited with persuading many European mathematicians to change to the new decimal system from the Roman numeral system. The problem contained in his book that led to a recursive sequence bearing his name and for which he has become famous states

A certain man put a pair of rabbits in a place surrounded by a wall. How many pairs of rabbits can be produced from that pair in a year if it is supposed that every month each pair begets a new pair, which from the second month on becomes productive?

The first month there is just one pair, assumed to be newborn. During the second month, the original pair is only one month old and therefore do not produce. Thus there is only 1 pair during the second month. In the third month, the original pair produces a new pair and so there are 2 pairs in the third month. In the fourth month the original pair produces a new pair, but their first offspring pair do not as they are only one month old. Thus there are 3 pairs in the fourth month. In the fifth month, both the original pair and their first offspring pair produce giving a total of 5 pairs. How many pairs are there in the sixth month?

Let us model the situation. We let

n = number of months
u_n = number of pairs of rabbits in the nth month.
$u_1 = 1$
$u_2 = 1.$

The number of pairs in month n is the number of pairs that existed in the previous month, $n-1$, plus the number of new pairs. The number of new pairs is equal to the number of pairs in month $n-2$, because each of these is at least two months old and thus productive. Therefore we have the famous recursive sequence called the Fibonacci sequence, which we express in subscript notation:

$$u_n = u_{n-1} + u_{n-2}.$$

Iterating the first few values of the Fibonacci sequence gives

$$1, 1, 2, 3, 5, 8, 13, 21, \ldots .$$

Two observations suggest a link between harmony scales in music and the Fibonacci sequence; and between the sequence of ratios of Fibonacci numbers and the Golden Ratio.

1. The number of keys in an octave, say from A to A, including both As is 13. Each octave contains a scale consisting of 8 keys. The most important grouping within a scale is the fifth, the first five keys beginning with the tonic or home key of the scale. Triads, groupings of three notes, and seconds, groupings of two notes, are along with fifths building blocks of composition. The signature note of a scale is the tonic or home key. These groupings within a scale occur naturally based on what is pleasing to the ear. Note that these numbers $1, 2, 3, 5, 8, 13$ are Fibonacci numbers.

2. Computing the ratio of a Fibonacci number to the preceding one gives the sequence $\frac{u_{n+1}}{u_n}: \frac{1}{1}, \frac{2}{1}, \frac{3}{2}, \frac{5}{3}, \frac{8}{5}, \frac{13}{8}, \frac{21}{13}, \ldots$. Writing this sequence in decimal form gives: 1, 2, 1.5, 1.6667, 1.6, 1.6250, 1.6154,... . Can it be that this sequence converges to the Golden Ratio?

Exercises 4.8

1. Assume the ratio of the frequencies of any two successive (higher to lower) notes on a piano is denoted by r.
   - **a.** Model the frequencies of the keys of a piano as a recursive sequence.
   - **b.** Using your model from Part a, compute the value of r.
   - **c.** Iterate your model to compute the frequency of the C note above middle A (440 hertz).
   - **d.** Determine an analytical solution to your model in part a.

e. Compute the frequency of the left-hand-most note of the piano keyboard. What is the name of this note?

f. Compute the frequency of the right-hand-most note of the piano keyboard. What is the name of this note?

2. The audible frequency range is 20 hertz to 20,000 hertz.
   a. What is the highest audible note?
   b. What is the lowest audible note?
   c. How many octaves are possible in the audible frequency range?

3. Measure the width and height of the vase in Figure 4.8.7. How does the ratio of the width to the height compare to the Golden Mean or Golden Ratio?

4. Consider the picture of the violin in Figure 4.8.8. How does the ratio of the length of the neck to the full length of the violin compare to the Golden Mean or Golden Ratio? Consider a double bass. How does the ratio of the length of the neck to the full length of a double bass compare to the Golden Mean or Golden Ratio?

5. Find five examples, outside this text, of the Golden Ratio.

6. The first eight terms of the Fibonacci sequence are: $1, 1, 2, 3, 5, 8, 13, 21$.Continue the sequence for another seven terms and then form the sequence of ratios of Fibonacci numbers for these terms. Does this sequence appear to be converging to the Golden Ratio?

Exercises 7–10 are discovery exercises based on the Fibonacci sequence. Hint: write down the first 12 terms of the Fibonacci sequence: $u_1, u_2, u_3, \ldots u_{12}$ and then use this sequence as a reference for Exercises 7-10.

7. Form the sequence $w_1, w_2, w_3, \ldots$ where $w_1 = u_1$, $w_2 = u_1 + u_3$, $w_3 = u_1 + u_3 + u_5$ and so on with w_n being the sum of the first n odd numbered elements in the Fibonacci sequence. Write down the numerical values of the first five terms of the w sequence. How are these numerical values related to particular Fibonacci numbers?

8. Form the sequence $z_1, z_2, z_3, \ldots$ where $z_1 = u_2$, $z_2 = u_2 + u_4$, $z_3 = u_2 + u_4 + u_6$ and so on with z_n being the sum of the first n even numbered elements in the Fibonacci sequence. Write down the numerical values of the first five terms of the z sequence. How are these numerical values related to particular Fibonacci numbers?

9. Form the sequence $q_1, q_2, q_3, \ldots$ where $q_1 = u_1^2$, $q_2 = u_1^2 + u_2^2$, $q_3 = u_1^2 + u_2^2 + u_3^2$ and so on with q_n being the sum of the squares of the first n elements in the Fibonacci sequence. Write down the numerical values of the first five terms of the q sequence. Factor each of these numbers being alert to the occurrence of Fibonacci numbers. How are the elements in the q sequence related to the elements in the Fibonacci sequence?

10. Form the sequence $p_1, p_2, p_3, \ldots$ where $p_1 = u_2^2 - u_1 u_3$, $p_2 = u_3^2 - u_2 u_4$, $p_3 = u_4^2 - u_3 u_5$ and so on with $p_n = u_{n+1}^2 - u_n u_{n+2}$. Write down the numerical values of the first five terms of the p sequence. What pattern do you recognize? What is the value of p_{20}?

4.9 Summary

Chapter 4 (Modeling) focuses on applying mathematics to understand real-life situations. This involves a three-step procedure as diagrammed in Figure 4.1.1:

(1) Develop a mathematical model (description) of a given situation.
(2) Solve the mathematical model.
(3) Interpret the solutions in terms of the original situation.

A mathematical model is a mathematical description of a situation that allows one to analyze the situation in greater depth and breadth than just answering a specific question. The first step of the modeling process almost always involves making simplifying assumptions in order to make the real-world situation tractable. Thus mathematical models are approximations of the real situation.

We focus primarily on recursive sequence models based on the paradigm

(New Situation) = (Old Situation) + (Change).

Investment-type problems in which interest is paid and deposits made at regular intervals of time illustrate this type of model. The approach is to assume that all activity takes place at regular periods of time and then to form a recursive sequence by writing an equation expressing the situation at the end of period n as equal to the situation at end of period $n-1$ plus the changes that take place during period n. The recursive sequence models in this chapter have the form

$$a(n) = ra(n-1) + d$$

and their general solution is

$$a(n) = r^n c + \frac{d}{1-r} \quad \text{where } c = a(0) - \frac{d}{1-r}.$$

This investment model is used as a paradigm for recursive sequence models in several different disciplines. An advantage of a recursive sequence model is that a numerical solution to the specific situation can be obtained by iterating the recursive sequence. Exponential expressions and geometric series arise naturally when iterating a recursive sequence. Sections 4.1 and 4.2 illustrate how to generalize a pattern of iterations into a symbolic model. Because iterations generate exponential terms, the symbolic model contains exponential terms. Therefore their solutions often involve using logarithms. A table of rules for exponents and logarithms is included in Section 4.2. A particularly important result developed in Section 4.2 is that the defining characteristic of an exponential function is that the amount of change is proportional to the amount of the quantity present.

Motion problems are addressed in Section 4.3. Newton's Second Law of Motion provides the basis for the models of motion problems. Modeling nonvertical motion requires the introduction of parametric and trigonometric functions. Parametric functions provide for modeling projectile motion by expressing both the x and y coordinates of a point as functions of a third variable, called the parameter. Thus both x and y are dependent variables and the parameter is the independent variable. Often the parameter represents time, but not always. An introduction to trigonometric functions is included for the purpose of decomposing force expressions into horizontal and vertical components. Because Newton's Second Law of Motion leads to models involving quadratic functions, there is a need to solve quadratic polynomials. The quadratic formula is derived to fill this need

Newton's Law of Cooling is introduced in Section 4.4 (Modeling in the Physical Sciences). Both discrete (recursive sequence) and continuous models are presented. Because the development and solution of continuous models requires calculus, the models and solutions are stated and then used. Concavity properties of curves, concave upward (curve bending upward) and concave downward (curve bending downward), are used to motivate aspects of a model. For example, the temperature curve of a warming can of soda is concave downward (although rising) and the growth curve of an investment is concave upward. Long-term behavior is discussed in terms of equilibrium and asymptotic behavior, such as the long-term temperature of a cooling object is the temperature of the surrounding medium. The section concludes with balancing chemical reactions using systems of equations and using equilibrium constants in analyzing long-term behavior of chemical reactions.

Blend problems are modeled in Section 4.5. An actual blend problem submitted by the EXXON Petroleum Corporation illustrates how a *Contemporary College Algebra* application saved the corporation considerable money.

Section 4.6 addresses situations in the life sciences, retention of medication in the body and population growth in particular. Recursive sequences are used to model the amount of medicine in a person's body as a result of taking medication over time. We show that marked changes in the level of medicine in a person's body can occur when medication is taken according to a schedule other than the prescribed schedule, even though the total amount of medicine remains unchanged. A second method of creating a recursive sequence model, one that relies on conjecture and verification rather than on iteration, is developed. Modeling the growth of a yeast culture with a recursive sequence is used to develop logistic models for population growth in constrained environments. The development of the logistic model involves applications of shifting and scaling graphs (Section 3.5) and graphical approximations (Section 3.7) as well as understanding the defining characteristic of the exponential function.

Two important aspects of decision making in economics, opportunity cost and multipliers, are presented in Section 4.7. Graphical, algebraic, and recursive sequence models are developed to illustrate both the spending and taxing multipliers in the economy.

In Section 4.8, recursive sequences are used to illustrate how music, art, and mathematics share a common goal of exhibiting beauty, harmony, and order within nature. We show how the musical scale for the piano, which evolved from Pythagoras's discovery of the beautiful numerical relation between the lengths of strings and musical intervals, and the Golden Mean in architecture have a connection through the Fibonacci (recursive) sequence.

Fun Projects

(See Fun Projects—Chapter 2, for suggestions on assigning these projects, formatting of the project reports, and assigning student responsibilities related to their groups.)

1. Consumer Price Index (CPI)

(Purposes: Raise awareness of an important economic index—what it measures, how it is used; provide a means to compare prices from different years; relate economic changes to major national and world events; provide a writing exercise; provide a small-group experience.)

In 1903, the Ford Motor Company introduced is first car, a Model A 2-person runabout. The car sold for $850. How much is this in today's dollars? In 1918, a first class stamp cost three cents. How much is this in today's dollars? To answer these and similar questions, we turn to the Consumer Price Index (CPI)—the primary economic index for measuring how prices change over time. The Index, maintained by the Bureau of Labor Statistics, tracks the prices of over 80,000 items, called the market basket, that include a broad array of consumer goods and services. The Bureau set the Index level at 100 to represent the average price of the market basket over three year period, 1982-84. This provides a base reference against which changes are measured. For example, an index of 120 means a 20% increase from the 1982-84 level. The CPI provides a way to compare the monetary value of an item (such as the cost of a stamp) in different time periods.

$$\text{monetary value at time A} = (\text{monetary value at time B}) * \frac{\text{CPI at time A}}{\text{CPI at time B}}$$

The following table gives CPI readings for a selected number of years.

Year	CPI	Year	CPI	Year	CPI
1913	9.8	1945	17.8	1980	77.8
1918	14	1960	29.3	1990	127.4
1925	17.3	1970	37.8	2000	168.8

Your tasks are:

1. Expand the preceding table to give the yearly January CPI readings from 1913 to the present. (See http://www.bls.gov/cpi/.htm)
2. Approximate the cost of Henry Ford's 1903 Model A in today's dollars. (Explain how you extrapolated the CPI value for 1903.)

3. Approximate the cost the 1918 first class letter stamp in today's dollars.
4. In 1970, gasoline costs $.50 per gallon. What would that amount be in today's dollars?
5. Create a scatter plot of the CPI from 1913 to 1970. Write a paragraph discussing the behavior of the CPI in terms of major national or world events during this period. (Suggestion: talk with a history professor.)
6. Create a scatter plot of the CPI from 1970 to 2003. Write a paragraph discussing the behavior of the CPI in terms of major national or world events during this period.(Suggestion: talk with a history professor.)
7. Form a multiplot of the CPI and the tuition at your school over the period 1970 to 2003. (That is, impose the graph of your school's tuition on the graph in Exercise 5.) What observation(s) can you make?
8. Write a paragraph discussing the CPI as measure of inflation.

2. Shingling a Roof

(Purposes: model a real-world situation; develop a bid on a roofing project, obtaining prices from two different stores; provide a writing exercise: provide a small-group work experience.)

Introduction: Eliza Key purchased an old truck garage to expand her Safe Keep storage company. She knows that hiring college students is always a good investment and thus when her building needed a new roof, she quickly offered your group an opportunity to bid on the job.

The garage is 120 feet long and 50 feet wide. The peak of the pitched roof is 30 feet high and the height of the edge of the roof is 20 feet. The roof is symmetrical in the sense that both sides of the roof are the same size. The old shingles must be removed, the roof covered with a layer of roofing paper, and then 25-year warranty "three-in-one" asphalt shingles stapled to the roof. Strips of drip-edge are installed along the roof edges after the roofing paper is put on and before the shingles are stapled. Drip-edge is a strip of aluminum with a cross-section in the shape of an L. The long side of the L is laid on the roof and the short side hangs over the edge. The purpose of the drip-edge is to cause rainwater to drip off the edge rather than flow up under the shingles.

A role of roofing paper is 3 feet wide and 250 feet long. Roofing paper is installed in strips parallel to the edge of the roof. Successive strips of roofing paper overlap the preceding strip by 4 inches. Three-in-one shingles are a foot wide and a yard long. (Each shingle has three tabs separated by 5-inch slits.) The shingles are laid end to end in rows parallel to the edge of the roof. Each row overlaps the preceding row by 7 inches and are offset by half a tab so that the tab slits are staggered to not line up.

A "square" of shingles (three bundles) covers 100 square feet. Drip-edges come in 10-foot lengths.

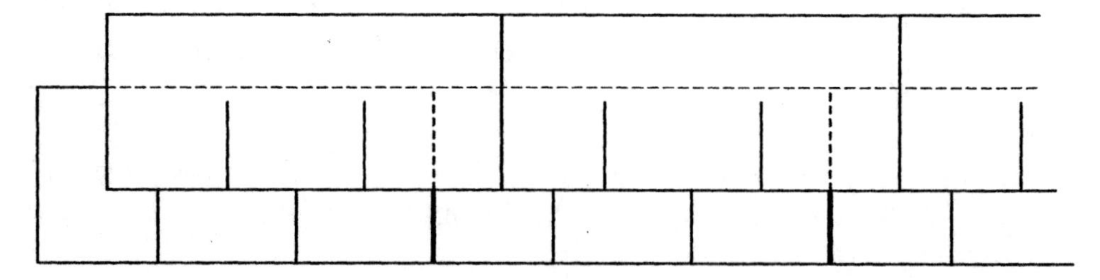

Figure 4.FunProject1.1

Ms. Key agrees to pay each worker $10 per hour provided your bid for materials is broken down by item and the total amount is within 10% of a professional company's material cost.

In order to prepare your bid, you need to determine the dimensions of the roof and the amount of materials. You then need to obtain prices from two stores (for example, lumber yard, roofing company, hardware store.) and formulate a cost sheet listing the amount and cost of each item. Be sure to include sales tax in your total cost.

1. Dimensions and amounts:
   a. The size of the roof (in square feet)
   b. Number of squares of shingles
   c. The number of lengths of drip-edge
   d. Number of rolls of roofing paper
   e. Number of pounds of nails for the staple (nail) guns

2. Costs/item (list prices from two stores for each item).

   Name of Store #1 ______________________

   Name of Store #2______________________

	Store #1	Store #2
a. Roll of roofing paper		
b. Length of drip-edge		
c. Square of shingles		
d. Pound of nails (staple gun)		

3. Cost Sheet

Item	No. of Item	Cost/Item	Total Cost
	Cost of Materials		________
	State Sales Tax		________
	Total Cost		________

3. Pollution—Walton Lake Cleanup

(Purposes: develop recursive models of a real-world pollution problem; experience using exponential algebra; provide a writing exercise; provide a small-group work experience.)

(This problem is abstracted from an article in The Times Herald-Record (NY) newspaper, October 1, 1999 and a draft report written by Leslie J. Surprenant, a Fisheries Biologist for the State of New York.) Walton Lake is a 120 acre natural lake that is the reservoir for drinking water for the Village of Chester, 50 miles north of New York City. The Lake is a popular recreational site, particularly for boating and fishing. However, large dense mats of vegetation severely restricted its recreational use. The Village formed a conservation weed district to get rid of the vegetation which, in 1987, was judged to cover a third of the lake's surface. After conducting a lengthy environmental impact study, the decision was made to stock the lake with grass carp, a sterile fish that feeds on the vegetation.

A study of the situation approximated the growth rate of the vegetation when left alone to be 33% per year, the death rate of carp to be 20% per year, and the average amount of vegetation eaten by a single carp in one year to be 1.522 g/m^2. In 1988, when the carp were first introduced into the Lake, the amount of vegetation was estimated to be 4,273 g/m^2.

Your tasks are:

1. Develop a recursive sequence model for the yearly number of carp in the Lake for the ten year time period 1988-1997, assuming that 1,000 carp were stocked on January 1, 1988. Iterate your model for the ten year period and then determine a symbolic solution to your model.
2. Develop a recursive sequence model for the yearly amount of vegetation in the Lake for the ten year time period 1988-1997. Iterate your model to determine the amount of vegetation in 1997.
3. Experiment using different initial amounts of carp in your model in #2 to determination the number of carp needed to be stocked in 1988 in order to eliminate the vegetation during the ten year time period.
4. Based on your answer to #3, record the year, number of carp, and amount of vegetation in a three column table for the period 1988-1997. Create a multiplot showing the plot of the number of carp versus time and the plot of the amount of vegetation versus time for the years 1988-1997.
5. Write an essay based on your study of the effects of using carp to control unwanted lake vegetation.

4. Sweepstakes

(Purposes: experience developing an amortization model for present value; understand the mathematics of paying off a large jackpot winning; provide a writing exercise; provide a small-group work experience.)

Introduction: J. D. Gryder, a thirty-one-year-old Houston rock band drummer, won the $11 million Family Publishers Sweepstakes in March 1997 (as reported by Tom Lowry in *USA Today*, March 20, 1998). Mr. Gryder was given an immediate bonus of $45,000 and promised 30 yearly payments of $366,666 each. His yearly payments are to be made on the anniversaries of his winning the sweepstakes.

Complete the following tasks.

1. Write out a dictionary definition of the word amortization.
2. Ask a finance officer (bank officer, college treasurer, investment officer, etc.) to estimate an average annual interest rate over the next thirty years for an investment of several million dollars. Also, ask the officer to estimate an average annual rate of inflation over the next thirty years. List the name, position, and institution of the person consulted. (If the person does not want to estimate rates of interest and inflation over the next thirty years, ask the person to estimate the rates over the past thirty years.)
3. Convert the annual interest rate obtained in Task 2 to a monthly rate.
4. Using the following outline, derive a present value formula for a monthly amortization.
5. Using the monthly rate found in Task 3 and the present value formula found in Task 4, determine how much Mr. Gryder's winnings cost American Family Publishers in terms of 1997 dollars, assuming no inflation.
6. Using the monthly rate found in Task 3, determine how much Mr. Gryder's winnings cost American Family Publishers in terms of 1997 dollars assuming the estimated rate of inflation found in Task 2. (Clearly explain your reasoning.)

Outline for deriving a present value formula for an amortization:

Terminology: i = monthly interest rate
A_k = present value of the kth payment of $366,666
(That is, A_k when invested for k years at i % per month yields $366,666.)

a. Show that $(1 + i)^{12}A_1$ = amount after 12 months (1 year)

b. Show that the 1997 value of the first payment (1998) is $A_1 = \frac{366,666}{(1+i)^{12}}$.

c. Show that $(1 + i)^{24}A_1$ = amount after 24 months (2 year).

d. Show that the 1997 value of the second payment (1999) is $A_2 = \frac{366,666}{(1+i)^{24}}$.

e. Show that $(1 + i)^{36}A_1$ = amount after 36 months (3 year).

f. Show that the 1997 value of the third payment (2000) is $A_3 = \frac{366,666}{(1+i)^{36}}$.

g. Conjecture the 1997 value of the fourth payment and verify your conjecture. (You should now be able to conjecture that the 1997 value of the kth payment is $A_k = \frac{366,666}{(1+i)^{12k}}$)

h. Sum the 1997 value of the 30 yearly payments: $A = A_1 + A_2 + A_3 + ... + A_{30}$.

i. Factor $\frac{366,666}{(1+i)^{12}}$ out of the expression for the sum in Part h. The remaining factor is a geometric series with ratio $\frac{1}{(1+i)^{12}}$.

j. Sum the geometric series and then simplify the resulting product.

5. Mercury in the Reservoirs: Water's OK, but Don't Eat the Fish

(Purposes: develop a recursive sequence model for a real-life mercury pollution problem; investigate long-term effects of the pollution; provide a writing exercise; provide a small-group work experience.)

Introduction: The amount of toxic mercury in the smallmouth bass population of New York's Neversink Reservoir is of serious concern to public officials. Neversink Reservoir supplies drinking water to New York City. The July 11, 1999, issue of the *Times Herald-Record*, a New York newspaper, carried a story about the elevated levels of toxic mercury pollution in New York City's drinking water. For more than a century, scientists have known about the adverse affects of mercury to the health of humans. The term "mad as a hatter" stems from the 19th century use of mercuric nitrate in the making of felt hats, causing insanity in hatters.

How does mercury get into the reservoir and thus into the smallmouth bass? Human activities are responsible for most of the mercury emitted into the environment. Mercury, a by-product of coal, comes from acid rain from smokestack emissions of old, coal-fired power plants in the Midwest and

South. Mercury particles rise on the smokestack plumes and are carried on prevailing winds, which often blow northeast. After colliding with the New York mountain range, the particles drop to earth. Once in the ecosystem, microorganisms in the soil and reservoir sediment break down the mercury and produce the toxic chemical methylmercury. This chemical is readily absorbed into the food chain. Once in the chain, the mercury works its way to the top as larger creatures eat smaller ones, finally concentrating at high levels in very long-lived fish, such as smallmouth bass, which may live for fifteen years. Thus the level of mercury concentration in the water may be safe, whereas it may be dangerous to eat smallmouth bass. The U.S. Environmental Protection Agency states that the water in Neversink Reservoir is safe to drink, but recommends eating no more than one meal of bass from the reservoir per month.

Mercury undergoes a process known as bioaccumulation. Bioaccumulation occurs when organisms including humans take in contaminants more rapidly than their bodies can eliminate them, thus the amount of mercury in their bodies accumulates over time. If for a period of time an organism does not ingest any more mercury, the body content of mercury will decline. If, however, an organism continues to ingest mercury, its body content can increase to toxic levels. Humans eliminate mercury in their system at a rate proportional to the amount in their system. Methylmercury decays about 25 percent per 30 days if no further mercury is ingested during that time.

Based on case studies and substantial human and animal data, the U.S. Environmental Protection Agency (USEPA) set the safe monthly dose for methylmercury at 3 micrograms per kilogram (μg/kg) of body weight. This monthly dose is intended to protect the average adult who weighs 70 kg. (The USEPA's safe dose is 0.1 micrograms per kg of body weight per day.)

Your group is tasked to analyze the situation of a fictitious character named Bubba who lives on the shore of Neversink Reservoir. Bubba eats one smallmouth bass taken from the reservoir each month on the first day of the month. Assume that he did not start this habit until he was 15 years old.

1. Construct a recursive sequence model using the paradigm (new situation) = (old situation) + (change) for the amount of methylmercury that will accumulate in Bubba's body. Suggestion: Review the solution to Sue's plan in Section 4.2. Then begin by defining variables and stating facts. For example, let $m(n)$ = number of grams of methylmercury in Bubba's system after the nth month. Assume each month has exactly 30 days. Note Bubba's intake of (0.09)(1000) g of mercury on first day of each month decays to (0.75)(0.09)(1000) g by the end of the month. Hint: The change that occurs during the month is (– decay of old situation + intake with decay).

2. Determine numerical, graphical, and analytical solutions to the following recursive sequence model and then conjecture the long-term behavior.

   $m(n) = m(n-1) - .25m(n-1) + 67.5$ with $m(0) = 0$.

   a. Compute a numerical solution by using the table function of your calculator to iterate the recursive sequence for $n = 0...20$. Copy the calculator table into your report.
   b. Conjecture the long-term value of $m(n)$ based on your calculator table from Part b. What can you do to gain more confidence in the truth of your conjecture?
   c. Plot the recursive sequence (graphical solution).
   d. Conjecture the long-term value of $m(n)$ from the graphical solution in Part c.
   e. Iterate $m(n)$ by hand for the purpose of recognizing a pattern. (See the solution to Sue's plan in Section 4.2.) Use exponential notation ($2*2*2=2^3$) to simplify expressions, but do not do any arithmetic (adding, multiplying, etc.) when trying to discover a pattern. Identify the geometric series pattern.

f. Sum the geometric series.
g. Formulate the general solution in the form $m(n) = (.75)^n c + b$, where c and b are constants and n denotes the nth input to the function m.
h. Use the initial condition to determine the value of c and obtain the particular solution (analytical solution).
i. Determine the long-term value of $m(n)$ from the analytical solution in Part h.

3. Determine the effects of changing the decay rate on the long-term value of $m(n)$.
   a. Determine Bubba's long-term situation if the decay rate is 40% rather than 25%.
   b. What decay rate would yield a long-term value of 100 g of methylmercury in Bubba's body?

6. The Future of the World's Oil Supply

(Purposes: Model the effect of a growing consumption rate on a fixed resource such as the world's oil resource; create an awarenessof a contentious, but very serious national issue, provide an application of the exponential function; provide a writing exercise; provide a small group experience.)

(The material for this Fun Project was adapted from *The New York Times* article, "The Mirage of a Growing Fuel Supply," by Evar D Nering, June 4, 2001.)

The history of using petroleum to enrich the quality of life dates back at least to the Mesopotamians of 3000 BC. Petroleum seeping to the surface was used to make a *natural* asphalt used to seal containers and boat bottoms. Around 1000 BC, Chinese found natural gas when drilling for salt and used it for fuel. However, the modern petroleum age dates from Edwin L. Drake's discovery of oil in August 1859 at Oil Creek in northwestern Pennsylvania. The oil boom spread rapidly and by 1900 oil was being commercially produced in more than a dozen countries. In the United States alone, production rose to 64,000,000 barrels per year. Petroleum became the dominant energy source in the United States in the mid 1900s due, in large part, to the automobile industry. Consumption of petroleum products in the United States has increased from approximately five million barrels per day to almost 20 million barrels per day during the past 50 years. The world-wide rate of petroleum consumption in 1999 was 26.7 billion barrels per year or approximately 73 million barrels per day.

Assume the amount of available oil in the earth is known and suppose the life-span of this oil will last n years at the present rate of consumption. The following questions lead you through an investigation of how changes in the consumption rate effect the present life-span.

1. Develop a consumption model for oil based on a constant rate of consumption.
   a. Suppose the world supply of oil is being consumed at the rate of 25 billion gallons per year. Suppose also that the present known supply will last 100 years. What is the present known supply of oil?
   b. If the consumption rate is r barrels per year, how long will the present known supply last?
   c. Determine a model (function) for the total consumption of oil in terms of years (t) at a rate of r barrels per year.
2. Using a recursive sequence, develop a consumption model for oil based on an exponential rate of growth in consumption. Assume the rate of consumption for the first year is r barrels per year.
   a. If the consumption rate increases 5% per year, determine the rate of consumption for the second, third, fourth, and fifth years.
   b. Generalize the results of (a) to determine the consumption rate for the nth year.
   c. Determine a model (function) for the total consumption of oil over t years if the rate increases 5% per year, beginning with a rate of r barrels per year. (Hint: Express the total as a geometric series and then sum the series.)
   d. Generalize the result of (c) to model the total consumption of oil if the rate increases p% per year beginning with a rate of r barrels per year.
3. Compare the two models. The known supply of oil, rt, will last t years at the present constant consumption rate of r barrels per year. How long will this known supply last if the consumption rate increases by p% each year?
   a. If the known oil supply is projected to last 100 years at the present rate of consumption (25 billion gallons per year), how long will it last if the rate increases 5% each year?
   b. Redo part (a), assuming that enough new oil has been discovered to last 200 years at the present rate of consumption.
   c. Redo part (a), assuming that enough new oil has been discovered to last 1,000 years at the present rate of consumption.
   d. Redo part (a), assuming that the consumption rate will only increase 1% per year.
   e. Redo part (a), assuming that the consumption rate will only increase 0.3% per year.
4. Verify Evar D. Nering's statement:

   "Doubling the size of the oil reserve will add at most 14 years to the life expectancy of the resource if

   we use it at the currently increasing rate, no matter how large it is currently. On the other hand, halving

   the growth of consumption will almost double the life expectancy of the supply, no matter what it is."
5. Write an essay reflecting on Evar D. Nering's statement and the impact it should have on the national energy debate of whether we should emphasize oil exploration and developing new oil fields or emphasize conservation measures.

7. College Tuition Plans

(Purposes: develop models to compare a prepaid tuition plan and a state savings plan; graphically fit curves to data in order to make predictions; provide a writing exercise; provide a small-group work experience.)

Introduction: Earning a college education is becoming more important every year. Employers are raising their education and skill requirements for new employees. Job markets in general are good and getting better for the college-educated person, but are poor and getting poorer for those without a college education. The cost of attending college is a major obstacle for many today, and the slope of the cost trend is definitely positive. The tuition at the University of Texas at Austin has increased 310% over 10 years from $375 per semester in 1988 to $1200 per semester in 1997. What will the tuition be when your children are ready to enroll in college, say 20 years from now?

The traditional forms of student aid—scholarships, grants, and work-study—do not meet the financial needs of the majority of students. As a result, many students are forced to compromise their educational opportunities by having to work part or fulltime while taking classes in addition to taking out student loans. The accumulated loan debt financially handicaps many graduates and cripples their potential careers. To help alleviate these problems, several states are offering some form of tuition savings plans. In this project, you will analyze the two most popular plans: (1) prepaid tuition plan and (2) state savings plan.

Prepaid Tuition Plan. This plan allows people to purchase tuition credits that can be redeemed at a future time with the plan paying for any increase that has occurred. Thus people can lock in future tuitions at the rate in effect when they purchase tuition credits. For example, a parent who had purchased $375 worth of tuition credits in 1988 (a semester's tuition at the University of Texas) would now have one semester of their son's or daughter's tuition paid for by the plan even though the current tuition rate is $1200 per semester. Under this plan, investors would lose if inflation increases faster than tuition, an unlikely event.

State Savings Plan. Under this plan people invest money in state-sponsored accounts. Federal tax on the interest received is postponed until the funds are used for college and are then taxed at the student rate. The performance of these state-sponsored accounts is usually subject to the fluctuations of the financial markets. Thus there is no assurance that the funds will grow as fast as tuition. On the other hand, invested funds may grow faster than tuition.

Complete the following tasks.

1. Graphically fit curves to data

   a. Fit a curve to the following tuition data from the University of Texas at Austin and then project the tuition cost for the year 2000.

   b. Fit a curve to the following room and board data from the University of Texas at Austin and then project the room and board cost for the year 2000.

Year	Tuition (Texas)	Room and Board
1988	375	3166
1989	410	3166
1990	440	3166
1991	510	3261
1992	590	3309
1993	715	3498
1994	770	3528
1995	930	3683
1996	1085	3901
1997	1200	4096

Table 4.FunProject 5.1

   Hint: plot the data with the horizontal axis scale being 1 through 10 and the vertical scale being 0 through 1200. Let year 1988 correspond to 1 on the horizontal scale and year 1997 correspond to 10.

   c. Write an interpretation of the accuracy of your projected values. Describe additional factors that need to be considered in projecting the costs of tuition and room and board.

2. Prepaid tuition plan

   Will and Pearl are considering prepaying their daughter's tuition at your college. They have asked your group to run a simulation under the following assumptions:

   (a) They purchase tuition credits each fall from 1988 through 1993.

   (b) A total of 120 credits, the amount needed for graduation, are purchased.

   (c) Their daughter entered your college as a first-year student in the fall of 1994 and graduated in the spring of 1997.

   Your task is to select the number of tuition credits Will and Pearl purchased each year and then to fill in the following table. You may need to consult the appropriate person (e.g., treasurer) at your school to find out the costs of tuition credits for each of the last 10 years. Assume that your school requires 120 credits to graduate (15 per semester). State the name of the college official whom you consulted.

Year	No. Credits Purchased	Price/Credit	Amount Prepaid
1988			
1989			
1990			
1991			
1992			
1993			
1994			
1995			
1996			
1997			

Table 4.FunProject5.2

3. State Savings Plan

   Will and Pearl are considering prepaying their daughter's tuition at your college. They have asked your group to run a simulation assuming for each of the years 1988-1997, the amounts that had been used to purchase tuition credits that year under the prepaid tuition plan are invested in a state-supported savings account at your local bank. Your task is to fill in the following table. You may need to consult someone at your local bank to determine the interest rate that the bank paid on its savings account for each of the years 1988 through 1996. Assume that only the amount of one year's tuition is withdrawn in the falls of 1994, 1995, 1996, and 1997. State the name of the bank and the name of person consulted.

Year	Amount Invested	Interest Rate	Interest Earned	Balance
1988				
1989				
1990				
1991				
1992				
1993				
1994				
1995				
1996				
1997				

Table 4.FunProject5.3

4. Write a recommendation to Will and Pearl addressing the following:
   a. Comparison of the two plans based on your analysis.
   b. Explanation of which plan carries the greater risk and the reasons for it.
   c. Additional factors that should be considered and your reasons for including each of the additional each factors.

8. Oxygen Levels in the Narraguagus River

(Purposes: provide research experience using the Internet; use of various regression programs in a graphing calculator; analysis to determine the best model; provide a writing exercise; provide for a a small-group work experience.)

Introduction: Environmentalists are growing concerned over the algae build-up in the Narraguagus River. A scientist assigned to look into the issue of algae build-up has asked that the oxygen levels at a designated location in the river be monitored for four successive days. Due to a mix-up in communications, the field worker assigned to do the monitoring understood that he should record the oxygen level in the river at four different times on the same day. He reported the following data. (The oxygen level is reported as the number of milligrams of oxygen for every 1000 grams of water.)

Time	Oxygen Level (ppm)
5:00AM	8
11:00AM	10
5:00PM	17
12:00PM	10

Table 4.FunProject6.1

The scientist, disappointed by having data for only one day rather than four days, has turned to your group for help. She comments that the day-to-day changes in the weather during the four-day period were minimal and asks your group to develop several models of the oxygen level as a function of time for the four day period. She then wants you to determine which of the models is best and to describe your reasoning in clear statements.

Your tasks are the following:

1. Research oxygen levels in a river
    a. List five factors that affect oxygen levels in a river.
    b. Determine the most important factor (explain your reasoning).
    c. Describe how oxygen gets into the water.
    d. Explain why oxygen levels may differ at different times of the day.
2. Plot the data.
3. Graphically model the oxygen level by fitting a curve to the data. List the major characteristics you would like the curve to have.
4. Use the regression capability of your graphing calculator, to determine the following regression models:
    a. Linear
    b. Quadratic
    c. Cubic
    d. Quartic
    e. Sine (remember that after you enter the data lists, you need to specify the period)
5. Determine which of the five regressions gives the best model for the four-day period. Describe your reasoning for each of the five regressions.
6. Superimpose the plot of your best model on the plot of the data.
7. Use your best model to predict the oxygen level at 2:00 PM on the third day.

9. Do Manatees have a Future?

(Purposes: provide a real-life modeling experience that emphasizes the role of assumptions;

provide for an inquiry and writing exercise; provide for a small-group work experience.)

West Indian manatees are large aquatic mammals whose closest relative is the elephant. An adult manatee is about 10 feet long and weighs approximately1,000 pounds. Their grayish-brown body tapers to a flat tail. Their two flippers have nails and their wrinkled face has whiskers on their snouts. Manatees are a calm, slow-moving mammal who like warm shallow water (1-2 meters). The majority of the manatees spend their winters in the estuaries, saltwater bays, and canals along the Florida coastline, migrating west and north during the summer as the water warms. They are completely herbivorous and spend most of their lives eating and resting.

Although the manatee has no natural enemy, many are killed each year by colliding with motor boats, getting caught in canal lock systems, becoming ensnared in fishing gear, and vandalism. Cold water also takes a toll. In 1990, cold water was responsible for the death of 47 manatees. Their greatest threat, however, is loss of habitat. Today, manatees are protected by the Marine Mammal Protection Act (1972), Endangered Species Act (1973), and the Florida Manatee Sanctuary Act (1978).

Although no scientific procedure exists for determining the age of a manatee, they are generally believed to live 50-60 years if left alone. Female manatees become sexually active between 5 and 9 years of age and males around 9 years of age. The gestation period is 13 months. Female manatees give birth to one calf every 2-5 years. Birth of twins is rare.

Reported deaths of manatees (Source: www.seaworld.org/manatee)

Year	**Deaths**	**Year**	**Deaths**	**Year**	**Deaths**
1979	78	1985	123	1991	181
1980	65	1986	125	1992	167
1981	117	1987	120	1993	147
1982	117	1988	135	1994	195
1983	81	1989	174	1995	203
1984	130	1990	218		

Table 4.FunProjects 7.1

Your task is to model the future of Manatees based on the data provided.

1. Develop a manatee death function by year:
   (a) Display the reported deaths data in a scatter plot.
   (b) Eyeball a line that appears to best fit the data and then approximate the equation of the line.
   (c) Fit a linear equation to the data using your calculator's linear regression program.

2. Develop a population function for manatees for the years 2000 through 2005. Assume the population in 1998 is 2,000. Clearly state:
   (a) Your assumptions and the reasons for them. For example, describe the population distribution by age in 1998.
   (b) Your projections for the number of deaths and the number of births per year.
   (c) Your reasoning in creating your function.

3. Write a half-page essay discussing the survival possibilities for the West Indian Manatee based on

Percy Lavon Julian

Percy Lavon Julian was born April 11, 1899, in Montgomery, Alabama. Although graduating with honors from public school, he was classified as a "slow learner" when he entered DePauw University and was required to take remedial courses for his first two years. In order to pay for his education, he washed dishes, waited on tables, scrubbed floors, and took on any other jobs that were open to him. His dedication, commitment, and perseverance to learn led him to graduation in 1923 as class valedictorian and member of Phi Beta Kappa, the highest academic honor society. His next goal was to pursue study for a Ph.D. in chemistry. Although easily earning a master's degree (Harvard University), he was temporarily rebuffed in his goal to study for a Ph.D. because of racial discrimination. In 1931, he realized his goal when he received his Ph.D. in chemistry from the University of Vienna. He taught at several schools before turning to industry in 1936, first with Institute of Paper Chemistry in Appleton, Wisconsin, and then with the Gliddon Company as Research Director for one of its Divisions.

Dr. Julian founded his own research laboratories, Julian Laboratories, in 1954, one in Chicago and one in Guatemala. Seven years later he sold his Chicago labora- to Smith, Kline, and French, and three years later he sold his Guatemalan laboratory to the Upjohn Company.

In addition to his earned degrees, Dr. Julian received fifteen honorary degrees. He was a member of the Board of Regents of the State of Illinois and a Trustee of DePauw, Fisk, Howard, and Roosevelt Universities and the Chicago Theological Seminary.

Faced with enumerable discrimination barriers, Dr. Julian never lost hope or his determination to succeed. Nor did he ever lose sight of his goal of helping youth to function at the very high levels of society. He served not only as a role model, but also as a mentor throughout his life. Speaking of his work with youth, he said

> Youth today are seeking self-identification to signify their struggle to find who they are. Properly explored and executed, this resolve could be the harbinger of the greatest emancipation vouchsaved to us in three-and-a half long centuries.

Appendix A Computational Skills and Basic Functions

Computational Skills

1. Arithmetic of Fractions

a. Addition and Subtraction of Fractions

i. Fractions with a common (same) denominator:
Two fractions are added by adding the numerators and dividing by the common denominator.

$$\frac{2}{3} + \frac{5}{3} = \frac{2+5}{3} = \frac{7}{3}$$

$$\frac{3}{5} - \frac{4}{5} = \frac{3-4}{5} = -\frac{1}{5}$$

ii. **Fractions with different denominators:**
The fractions must be transformed to have a common denominator. This is done by multiplying and dividing each fraction by a suitable number to form a common denominator as illustrated by the following examples. Note that the lowest common denominator is not used in the fourth example as it is in the first three examples.

$$\frac{2}{3} + \frac{3}{4} = \frac{2}{3}\left(\frac{4}{4}\right) + \frac{3}{4}\left(\frac{3}{3}\right) = \frac{8}{12} + \frac{9}{12} = \frac{8+9}{12} = \frac{17}{12}$$

$$\frac{3}{7} - \frac{5}{2} = \frac{3}{7}\left(\frac{2}{2}\right) - \frac{5}{2}\left(\frac{7}{7}\right) = \frac{6}{14} - \frac{35}{14} = \frac{6-35}{14} = \frac{29}{14}$$

$$\frac{3}{4} - \frac{5}{6} = \frac{3}{4}\left(\frac{6}{6}\right) - \frac{5}{6}\left(\frac{4}{4}\right) = \frac{18}{24} - \frac{20}{24} = \frac{18-20}{24} = \frac{-2}{24} = -\frac{1}{12}$$

$$\frac{1}{2} + \frac{3}{4} = \frac{1}{2}\left(\frac{2}{2}\right) + \frac{3}{4}\left(\frac{1}{1}\right) = \frac{2}{4} + \frac{3}{4} = \frac{2+3}{4} = \frac{5}{4}$$

b. Multiplication of Fractions

Two fractions are multiplied by multiplying the numerators and dividing by the product of the denominators.

$$\frac{2}{3} * \frac{4}{5} = \frac{2*4}{3*5} = \frac{8}{15}$$

$$\frac{1}{4} * \frac{-2}{7} = \frac{1*(-2)}{4*7} = \frac{-2}{28} = -\frac{1}{14}$$

c. Division of Fractions

Two fractions are divided by inverting the denominator fraction and multiplying it by the numerator fraction. (Explain the rationale for this rule.)

$$\frac{\frac{2}{3}}{\frac{4}{5}} = \frac{2}{3} * \frac{5}{4} = \frac{2*5}{3*4} = \frac{10}{12} = \frac{5}{6}$$

$$\frac{3}{\frac{1}{4}} = \frac{3}{1} * \frac{4}{1} = \frac{3*4}{1*1} = 12$$

$$\frac{\frac{2}{3}}{5} = \frac{2}{3} * \frac{5}{1} = \frac{2*5}{3*1} = \frac{10}{3}$$

2. Rules of Exponents

Let r, s, m, and n be real numbers. The following rules apply.

a. $r^{-n} = \frac{1}{r^n}$

b. $r^0 = 1$

c. $r^1 = r$

d. $r^n = r * r * r * r \ldots r$ for n a positive integer

e. $r^n r^m = r^{n+m}$ (e.g., $2^2 2^3 = 2^{2+3} = 2^5$)

f. $(r^n)^m = r^{nm}$ (e.g., $(2^2)^3 = 2^6$)

g. $(\frac{r}{s})^n = \frac{r^n}{s^n}$ (e.g., $(\frac{2}{3})^2 = \frac{2^2}{3^2} = \frac{4}{9}$)

3. The quadratic formula

If a, b, and c are real numbers with a not equal to zero, then

$$ax^2 + bx + c = 0$$

has two solutions: $x_1 = \frac{-b+\sqrt{b^2-4ac}}{2a}$ and $x_2 = \frac{-b-\sqrt{b^2-4ac}}{2a}$.

If $b^2 - 4ac \geq 0$, then both solutions are real numbers. Otherwise, both solutions are complex numbers.

If $2x^2 + 4x - 1 = 0$, then $x = \frac{-4\pm\sqrt{16+8}}{4} = \frac{-4\pm\sqrt{24}}{4} = \frac{-4\pm2\sqrt{6}}{4} = \frac{-2\pm\sqrt{6}}{2}$

and so: $x = \frac{-2+\sqrt{6}}{2}$ and $x = \frac{-2-\sqrt{6}}{2}$.

4. **Factoring**

a. **Common Factor**

Each term contains a factor common to all the terms.

$$3x^2 + 6x + 12 = 3(x^2 + 2x + 6)$$

$$ax + ay + az = a(x + y + z)$$

b. **Difference of Squares**

$$x^2 - 4 = (x - 2)(x + 2)$$

$$a^2 - b^2 = (a - b)(a + b)$$

c. **Difference of Cubes**

$$x^3 - 8 = (x - 2)(x^2 + 2x + 4)$$

$$a^3 - b^3(a - b)(a^2 + ab + b^2)$$

5. **Roots and Factors of Polynomials**

A root of a polynomial is a number that when substituted for the variable in the polynomial yields zero.

If $x = r$ is a root of a polynomial than $(x - r)$ is a factor of the polynomial.

If $(x - r)$ is a factor of a polynomial, then $x = r$ is a root of the polynomial.

$(x - 2)$ and $(x - 1)$ are factors of $x^2 - 3x + 2 = (x - 2)(x - 1)$ and 2, 1 are roots of $x^2 - 3x + 2 = 0$.

6. **Binomial Theorem**

$$(x + a)^2 = (x + a)(x + a) = x^2 + 2ax + a^2$$

$$(2x + 3y)^2 = (2x)^2 + 2(2x)(3y) + (3y)^2 = 2x^2 + 12xy + 9y^2$$

7. **Pythagorean Theorem**

The length of the hypotenuse of a right triangle is the square root of the sum of the squares of the two legs of the triangle.

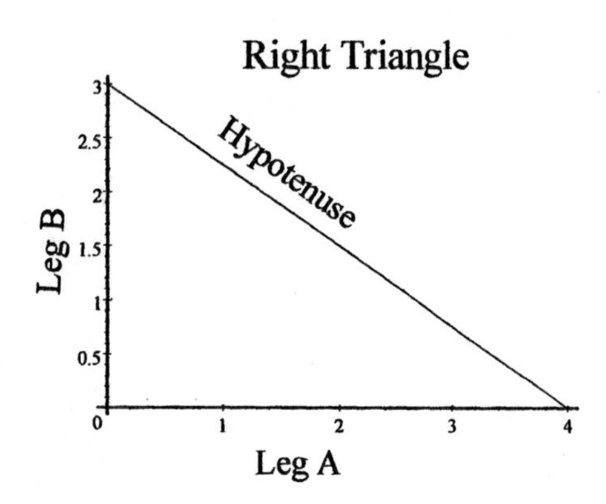

Length of Leg A = 4

Length of Leg B = 3

Length of Hypotenuse = $\sqrt{4^2 + 3^2} = 5$

8. Distance between Two Points.

Form a right triangle with the two given points being the endpoints of the hypotenuse and then use the Pythagorean theorem.

The distance between points (2, 5) and (4, 1) is $\sqrt{(4-2)^2 + (5-1)^2} = \sqrt{20} = 2\sqrt{5}$

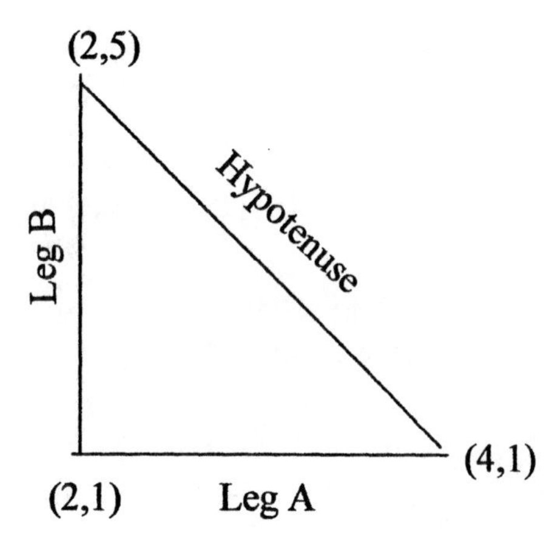

Length of Leg A = $(4 - 2)$

Length of Leg B = $(5 - 1)$

Length of Hypotenuse = $\sqrt{(4-2)^2 + (5-1)^2}$

Basic Functions

1. Polynomial Function

Each term of the range expression is the product of a coefficient and the independent variable raised to a <u>nonnegative integer</u> power.

$p(x) = -x^4 - 2x^3 + 12x^2 + x - 10$

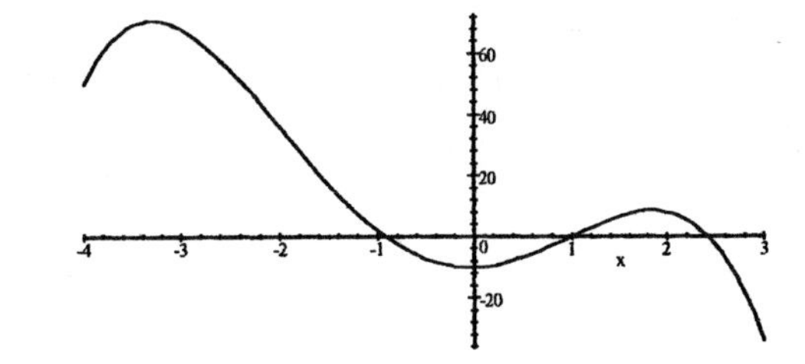

Definition: Any number raised to the zero power is one: $a^0 = 1$. (Thus $-10x^0 = -10$.)

Definition: The degree of a polynomial is the value of the largest exponent in the polynomial. (The preceding polynomial has degree 4.)

2. **Exponential Function**

A function whose range expression contains an exponential term. That is, a term that is the product of a coefficient and a number with an exponent which contains the independent variable.

$f(x) = 3 * b^{5x}$ The number b is called the base of the exponential. When $b = e \sim 2.71828...$, the exponential is called the natural exponential.

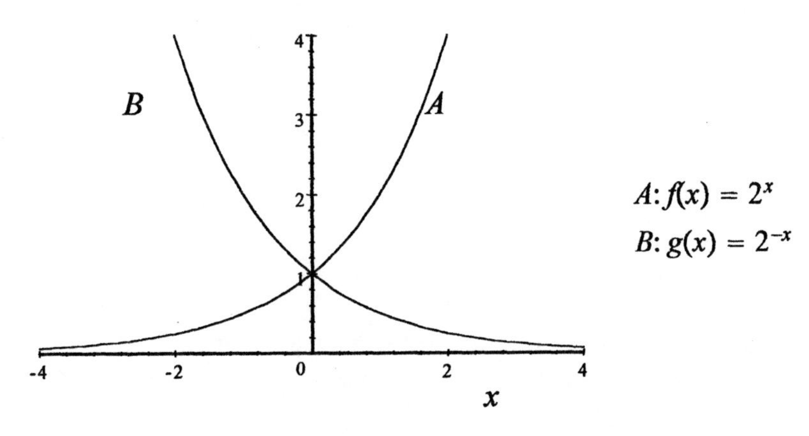

$A: f(x) = 2^x$

$B: g(x) = 2^{-x}$

3. **Logarithm Function** (inverse of the exponential function)

Defining property: $\log_b(x) = y$ if and only if $x = b^y$

b is called the base of the logarithm.

When $b = e \sim 2.71828...$, the logarithm is called the natural logarithm.

When $b = 10$, the logarithm is called the common logarithm.

Because the logarithm function is the inverse of the exponential function, the graph of the logarithm function is found by reflecting the graph of the exponential function in the line $y = x$ as shown in the following multiplot,

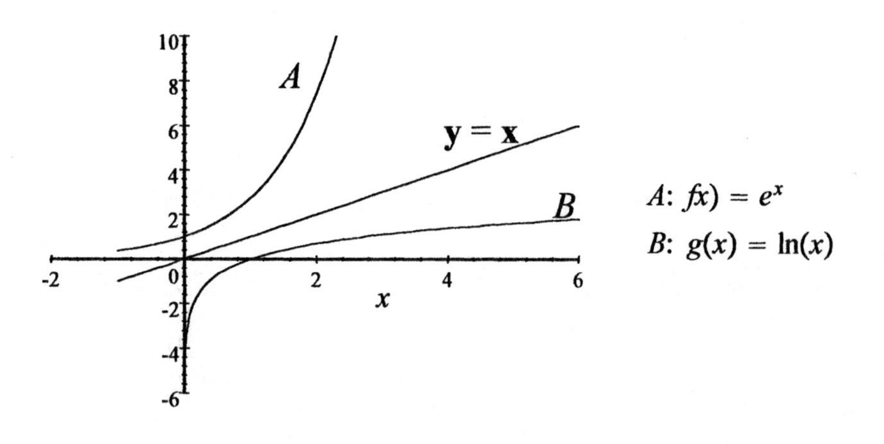

Rules of Logarithms

Let b, r, s, and n be real numbers with $b > 0$. The following rules apply

a. If $r = b^s$, then $s = \log_b(r)$, b is called the **base** of the logarithm.
b. $\log_b(r^{-1}) = -\log_b(r)$
c. $\log_b(1) = 0$
d. $\log_b(b) = 1$
e. $\log_b(r^n) = n\log_b(r)$
f. $\log_b(r * s) = \log_b(r) + \log_b(s)$
g. $\log_b(\frac{r}{s}) = \log_b(r) - \log_b(s)$

4. Trigonometric Functions

Trigonometric functions are periodic functions and thus are often used to model wave motion and cyclic phenomena. A function f is periodic if there exists a number p such that $f(x+p) = f(x)$. The smallest positive number p for which this is true is called the period of the function. The six Trigonometric functions are defined in terms of ratios of sides of a triangle formed from the center point of a circle of radius r and a point, (a, b), on a circle circumference, as shown in the following diagram.

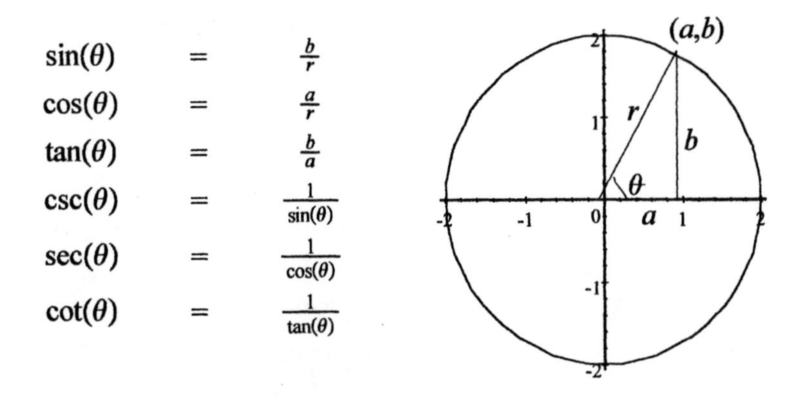

$$\sin(\theta) = \frac{b}{r}$$
$$\cos(\theta) = \frac{a}{r}$$
$$\tan(\theta) = \frac{b}{a}$$
$$\csc(\theta) = \frac{1}{\sin(\theta)}$$
$$\sec(\theta) = \frac{1}{\cos(\theta)}$$
$$\cot(\theta) = \frac{1}{\tan(\theta)}$$

Reference angles and values for $\sin(\theta)$ and $\cos(\theta)$

$\theta = 0$	$(a,b) = (r,0)$	and	$\sin(0) = 0$	$\cos(0) = 1$
$\theta = \frac{\pi}{4}$	$(a,b) = (\frac{r}{\sqrt{2}}, \frac{r}{\sqrt{2}})$	and	$\sin(\frac{\pi}{4}) = \frac{1}{\sqrt{2}}$	$\cos(\frac{\pi}{4}) = \frac{1}{\sqrt{2}}$
$\theta = \frac{\pi}{2}$	$(a,b) = (0,r)$	and	$\sin(\frac{\pi}{2}) = 1$	$\cos(\frac{\pi}{2}) = 0$
$\theta = \frac{3\pi}{4}$	$(a,b) = (\frac{-r}{\sqrt{2}}, \frac{r}{\sqrt{2}})$	and	$\sin(\frac{3\pi}{4}) = \frac{1}{\sqrt{2}}$	$\cos(\frac{3\pi}{4}) = -\frac{1}{\sqrt{2}}$
$\theta = \pi$	$(a,b) = (-r,0)$	and	$\sin(\pi) = 0$	$\cos(\pi) = -1$
$\theta = \frac{5\pi}{4}$	$(a,b) = (\frac{-r}{\sqrt{2}}, \frac{-r}{\sqrt{2}})$	and	$\sin(\frac{5\pi}{4}) = \frac{-1}{\sqrt{2}}$	$\cos(\frac{5\pi}{4}) = -\frac{1}{\sqrt{2}}$
$\theta = \frac{3\pi}{2}$	$(a,b) = (0,-r)$	and	$\sin(\frac{3\pi}{2}) = -1$	$\cos(\frac{3\pi}{2}) = 0$
$\theta = \frac{7\pi}{4}$	$(a,b) = (\frac{r}{\sqrt{2}}, \frac{-r}{\sqrt{2}})$	and	$\sin(\frac{7\pi}{4}) = \frac{-1}{\sqrt{2}}$	$\cos(\frac{7\pi}{4}) = \frac{1}{\sqrt{2}}$

Plots of $\sin(\theta)$ and $\cos(\theta)$

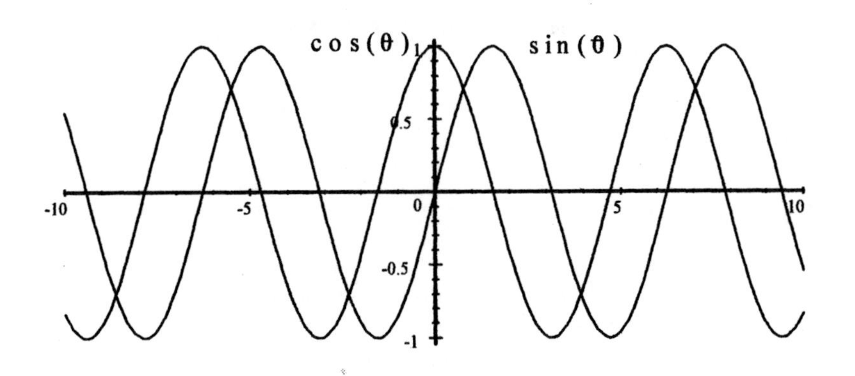

Appendix B Solutions to Selected Exercises

Chapter 2

Section 2.1:

1a. 52 1b. 263.90 1c. 103.50 2a. 20% 2b. 5% 2c. 40.58%

5.

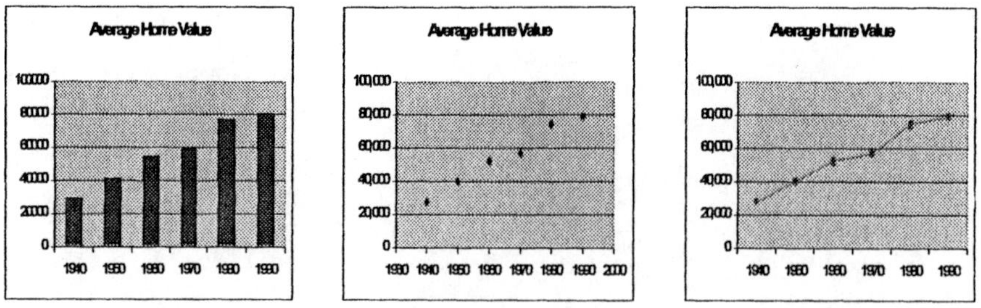

7.

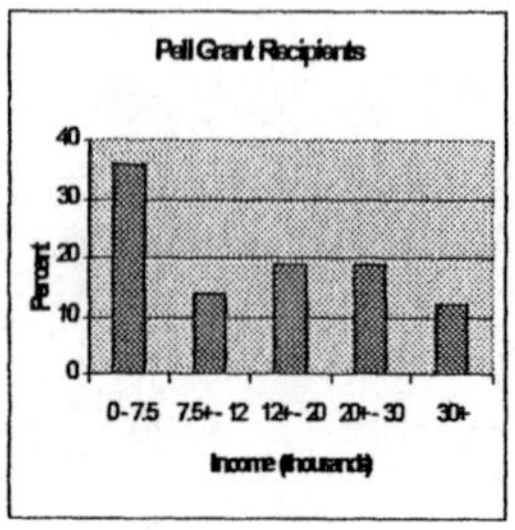

Income (thousands)	Percent
0 – 7,5	36
7.5+ - 12	14
12+ - 20	19
20+ - 30	19
30+	12

Section 2.2:

1a. 5 1b. 6.5 1c. 26

2a. 7 2b. $\frac{6+10+x}{3} = 5 \Rightarrow 16 + x = 15 \Rightarrow x = -1$ 2c. $\frac{5+4+6+7+x}{5} = 6 \Rightarrow x = 8$

5. $\frac{72+77+81+86+2x}{6} = 80 \Rightarrow \frac{316+2x}{6} = 80 \Rightarrow 316 + 2x = 480 \Rightarrow 2x = 164 \Rightarrow x = 82$

8. $\frac{25+2+3}{100+3+4} = \frac{30}{107} = 280$ 11. $\frac{23+32+33+43+57+70}{6} = 43$

13a. Set has an odd number of elements 13b. Always true: $\frac{\text{sum}}{\text{no.elements}}$ =average

17. Classes have the same number of students. 19. $22,272.73

Section 2.3:

1a. 3, every element 1b.3.5, 5 1c. 5, every element

2a. 5 2b. 7 2c. 9 4-14: several correct answers

Section 2.4:

1a. $2 + 2x = 3x \Rightarrow x = 2$ 1b. $x^2 - 1 = x^2 + x \Rightarrow x = 1$

1c. $x^2 + 2x + 1 = x \Rightarrow x = -\frac{1}{2}$

2a. $2x + 2b = 5x \Rightarrow 2b = 3x \Rightarrow b = \frac{3}{2}x \Rightarrow b = \frac{3}{2}(4) = 6$

2b. $x^2 + 2x + 1 = 2b \Rightarrow b = \frac{19}{2}$ 2c. $x^2 - b^2 = -8 \Rightarrow b = x^2 + 8 \Rightarrow b = \pm 3$

4a. $x = 1,5$ 4b. $1,-2.5$ 4c. ~2.0945

7a. 1,2,3,5,8 7c. Let x be the second element. Sequence is

$2, x, 2 + x, 4 + x, 6 + 2x, 10 + 3x \Rightarrow 10 + 3x = 7 \Rightarrow x = -1$. Thus 2, 1, 3, 4, 7

7e. Let x be the first element and y the second element.

Sequence: $x, y, x + y, x + 2y, 2x + 3y, 3x + 5y$. Thus $\begin{cases} 2x+3y=1 \\ 3x+5y=2 \end{cases}$

Solve for x in 1st eq. and substitute into 2nd eq. $x = \frac{1-3y}{2}$.Substitute

$3\frac{1-3y}{2} + 5y = 2 \Rightarrow y = 1, x = -1$ sequence: $-1, 1, 0, 1, 1, 2$

9. Fibonacci sequence: 1, 1, 2, 3, 5, 8, 13, 21, 34, 55, 89, 144

11. Let x =Terry's time=>$\frac{x}{2}$ = Sandra's time and $1.5x$ = Doris' time (4 hrs.)
Thus $1.5x = 4 \Rightarrow x = \frac{8}{3}$. Sandra's time=$\frac{x}{2} = \frac{4}{3}$. 13 a.1,3, 13b. ~4.5,~3.3

Section 2.5:

1a. $90(\frac{\pi}{180}) = \frac{\pi}{2}$ 1b. $180(\frac{\pi}{180}) = \pi$ 1c. $(\frac{2}{3}\pi)(\frac{180}{\pi}) = 120°$

2a. area = $\frac{1}{2}r^2\theta = \frac{1}{2}(3^2)[30(\frac{\pi}{180})] = \frac{3}{4}\pi$ in^2

2b. area = $\frac{1}{2}r^2\theta = \frac{1}{2}(3^2)(\frac{4}{3}) = 6$ in^2 $(\theta = \frac{\text{arc length}}{\text{radius}} = \frac{4}{3})$

2c. area = $\frac{1}{2}r^2\theta = 10$ and $r = 3$. Thus $\frac{1}{2}(3^2)\theta = 10 \Rightarrow \theta = \frac{20}{9}$ radians.

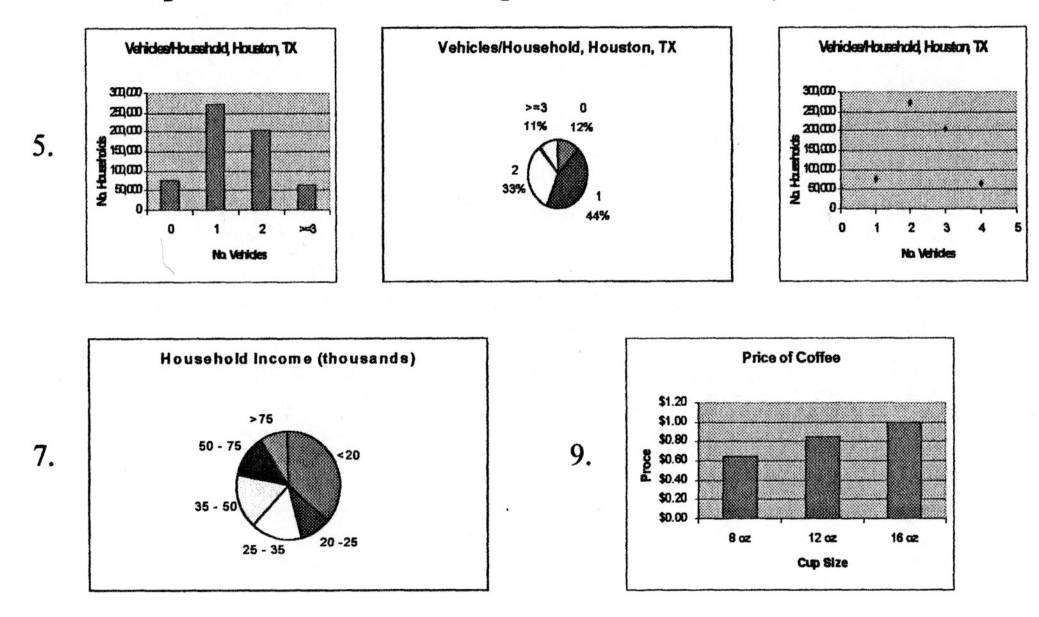

Section 2.6:

1a. $\frac{2}{3}(\frac{4}{4}) + \frac{3}{4}(\frac{3}{3}) = \frac{8}{12} + \frac{9}{12} = \frac{17}{12}$ 1b. $\frac{23}{6}$ 1c. $-\frac{211}{100}$

2a. $-\frac{15}{2}$ 2b. $\frac{63}{4}$ 2c. $-\frac{2x}{5} + \frac{3x}{4} = -1 \Rightarrow \frac{-8x+15x}{20} = -1 \Rightarrow x = -\frac{20}{7}$

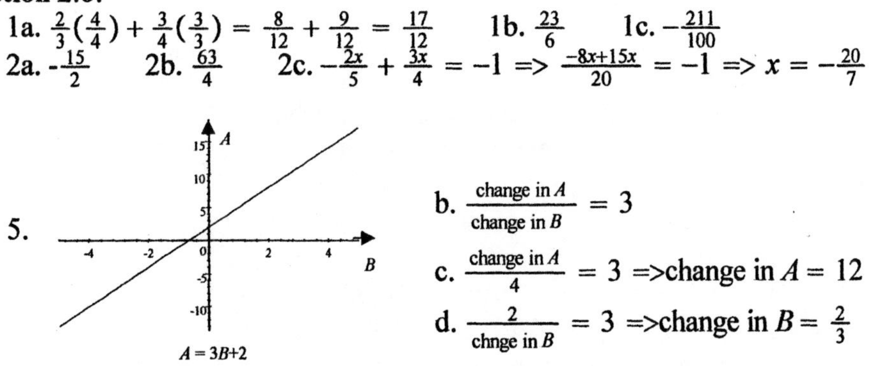

b. $\frac{\text{change in } A}{\text{change in } B} = 3$

c. $\frac{\text{change in } A}{4} = 3 \Rightarrow$change in $A = 12$

d. $\frac{2}{\text{chnge in } B} = 3 \Rightarrow$change in $B = \frac{2}{3}$

7. $\frac{y+3}{x+2} = \frac{1+3}{4+2} \Rightarrow 6y + 18 = 4x + 8 \Rightarrow y = \frac{2}{3}x - \frac{5}{3}$, y-intercept = $-\frac{5}{3}$

10. a<->B, b<->C, c<->D, no match for d (no plot has a y-intercept = -1)

13. Let $p(t)$ = pop. at year t. a. $p(t+20) = 2p(t)$, b. No, slope over 1910–30 is not the same as slope over 1910–70.

15. Let p = price and w = weight. Data points line on line $p = .06w + .84$
Price of Medium = $1.05, Price for Extra Large = $1.41.

17. D.

Section 2.7:

1a. 120 mi 1b. 100 =rate x $\frac{90}{60}$ => 66.67 mph 1c. $\frac{90}{40}$ = 2 hr and 15 min

2a. $y = -\frac{5}{4}x + \frac{13}{4}$ 2b. $y = 2x - 7$ 2c. $y = -2x - 3$

5. See Problem #1. 7. $x + y = 27$ and $y = 2x \Rightarrow x = 9$, $y = 18$
9. \$155.56 13. Equivalent exercise: How long does it take Doris to go 86 ft. when traveling 5 mph? Answer 11.73 seconds 15. $\$1.189\frac{1200}{27} = \$52.84$

17. 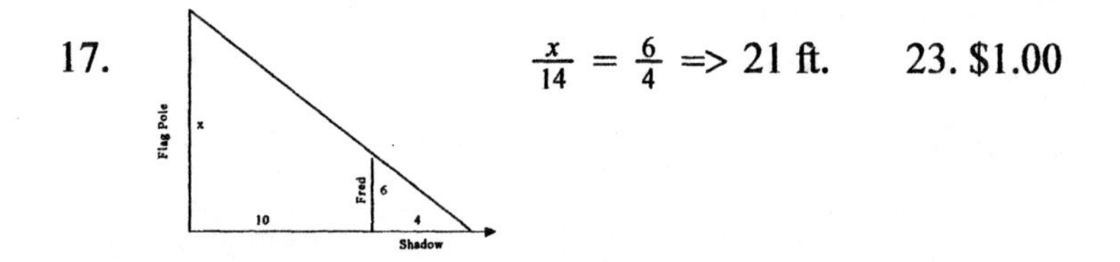

$\frac{x}{14} = \frac{6}{4} \Rightarrow 21$ ft. 23. \$1.00

Section 2.8:

1a. Yes 1b. No 1c. Yes 2a. $x < \frac{6}{5}$ 2b. $x < 2$

2c. There are two cases to consider when multiplying both sides by $x + 2$, one where $x + 2$ is positive and one where $x + 2$ is negative. The original problem is equivalent to $\begin{cases} 3(x+2)<4 & \text{for } x+2>0 \\ -3(x+2)<4 & \text{for } x+2<0 \end{cases}$ or $\begin{cases} 3x+6<4 & \text{for } x>-2 \\ -3x-6<4 & \text{for } x<-2 \end{cases}$ or $\begin{cases} x<-2/3 & \text{for } x>-2 \\ x>10/3 & \text{for } x<-2 \end{cases}$

Because there are no solutions for $x > 10/3$ and $x < -2$, the only solutions are for $x < -2/3$ and $x > -2$ or $-2 < x < -2/3$.

5a. $-5 < x < 1$ 5d. $x < -2$ or $x > 8$ 5e. $x < 1$ (graphical solution)

13c. $3Na_2CO_3 + 3Br_2 = 5NaBr + NaBr0_3 + 3CO_2$

14a. $4HCL + MnO_2 = MnCL_2 + 2H_2O + CL_2$ 14b. $5Fe + MnO_4 + 8H = Mn + 5Fe + 4H_2O$

16. $y = -2x + 5$, $z = z$ 17. Form the system of equations $xy + z = 15$, $u + v + w = 15$, $x + u = 7$, $y + v = 14$, $z + w = 9$, $y + w = 7$, $y + u = 9$. Magic square

8 1 6

3 5 7

4 9 2

Section 2.9:

1a. Yes 1b. No 1c. Yes

5. The slope of the graph of the objective function must be greater than the slope of the graph of the land constraint, that is greater than -1. Let x be the wheat subsidy. The slope of the objective function is $-\frac{60}{40+x}$. Thus $-\frac{60}{40+x} > -1$ or $x > 20$.

7. Objective function: $15ST + 30RC$. Corner points $(RC, ST) = (\frac{5}{4}, 0), (1, \frac{1}{2}), (0, \frac{3}{2})$
Just make rocking chairs.

9a. Minimize $50M + 40F$ subject to M and $F \geqslant 0, 20M + 50F \geqslant 500$,
$30M + 100F \geqslant 1000, 10M + F \geqslant 200, 15M + 2F \geqslant 50$

9b. Feasible region is the unbounded region **not** containing origin.

9c. minimum of \$11.44 at $M = 19.6, F = 4.1$.

10. 5 oz. of Product 1 and 1 oz. of Product 2 cost \$.19

11. 13 experienced and 3 apprentices seamstresses

Chapter 3

Section 3.1:

1a. $\frac{2}{5}$ 1b. 6 1c. $-\frac{3}{52}$ 2a. $\frac{49}{36}$ 2b. $\frac{49}{36}$ 2c. ~16.67

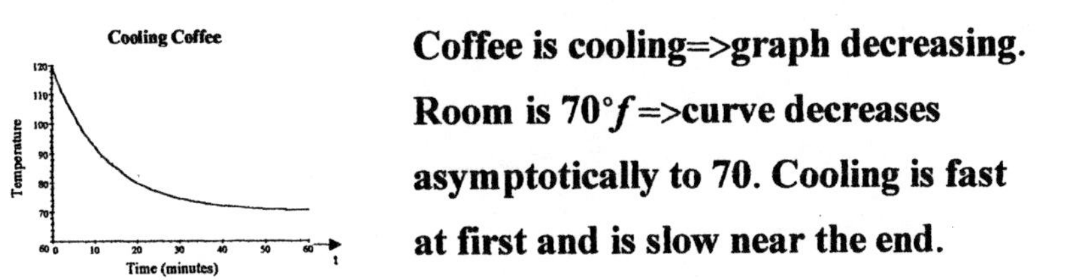

Coffee is cooling=>graph decreasing. Room is 70°f=>curve decreases asymptotically to 70. Cooling is fast at first and is slow near the end.

Section 3.2:

1a. 5 1b. $f(4) = 2^4 + 3(4^2) = 16 + 48 = 64$ 1c. $\sqrt{2} + \frac{3}{4}$

5a. nonnegative real numbers,

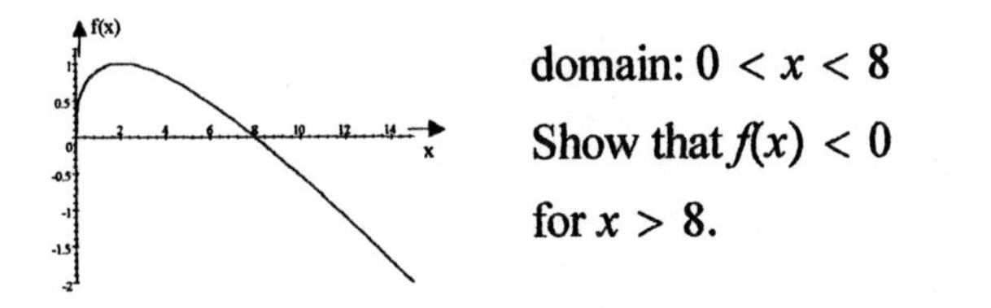

domain: $0 < x < 8$

Show that $f(x) < 0$ for $x > 8$.

7. Domain: all real numbers, Range: non-negative real numbers
9. Domain: all real numbers, Range: all real numbers 11. No, there are two different ordered pairs having the same first element.
13. No, if a person has two different phone numbers. 15. Mother to child is not a function relation; child to mother is a function relation
17. Temperature to time is not a function relation; time to function is a function relation. 19. Domain = $\{1, -2, 4, 2, 5\}$, Range = $\{3, 0, -1, 2\}$; Plot of f consists of 5 points. 21. No, ordered pairs $(1,2)$ and $(1,4)$ have the same 1st elements, but different 2th elements—violates the uniqueness prop.
23. Yes 25. Not a function, uniqueness property violated.

Section 3.3:

1a. 8 1b. 25 1c. 3 2a. B 2b. C 2c. A 2d. D

19. A: $y = x^2$, B: $y = -\cos(x)$, C: $y = e^{-x}$, D: $y = -x^3$

Section 3.4:

1a. Shift upward 10 units, $f(x) = x^2 + 10$ 1b. Shift upward 4 units and to the right 2 units, $f(x) = (x-2)^2 + 4$ 1c. $f(x) = (x+1)^2 + 5$

2a. $g(x) = (x-1)^3 + 0.5$ 2b. $g(x) = (x+1)^3 + 1$ 2c. $g(x) = (x-1)^3 + 8$

5a. $f(x) = |x-3| + 2$ 5b. $f(x) = |x+1| - 2$ 5c. $f(x) = |x+2| + 3$

7. $f(x) = (x-3)^2 + 2$; 9. $h(x) = \frac{1}{x+3} + 4$

11. #1 conjecture: $f(x) = -x^3 + 1$, correct basic shape, needs small shift left
#2 conjecture: $f(x) = -(x + .15)^3 + 1$, better, needs to be scaled
#3 conjecture: $f(x) = -2(x + .15)^3 + 1$, O.K.

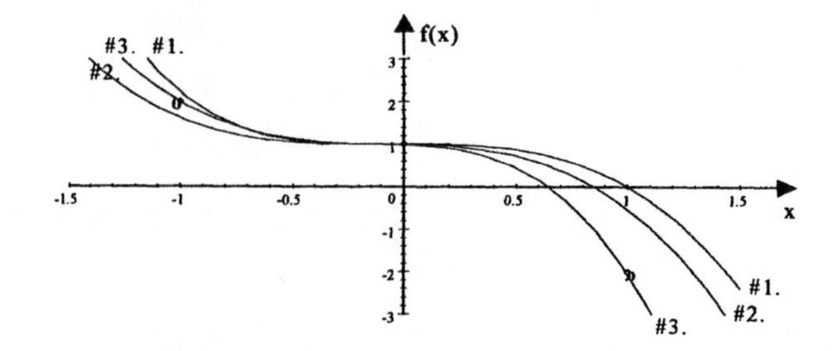

13.#1 conjecture: $f(x) = e^x - 4$, need to scale to flatten curve
#2 conjecture: $f(x) = e^{x/2} - 4$, need to lessen scale factor
#3 conjecture: $f(x) = e^{x/1.8} - 4$, O.K.

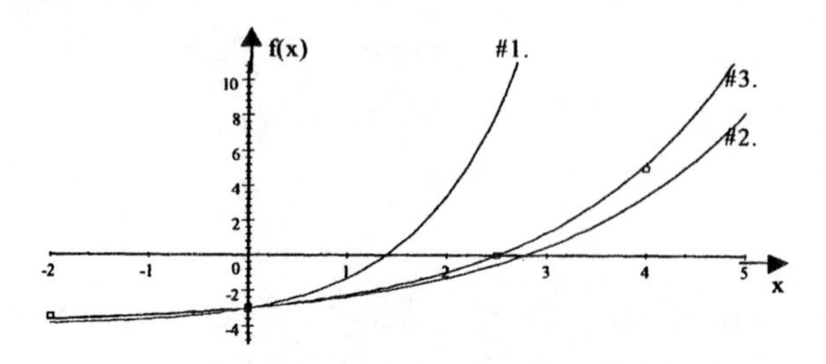

15. #1. conjecture: $f(x) = \sin(x+3) + 1.5$, need to shift upward
#2. conjecture: $f(x) = \sin(x+3) + 2$, should have scaled instead of shifting. Shift a little to left and scale
#3. conjecture: $f(x) = 2\sin(x+\pi) + 1$, O.K.

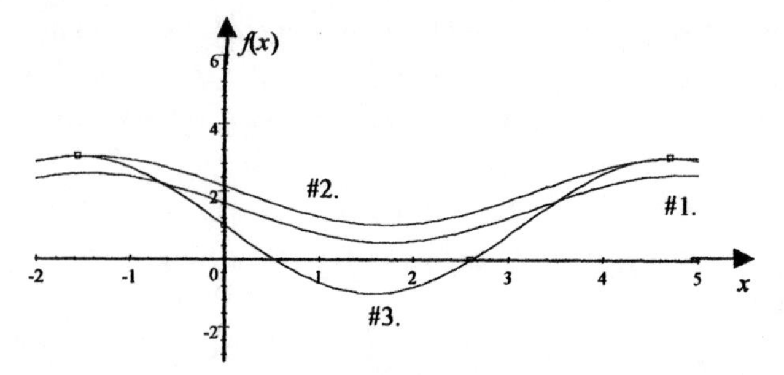

17. Matching: A-e, D-b; 19. Matching: B-a, A-b, C-c, D-d

Section 3.5:

1a. $12x + 12$ 1b. $\frac{x^2 - 4}{x - 3}$ 1c. -96

2.

x	f(x)	g(x)	f(x)g(x)	f(x)/g(x)	f(g(x))	g(f(x))
−4		−2				
−2		5			0	
−1	3					3
0	1	2	2	1/2	2	
2	2					
3		3				
4	−1	−1	1	1	3	
5	0					2

(Blank spaces means function is undefined.)

11.

x	$f(x)$	$g(x)$	$f(x)+g(x)$	$f(x)-g(x)$	$f(x)g(x)$	$f(g(x))$
0	4	3	7	1	12	1
1	2	0	2	2	0	4
2	4	1	5	3	4	2
3	1	4	5	−3	20	

(Blank space means function is undefined.)

13.

15a. $C(F) = \frac{5}{9}(F-32)$

17a. Is not a function   17b. Is a function

Section 3.6:

1a. $x^2 + (3-2)x + (-2)(3) = x^2 + x - 6$

1b. $x(x-2)^2 = x(x^2 - 4x + 4) = x^3 - 4x^2 + 4x$

1c. $x(x-3)^2 = x(x^2 - 6x + 9) = x^3 - 6x^2 + 9x$

2a. $x^3(x+7)$ 2b. $(x-2)(x+2)$ 2c. $x(x^2 - 5x + 6) = x(x-2)(x-3)$

5.

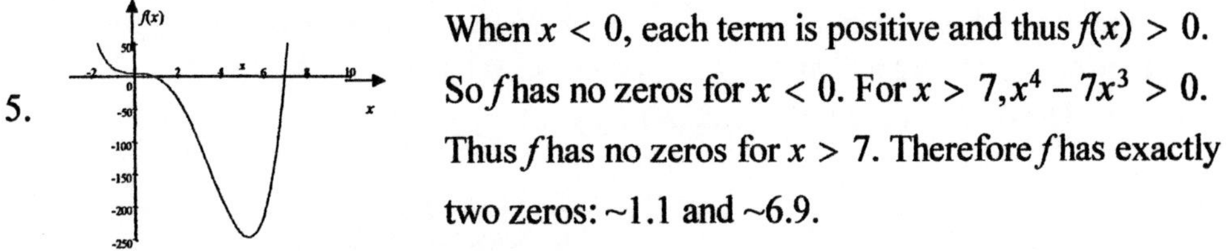

When $x < 0$, each term is positive and thus $f(x) > 0$. So f has no zeros for $x < 0$. For $x > 7, x^4 - 7x^3 > 0$. Thus f has no zeros for $x > 7$. Therefore f has exactly two zeros: ~1.1 and ~6.9.

9. Plot the data using 82 for 1982, 85 for 1985, and so on. The data appears to be "almost" linear. Using the first and last data points to determine the equation of a line gives $f(x) = \frac{10.7}{12}x - 34.82$. $f(97) = 51.67$ mph.

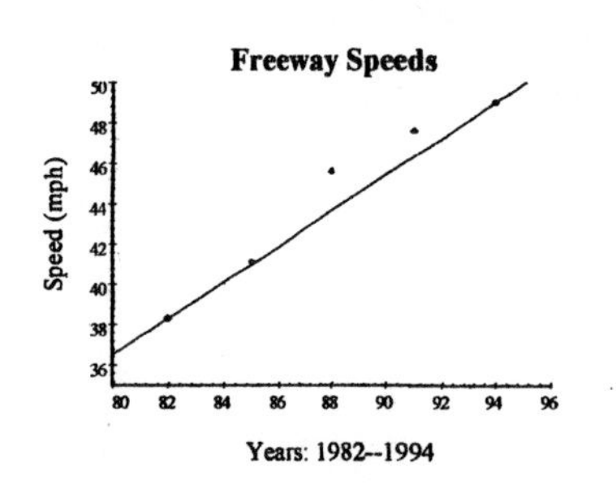

11. Plot the data using 90 for 1990, 91 for 1991, and so on. The basic shape is quadratic. Thus begin with $f(x) = x^2$ and shift right (90) and up (10), *app*#1: $f(x) = (x-90)^2 + 10$. Need to flatten the curve and shift left just a little. *app*#2: $f(x) = .9(x-88)^2 + 9$. Need to lower (decrease the 9) and flatten more (decrease the .9). After several attempts, *app*#3: $f(x) = .53(x-88)^2 + 6.4$

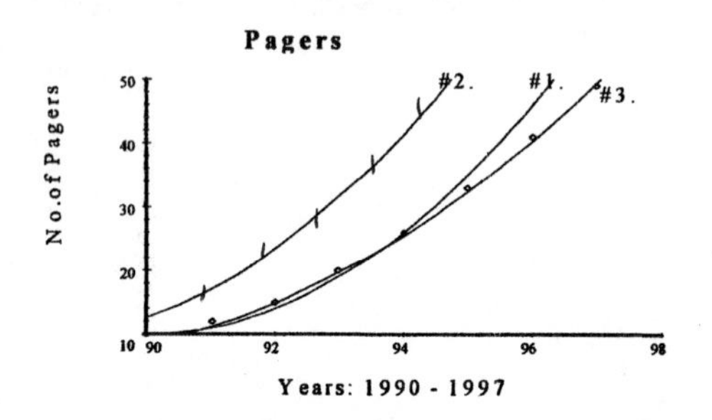

13. Plot the data using 10 for 1910, 30 for 1930, and so on. Because the data increases faster than 2^t, we begin by scaling t. Let *app*#1 $f(t) = 2^{t/5}$. Need to flatten the curve, thus need to scale t further. Let *app*#1 $f(t) = 2^{t/10}$. The shape is looking good, but needs to be shifted left and "rotated." To shift left, replace t by $t + 3$. To rotate, scale (up) the exponential term and subtract a constant. After several modifications, we have *app*#3 $f(t) = 28 * 2^{\frac{x+3}{23}} - 10$.

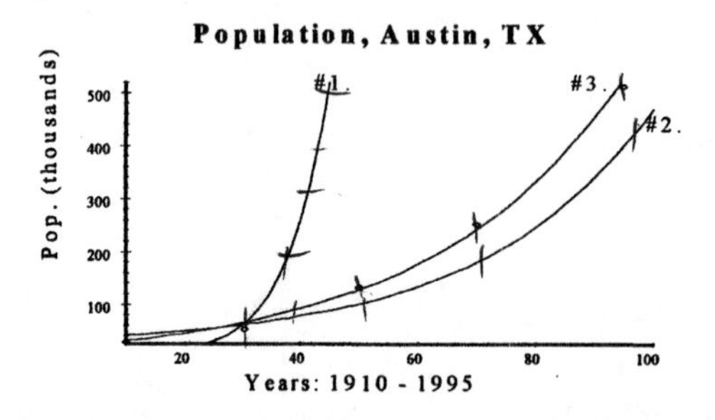

15. Plot $f(x) = \cos(x)$ (dark curve). Shape appears to be a parabola, opening down. Thus we let *app*#1 $g(x) = -x^2 + 1$. (Why?) The shape looks correct, but we need to scale x in order to "spread out" the curve. Thus let *app*#2. $g(x) = -\frac{1}{2}x^2 + 1$. Need to scale a little more. Let *app*#3 $g(x) = -\frac{1}{2.4}x^2 + 1$.

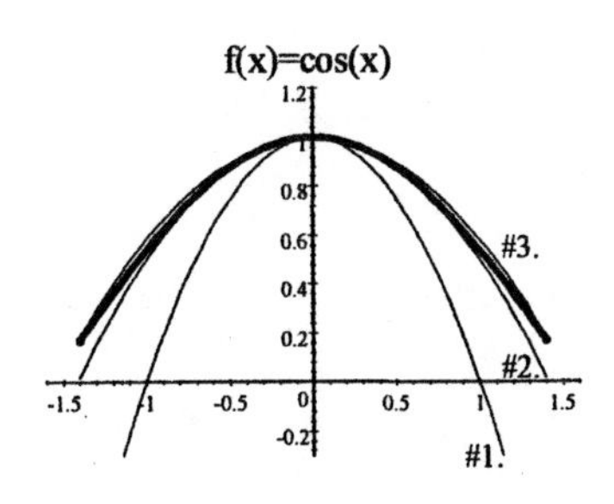

Section 3.7:

1a. $a = -\frac{1}{7}$, $b = \frac{5}{7}$ 1b. $a=\frac{8}{7}$, $b = \frac{2}{7}$ 1c. Both equations represent the same line.
2a. $2\sqrt{2}$ 2b. $\sqrt{26}$ 2c. $5\sqrt{2}$ 5. Let $y = mx$ be the equation of a line passing through the first data point, (0, 0). There are three errors, one for each of the other three data points. Their sum is $error(m) = \sqrt{(-.2-5m)^2} + \sqrt{(-.8-15m)^2} + \sqrt{(-1.3-25m)^2}$. The minimum of the error function occurs at $m = -0.054$ (approximately). Thus the linear function that best fits the data is: $y = -0.054x$.

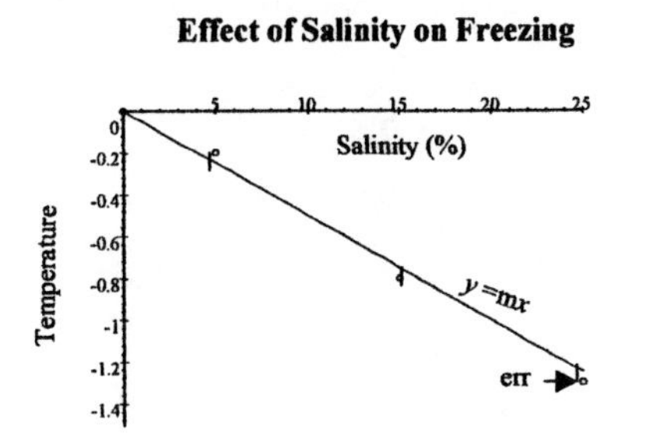

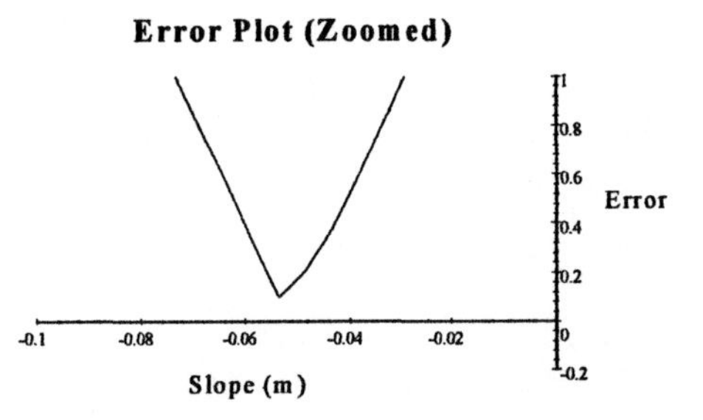

7.

Data Point		$y = ax^2 + bx + c$
$(0,0)$	=>	$c = 0$
$(\frac{\pi}{2}, 1)$	=>	$a(\frac{\pi}{2})^2 + b(\frac{\pi}{2}) = 1$
$(\pi, 0)$	=>	$a\pi^2 + b\pi = 0$

Solve the System of 2 Equations in a and b
Multiply 1st equation by 4. Then
$\begin{cases} a\pi^2+2b\pi=4 \\ a\pi^2+b\pi=0 \end{cases}$ Subtract to get $b\pi = 4, b = \frac{4}{\pi}$
Substitute for b and solve for a. $a = -\frac{4}{\pi^2}$

Thus $p(x) = -\frac{4}{\pi^2}x^2 + \frac{4}{\pi^2}x$. The error plot indicates the error is < 0.05.

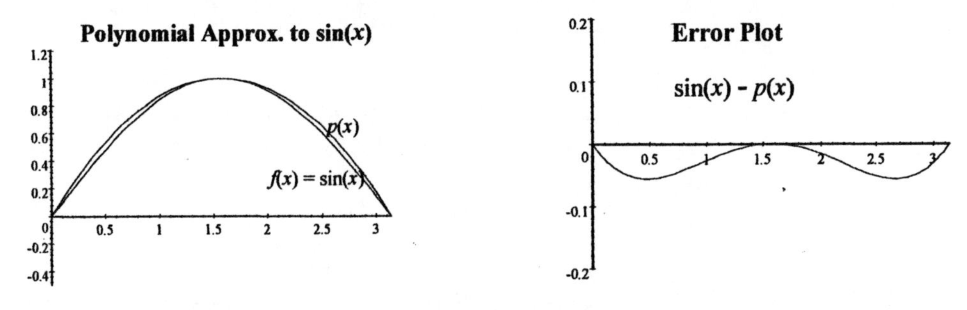

13. $y = -0.0145x^4 + 0.1932x^3 - 0.6853x^2 + 0.2944x + 3$

Section 3.8:

1. max =2 (due to the constraint on the range)
   max = 2.1773 without the constrtaint on the range
2. min=-.8056 (due to restriction on the range)
5. Conversion factor $= \frac{3600 \text{ sec/hr}}{x \text{ mi/hr}}$ (gr/sec)(sec/hr)(1/mi/hr)=gr/mi

speed = 35.7143 mi/hr

7. Obj. function: $A(x,y) = xy$, Constraint eq. $2x + 2y = 100$; ans. $x = y = 25$ m

9. Obj. function: $V(h,w,l) = hwl$, Constraint eq.s $w = 16 - 2h$, $l = 20 - 2h$
ans. $h = 2.9449$ in.

11.(line: $y = -x + 2$), Obj. function: $A(a,b) = ab$, Constraint eq.: $a - b + 2$, ans.: $a = b = 1$

13. Obj. function: $A(w,h) = wh$, Constraint eq. $h = 16 - w^2/20$, ans.: w = 10 ft, $h = 11$ ft

15. Obj. function: $V(x,y) = xy + \frac{1}{2}(x\cos(30°))(x\sin(30°))$, Constraint eq. $2x + y = 10$, ans. = 4.39 in.

Chapter 4.

Section 4.1:

1a. ~1.250 1b.~1.333 1c.~1.333 2a. 219.615 2b. $\frac{91}{8}$ 2c. $\frac{255}{1012}$

5. Rose has stepped (10 steps): $\frac{1}{2} + (\frac{1}{2})^2 + (\frac{1}{2})^3 + \ldots + (\frac{1}{2})^9$
$= \frac{1}{2}[1 + \frac{1}{2} + (\frac{1}{2})^2 + \ldots + (\frac{1}{2})^8] = \frac{1}{2}[\frac{1-(\frac{1}{2})^9}{1-\frac{1}{2}}] = 1 - (\frac{1}{2})^9 = 0.998$ yds.
So Rose is $1 - 0.998 = 0.002$ yds from the door.

7. Model: $f(n)$ = # pairs of fruit flies after n hour.
$f(n) = f(n-1) + 3f(n-1) = 4f(n-1)$,
$f(0) = 1$. $f(n) = 4^n(1) = 4^n$, $f(24) = 4^{24} = 2.815x10^{14}$. (Lots of flies!)

9. For very large k : $0.11111... \sim \frac{1}{10} + (\frac{1}{10})^2 + (\frac{1}{10})^3 + ...(\frac{1}{10})^k$
$= \frac{1}{10}[1 + \frac{1}{10} + (\frac{1}{10})^2 + ... + (\frac{1}{10})^{k-1}] = \frac{1}{10}[\frac{1-(\frac{1}{10})^k}{1-\frac{1}{10}}] = \frac{1}{9}[1 - (\frac{1}{10})^k] = \frac{1}{9}$ when k becomes infinitely large.

11. For very large k : $0.34111... = \frac{3}{10} + \frac{4}{10^2} + \frac{1}{10^3} + \frac{1}{10^4} + ...$
$= \frac{3}{10} + \frac{4}{10^2} + \frac{1}{10^3}[1 + \frac{1}{10} + \frac{1}{10^2} + ... + \frac{1}{10^{k-2}}] = \frac{3}{10} + \frac{4}{10^2} + \frac{1}{10^3}[\frac{1-\frac{1}{10^{k-1}}}{1-\frac{1}{10^3}}]$
$= \frac{34}{100} + \frac{1}{999}[1 - \frac{1}{10^{k-1}}] = \frac{34}{100} + \frac{1}{999} = \frac{341}{1,000}$ when k becomes infinitely large.

13. Let v = volume, h =height, w =width, 3 = thickness. Let h^*, w^* denote the new height and width. $v = 3hw$ and $2v = 3h^*w^*$. Also $\frac{h}{w} = \frac{h^*}{w^*} = \frac{11}{8}$. Thus $h^* = \frac{11}{8}w^*$, so $2v = 3(\frac{11}{8}w^*)w^*$ or $2v = \frac{33}{8}(w^*)^2$. $w^* = \sqrt{\frac{16}{33}v} = \sqrt{\frac{16}{33}(264)} = 11.31$ in.

15. Let $a(n)$ = the balance in the account after n 3-month periods, $a(0) = 1,000, 0.03$ quarterly interest rate. Model: $a(n) = a(n-1) + 0.03a(n-1)$ and $a(0) = 1,000$.
So $a(n) = 1.03a(0)$. Solution: $a(k) = 1,000(1.03)^k$
Value: $a(4) = 1,000(1.03)^4 = 1125.51$.

17. Let $a(n)$ = the balance in the account after n days, $a(0) = 1,000, 0.0003$ daily interest rate.Model: $a(n) = a(n-1) + 0.000333a(n-1)$ and $a(0) = 1,000$.
So $a(n) = 1.000333a(0)$. Solution: $a(k) = 1,000(1.000333)^k$.
Value: $a(360) = 1,000(1.000333)^{360} = 1127.47$.

Section 4.2:

1a. Note: $\log(2x) = \log(2) + \log(x)$. 1.082 1b. 0.707 1c. –0.322

2a. 0.893 2b. 5.585 2c. -3.969 5. $a(k) = .85^k c + \frac{50}{15}, a(17) = 655.10$

7.a. $\frac{.12}{2} = .06$ 7b. $\frac{.12}{4} = .03$ 7c. $\frac{.12}{12} = .01$ 7d. $\frac{.12}{365} = 0.000329$

9. $p(t)$ = pop. at time t years. $p(t) = p(t-1) + 0.04p(t-1) = 1.04p(t-1)$
=> $p(k) = 1.04^k p(0)$. Solve for k when $p(k) = 2p(0)$ or when $2 = 1.04^k$.
$k = \frac{\log(2)}{\log(1.04)} = 17.67$ years. For 2% rowth rate, $k = \frac{\log(2)}{\log(1.02)} = 35$ years.

15. Let $a(n)$ = balance after n months, r = monthly interest rate, d = monthly deposit

Model: $a(n) = a(n-1) + ra(n-1) + d = (1+r)a(n-1) + d$.
Solution (see Sue's Plan) $a(k) = (1+r)^k c - \frac{d}{r}$ where $c = a(0) + \frac{d}{r}$.
17. (Reference Sue's Plan) $a(k) = (1+r)^k c - \frac{d}{r}$ where $c = a(0) + \frac{d}{r}$.
$a(18*12) = 1.005^{18*12}(100 + \frac{d}{.005}) - \frac{d}{.005} = 40,000$
$=> 1.005^{216}(100) + (1.005^{216} - 1)\frac{d}{.005} = 40,000$
$=> \frac{d}{.005} = \frac{40,000 - 1.005^{216}(100)}{1.005^{216} - 1} => d = 102.51$
19. Let $a(n)$ = balance after n months, r = monthly interest rate, d = monthly withdrawal
Model: $a(n) = a(n-1) + ra(n-1) - d = (1+r)a(n-1) - d$.
Solution (see Sue's Plan) $a(k) = (1+r)^k c + \frac{d}{r}$ where $c = a(0) - \frac{d}{r}$.
Maintaining her initial investment, means that $a(1) = a(0)$. Thus
$a(0) = (1.01)^1(a(0) - \frac{2,000}{.01}) + \frac{2,000}{.01} => (1 - 1.01)a(0) = -1.01\frac{2,000}{.01} + \frac{2,000}{.01}$
$=> a(0) = \frac{2,000}{.01} = 200,000$.

Section 4.3:

1a. $x = \frac{-3 \pm \sqrt{9+4}}{2} = \frac{-3 \pm \sqrt{13}}{2}$ 1b. $x = \frac{1 \pm \sqrt{1+16}}{4} = \frac{1 \pm \sqrt{17}}{4}$ 1c. $x = \frac{4 \pm \sqrt{20}}{2} = 2 \pm \sqrt{5}$
2a. $44° = 44(\frac{\pi}{180}) = \frac{11\pi}{45}$ rad 2b. $\frac{3}{5}$rad $= \frac{3}{5}(\frac{180}{\pi}) = \frac{108}{\pi}°$
2c. 60 mph= $\frac{60 \text{ mi}}{1 \text{ hr}} * \frac{\frac{5280 \text{ ft}}{1 \text{ hr}}}{\frac{3600 \text{ sec}}{1 \text{ hr}}} = 88$ ft/sec.
5. (Reference Problem #1) $s(t) = -16t^2 - 25t + 1,000$. Set $s(t) = 0$ and solve for t.
$t = \frac{25 \pm \sqrt{625+64000}}{-32} = -8.725$ or 7.163 seconds. The negative solution does not make any sense in the setting, just answer is 7.163 seconds.
7. (Reference Problem #2) $s(t) = (50(\frac{1}{2})t, -16t^2 + 50(\frac{\sqrt{3}}{2})t + 4)$, javelin hits the ground when $-16t^2 + 25\sqrt{3}t + 4 = 0$ or $t = 2.796$ seconds.
Distance = (25)(2.796) = 182.4 ft.
9. $\tan(0) = 0, \tan(\frac{\pi}{4}) = 1$.
13. $p(t) = (\frac{v}{\sqrt{2}}t, -16t^2 + \frac{v}{\sqrt{2}}t)$; set x-component =100 and solve for t in terms of v, substitute for t in the y-component and then set the component = 0. Solve for v.
$v = 40\sqrt{2}$ ft/sec.

Section 4.4:

1a. $\frac{1}{16}$ 1b. geometric series with $r = 2$. Sum= $\frac{1-2^{11}}{1-2} = -(1 - 2^{11}) = 2,047$
1c. geoemtric series with $r = -2$. Sum=$\frac{1-(-2)^{11}}{1-(-2)} = \frac{1+2^{11}}{3} = 683$
2a. $x = \frac{\ln(17)}{\ln(4)} = 2.044$ 2b.$3x = \frac{\ln(30)}{\ln(12)} => x = 0.456$
2c. $2^{3x} + 8^x = (2^3)^x + 8^x = 8^x + 8^x = 2 * 8^x$
So $2 * 8^x = 47 => 8^x = \frac{47}{2} => x = \frac{\ln(\frac{47}{2})}{8} = 0.395$
7. Solution format: $T(t) = ae^{-pt} + RM$, where $p = 1.7, RM = 350$, and $T(0) = 70$
$T(60) = 350°$ 9a. $f(x) = e^x$ 9b. $f(x) = -e^{-x}$ 9c. $fx) = e^{-x}$ 9d. $f(x) = -e^x$
11a. $2NaCL + H_2SO_4 \to Na_2SO_4 + 2HCL$, 11b. $Na_2CO_3 + Ca(OH)_2 = 2NaOH + CaCO_3$

Section 4.5:

1a. $\frac{3}{16}$ 1b. 1.6 1c.10 gallons 2a. $a(k) = 10(.7^k)$
2b. $a(k) = 10(.7^k) + \frac{5}{1-.7}$ 2c. $a(k) = 10(2^k) + \frac{-4}{1-2} = 10(2^k) + 4$
5. $\frac{1*90° + 1.25*60°}{2.25} = 73.33°$ 9. Form a recursive sequence model. Let $c(n)$ = the grapefruit concentration after the nth serving. $c(0) = \frac{12}{40}$.
$39c(n)$ = amount of grapefruit concentrate after nth cup is served. Thus
$c(n) = \frac{39}{40}c(n-1)$ and $c(k) = (\frac{12}{40})(\frac{39}{40})^k$. Solve for k such that $c(k) < \frac{6}{40}$

Section 4.6:

1a. Compute the numerical value of both sides without using a formula.

1c. Factor $\frac{1}{2}$ out of the left-hand side. $\frac{1}{2}[1 + (\frac{1}{2}) + (\frac{1}{2})^2 + \ldots + (\frac{1}{2})^{k-1}]$
$= (\frac{1}{2})[\frac{1-(\frac{1}{2})^k}{1-\frac{1}{2}}] = 1 - (\frac{1}{2})^k \text{ -> } 1$ as k becomes infinitely large.

2a. $b(1) = 9.6 + 0.00083(9.6)[675 - 9.6] = 14.90$
$b(2) = 14.90 + 0.00083(14.90)[675 - 14.90] = 23.06$

2c. $b(12) = \frac{0.00083}{1+69.3125(e^{675*0.00083*12})} = 1.44$

7. Model: $m(n) = m(n-1) - .15m(n-1) = .85m(n-1)$ with $m(0) = 900$
$m(k) = .85^k m(0) = 900(.85^k)$. $m(k) = 5 \Rightarrow 900(.85^k) = 5 \Rightarrow k = 31.95$ hrs.

9. 7 days; 13a. 100, 13b. $\frac{100}{1+4e^{-t}} = 50 \Rightarrow 2 = 1 + 4e^{-t} \Rightarrow e^{-t} = \frac{1}{4} \Rightarrow t = 1.386$

Section 4.7:

1a. (2.2,2.4) 1b. (-3.26,0.11) 1c. (-0.797,0.576), (2,4)

3a. 1.25 3b. 1.667 3c. 6.25 5. Opportunity cost: Bianca $\frac{1}{2}$, Jim $\frac{5}{12}$

7. 550 9. 740

Section 4.8:

1a. $f(n)$ = frequency of the nth note from the left-hand-most note of the piano keyboard;
$f(n+1) = r * f(n)$

1b. $r = 2^{1/12}$ 1c. C note—frequency = $440*2^{3/12} = 523.25$

1d. $f(k) = r^k f(0)$ 1e. $f(0) = 55$ A 1f. $f(87) = 55(2^{87/12}) = 8372.02$

7. 1,3,8,21,55 $u_1 + u_3 + u_5 + \ldots u_{2n+1} = u_{2n+2}$

9. $u_1^2 + u_2^2 + u_3^2 + \ldots + u_n^2 = u_n * u_{n+1}$

Index

G

H

I

J

K

L

M

T

U

V

W

Z